Current Topics in Microbiology and Immunology

Volume 334

Current Topics in Microbiology and Immunology

Previously published volumes

Further volumes can be found at springer.com

Vol. 309: **Polly Roy (Ed.):**
Reoviruses: Entry, Assembly and Morphogenesis. 2006. 43 figs. XX, 261 pp. ISBN 3-540-30772-9

Vol. 310: **Doerfler, Walter; Böhm, Petra (Eds.):**
DNA Methylation: Development, Genetic Disease and Cancer. 2006. 25 figs. X, 284 pp. ISBN 3-540-31180-7

Vol. 311: **Pulendran, Bali; Ahmed, Rafi (Eds.):**
From Innate Immunity to Immunological Memory. 2006. 13 figs. X, 177 pp. ISBN 3-540-32635-9

Vol. 312: **Boshoff, Chris; Weiss, Robin A. (Eds.):**
Kaposi Sarcoma Herpesvirus: New Perspectives. 2006. 29 figs. XVI, 330 pp. ISBN 3-540-34343-1

Vol. 313: **Pandolfi, Pier P.; Vogt, Peter K.(Eds.):**
Acute Promyelocytic Leukemia. 2007. 16 figs. VIII, 273 pp. ISBN 3-540-34592-2

Vol. 314: **Moody, Branch D. (Ed.):**
T Cell Activation by CD1 and Lipid Antigens, 2007, 25 figs. VIII, 348 pp. ISBN 978-3-540-69510-3

Vol. 315: **Childs, James, E.; Mackenzie, John S.; Richt, Jürgen A. (Eds.):**
Wildlife and Emerging Zoonotic Diseases: The Biology, Circumstances and Consequences of Cross-Species Transmission. 2007. 49 figs. VII, 524 pp. ISBN 978-3-540-70961-9

Vol. 316: **Pitha, Paula M. (Ed.):**
Interferon: The 50th Anniversary. 2007. VII, 391 pp. ISBN 978-3-540-71328-9

Vol. 317: **Dessain, Scott K. (Ed.):**
Human Antibody Therapeutics for Viral Disease. 2007. XI, 202 pp. ISBN 978-3-540-72144-4

Vol. 318: **Rodriguez, Moses (Ed.):**
Advances in Multiple Sclerosis and Experimental Demyelinating Diseases. 2008. XIV, 376 pp. ISBN 978-3-540-73679-9

Vol. 319: **Manser, Tim (Ed.):**
Specialization and Complementation of Humoral Immune Responses to Infection. 2008. XII, 174 pp. ISBN 978-3-540-73899-2

Vol. 320: **Paddison, Patrick J.; Vogt, Peter K.(Eds.):** RNA Interference. 2008. VIII, 273 pp. ISBN 978-3-540-75156-4

Vol. 321: **Beutler, Bruce (Ed.):**
Immunology, Phenotype First: How Mutations Have Established New Principles and Pathways in Immunology. 2008. XIV, 221 pp. ISBN 978-3-540-75202-8

Vol. 322: **Romeo, Tony (Ed.):**
Bacterial Biofilms. 2008. XII, 299. ISBN 978-3-540-75417-6

Vol. 323: **Tracy, Steven; Oberste, M. Steven; Drescher, Kristen M. (Eds.):**
Group B Coxsackieviruses. 2008. ISBN 978-3-540-75545-6

Vol. 324: **Nomura, Tatsuji; Watanabe, Takeshi; Habu, Sonoko (Eds.):**
Humanized Mice. 2008. ISBN 978-3-540-75646-0

Vol. 325: **Shenk, Thomas E.; Stinski, Mark F. (Eds.):**
Human Cytomegalovirus. 2008. ISBN 978-3-540-77348-1

Vol. 326: **Reddy, Anireddy S.N; Golovkin, Maxim (Eds.):**
Nuclear pre-mRNA processing in plants. 2008. ISBN 978-3-540-76775-6

Vol. 327: **Manchester, Marianne; Steinmetz, Nicole F. (Eds.):**
Viruses and Nanotechnology. 2008. ISBN 978-3-540-69376-5

Vol. 328: **van Etten, (Ed.):**
Lesser Known Large dsDNA Viruses. 2008. ISBN 978-3-540-68617-0

Vol. 329: **Diane E. Griffin; Michael B.A. Oldstone (Eds.):** Measles 2009. ISBN 978-3-540-70522-2

Vol. 330: **Diane E. Griffin; Michael B.A. Oldstone (Eds.):** Measles 2009. ISBN 978-3-540-70616-8

Vol. 331 **Villiers, E. M. de (Eds):**
TT Viruses. 2009. ISBN 978-3-540-70917-8

Vol. 332 **Karasev A.(Ed.):**
Plant produced Microbial Vaccines. 2009. ISBN 978-3-540- 70857-5

Vol. 333 **Richard W. Compans; Walter A. Orenstein (Eds):** Vaccines for Pandemic Influenza. 2009. ISBN 978-3-540-92164-6

Dorian McGavern · Michael Dustin
Editors

Visualizing Immunity

Springer

Editors
Dr. Dorian McGavern
National Institutes of Neurological Disorders and Stroke
National Institutes of Health (NIH)
10 Center Drive
Bldg 10, Rm 7C213
Bethesda, MD 20892
USA
mcgavernd@mail.nih.gov

Dr. Michael Dustin
New York University
School of Medicine
Skirball Institute of Biomolecular Medicine
540 First Ave.
New York NY 10016
USA
dustin@saturn.med.nyu.edu

ISBN 978-3-540-93862-0 e-ISBN 978-3-540-93864-4
DOI 10.1007/978-3-540-93864-4
Springer Dordrecht Heidelberg London New York

Current Topics in Microbiology and Immunology ISSN 0070-217x

Library of Congress Catalog Number: 2008944027

Cover legend: ***Localization of cytotoxic lymphocytes (CTL) during fatal LCMV-induced meningitis.***

This low magnification panel shows LCMV-specific CTL in green, meningeal blood vessels in red, and skull in blue. The 3D image was captured from a symptomatic mouse at day 6 post-infection using a two photon microscope. Virus-specific CTL tend to aggregate along meningeal blood vessels, because the stromal network that surrounds these vessels is infected by LCMV. This image was captured by Dr. Jiyun Kim at New York University.

Cover design: WMX Design GmbH, Heidelberg, Germany

Printed on acid-free paper

Springer is part of Springer Science+Business Media (www.springer.com)

Preface

Researchers have used a variety of techniques over the past century to gain fundamental insights in the field of immunology and, as technology has advanced, so too has the ability of researchers to delve deeper into the biological mechanics of immunity. The immune system is exceedingly complex and must patrol the entire body to protect us from foreign invaders. This requires the immune system to be highly mobile and adaptable - able to respond to diverse microbial challenges while maintaining the ability to distinguish self from a foreign invader. This latter feature is of great importance because the immune system is equipped with toxic mediators, and a failure in self/non-self discrimination can result in serious diseases. Fortunately, in most cases, the immune system operates within the framework of its elegant design and protects us from diverse microbial challenges without initiating disease.

Because the immune system is not confined to a single tissue, a comprehensive understanding of immunity requires that research be conducted at the molecular, cellular, and systems level. Immune cells often find customized solutions to handling microbial insults that depend on the tissue(s) in which the pathogen is found. Removal of immune cells from their natural environment is one common means by which immunity is studied; however, this approach comes with the caveat that immune cells interact uniquely with the microenvironments and tissue architecture they encounter. Because no two tissues are alike, immune cells will often adapt and respond based on the unique microenvironment in which they reside. Given this fact, it is of great importance to consider cellular context when deciphering the mysteries of the immune system. Lessons learned in one tissue may not necessarily apply to the entire body.

Understanding the contextual side of immunity necessitates study of immune cells in vivo. However, those that pursue in vivo research immediately encounter the obstacle of how best to study immune cells in their natural environments. This obstacle is not a trivial one, as it is far easier to remove cells from their natural environments and study them ex vivo or in vitro. Fortunately, scientists in other disciplines have come to the rescue with exciting advances in imaging techniques. These advances have enabled immunologists to quite literally "see" how immune cells respond to diverse challenges. Using static imaging approaches, researchers have captured snapshots in time, which are then pieced together with corroborating datasets to assemble a sequence of events. More recently, researchers have instituted an exciting upgrade, transitioning from static to dynamic imaging techniques such as two-photon laser scanning microscopy. These dynamic approaches are advantageous because researchers can use them to study immune cells in their natural environments in real time. Thus, it is no longer necessary to extrapolate from in vitro observations how immune cells operate in vivo. Immune cells operating in states of health and disease can now be filmed and studied afterwards in great detail.

The field of visualizing immunity has moved rapidly over recent years and has carved out an important niche within the broader discipline of immunology. This issue was assembled to pay tribute to those who have gleaned fundamental insights in immunology using imaging approaches. The reviews within span a breadth of knowledge that covers certain technical, molecular, cellular, and systems aspects of visualizing immunity. The issue begins with what should be considered when assembling a custom imaging platform and then progresses to the mechanics of T cell interactions and activation. From there, the issue moves on to lymphocyte motility/migration, B lymphocyte activation, and finally to visualizations of some challenges that immune cells face (e.g., pathogens and tumors). We felt that these topics have a natural flow in the order presented and allow the reader to "see" the development of immune responses at all levels. The visualizations within are aesthetically pleasing, and it is gratifying to know that many novel insights have been extracted from such stunning imagery. Now that the field has been set ablaze with enthusiasm, it is certain that exciting new immunological discoveries lie just around the corner. This issue is merely a snapshot in time for a field that should grow exponentially in the years to come.

Dorian B. McGavern

Contents

Contributors

Sebastian Amigorena
Institut National de la Santé et de la Recherche Médicale U653, Immunité et Cancer, Pavillon Pasteur, Institut Curie, 26 rue d'Ulm, 75245 Paris Cedex 05, France
sebastian.amigorena@curie.fr

Jeanette Ampudia
Department of Immunology and Microbial Science, The Scripps Research Institute, 10550 North Torrey Pines Rd., La Jolla, CA 92037, USA

Facundo D. Batista
Lymphocyte Interaction Laboratory, Cancer Research UK, London Research Institute, Lincoln's Inn Fields Laboratories, 44 Lincoln's Inn Fields, London WC2A 3PX, UK
facundo.batista@cancer.org.uk

Alexandre Boissonnas
Institut National de la Santé et de la Recherche Médicale U653, Immunité et Cancer, Pavillon Pasteur, Institut Curie, 26 rue d'Ulm, 75245 Paris Cedex 05, France

Andrew Bullen
Department of Pathology and Biological Imaging Development Center, University of California, 513 Parnassus Avenue, San Francisco, CA 94143-0511, USA
Andrew.Bullen@ucsf.edu

Jean-Pierre Clamme
Department of Immunology and Microbial Science, The Scripps Research Institute, 10550 North Torrey Pines Rd., La Jolla, CA 92037, USA

Michael L Dustin
Department of Pathology, Program of Molecular Pathogenesis, Skirball Institute of BioMolecular Medicine, NYU School of Medicine, 540 First Avenue, 2nd Floor, New York 10016, USA

Luc Fetler
Centre National de la Recherche Scientifique UMR 168, Laboratoire de Physico-Chimie Curie, Institut Curie, 26 rue d'Ulm, 75245 Paris Cedex 05 France

Reinhold Förster
Institute of Immunology, Hannover Medical School, Carl-Neuberg-Strasse 1, 30625 Hannover, Germany

Rachel S. Friedman
Department of Pathology and Biological Imaging Development Center, University of California, 513 Parnassus Ave, San Francisco, CA 94143-0511, USA

Guo Fu
Department of Immunology and Microbial Science, The Scripps Research Institute, 10550 North Torrey Pines Rd., La Jolla, CA 92037, USA

Nicholas R.J. Gascoigne
Department of Immunology and Microbial Science, The Scripps Research Institute, 10550 North Torrey Pines Rd., La Jolla, CA 92037, USA
Gascoigne@scripps.edu

Naomi E. Harwood
Lymphocyte Interaction Laboratory, Cancer Research UK, London Research Institute, Lincoln's Inn Fields Laboratories, 44 Lincoln's Inn Fields, London WC2A 3PX, UK

Il-Young Hwang
B-cell Molecular Immunology Section, Laboratory of Immunoregulation, NIAID, NIH, 10 Center Dr. Bethesda, MD 20892-1892, USA
jkehrl@niaid.nih.gov

Silvia S. Kang
National Institutes of Neurological Disorders and Stroke, National Institutes of Health, 10 Center Drive, Bldg 10, Rm 7C213, Bethesda, MD 20892, USA

John H. Kehrl
B-cell Molecular Immunology Section, Laboratory of Immunoregulation, NIAID, NIH, 10 Center Dr. Bethesda, MD 20892-1892, USA
jkehrl@niaid.nih.gov

Vjollca Konjufca
Washington University School of Medicine, Department of Pathology and Immunology, 660 South Euclid Avenue, St. Louis, MO 63110-1093 USA

Matthew F. Krummel
Department of Pathology and Biological Imaging Development Center, University of California, 513 Parnassus Avenue, San Francisco, CA 94143-0511, USA

Klaus Ley
La Jolla Institute for Allergy and Immunology, La Jolla, CA, USA

Carina Lotz
Department of Hematology, University Medical Center Utrecht,
Lundlaan 6, 3584EA Utrecht, The Netherlands

Michel Mallaun
Department of Biomedicine, University Hospital, Basel, Switzerland

Dorian B. McGavern
National Institutes of Neurological Disorders and Stroke, National Institutes of Health (NIH), 10 Center Drive, Bldg 10, Rm 7C213, Bethesda, MD 20892, USA
mcgavernd@mail.nih.gov

Stephen J. McSorley
Medicine, University of Minnesota Medical School, Minneapolis, MN 55455, USA
and
Center for Immunology, University of Minnesota Medical School, Minneapolis, MN 55455, USA
and
Center for Infectious Diseases and Microbiology Translational Research, University of Minnesota Medical School, Minneapolis, MN 55455, USA
mcsor002@umn.edu

Mark J. Miller
Washington University School of Medicine, Department of Pathology and Immunology, 660 South Euclid Avenue, St. Louis, MO 63110-1093, USA
miller@pathology.wustl.edu

James J. Moon
Center for Immunology, University of Minnesota Medical School, Minneapolis, MN 55455, USA
and
Department of Microbiology, University of Minnesota Medical School, Minneapolis, MN 55455, USA

Nathalie Niederberger
Serono Pharmaceutical Research Institute, Geneva, Switzerland

Ed Palmer
Department of Biomedicine, University Hospital Basel, Switzerland

Chung Park
B-cell Molecular Immunology Section, Laboratory of Immunoregulation, NIAID, NIH, 10 Center Dr. Bethesda, MD 20892-1892, USA
jkehrl@niaid.nih.gov

Vasily Rybakin
Department of Immunology and Microbial Science, The Scripps Research Institute, 10550 North Torrey Pines Rd., La Jolla, CA 92037, USA

Alix Scholer-Dahirel
Institut National de la Santé et de la Recherche Médicale U653, Immunité et Cancer, Pavillon Pasteur, Institut Curie, 26 rue d'Ulm, 75245 Paris Cedex 05, France

Tim Worbs
Institute of Immunology, Hannover Medical School, Carl-Neuberg-Strasse 1, 30625 Hannover, Germany
worbs.tim@mh-hannover.de

Pia P. Yachi
Applied Molecular Evolution, 3520 Dunhill Street, San Diego, CA 92121, USA

Tomasz Zal
Department of Immunology, University of Texas, MD Anderson Cancer Center, 7455 Fannin, Houston, TX 77030, USA

Alexander Zarbock
Department of Anesthesiology and Intensive Care Medicine, University of Münster, Münster, Germany
zarbock@uni-muenster.de

Two-Photon Imaging of the Immune System: A Custom Technology Platform for High-Speed, Multicolor Tissue Imaging of Immune Responses

Andrew Bullen, Rachel S. Friedman, and Matthew F. Krummel

Contents

Abstract Modern imaging approaches are proving important for addressing contemporary issues in the immune system. These approaches are particularly useful for characterizing the complex orchestration of immune responses *in vivo*. Multicolor, two-photon imaging has been proven to be especially enabling for such studies because of its superior tissue penetration, reduced image degradation by light scattering leading to better resolution and its high image quality deep inside tissues. Here, we examine the functional requirements of two-photon imaging instruments necessary for such immune studies. These requirements include frame rate, spatial resolution and the number of emission channels. We use this discussion as a starting point to compare commercial systems and to introduce a custom technology platform that meets these requirements. This platform is noteworthy because it is very cost-effective, flexible and experimentally useful. Representative data collected

A. Bullen (✉), R.S. Friedman, and M.F. Krummel
Department of Pathology and Biological Imaging Development Center, University of California San Francisco, 513 Parnassus Ave, San Francisco, CA 94143-0511, USA
e-mail: Andrew.Bullen@ucsf.edu or Matthew.Krummel@ucsf.edu

D. McGavern and M. Dustin (eds.), *Visualizing Immunity,* Current Topics in Microbiology and Immunology 334,
DOI: 10.1007/978-3-540-93864-4_1,

with this instrument is used to demonstrate the utility of this platform. Finally, as the field is rapidly evolving, consideration is given to some of the cutting-edge developments in multiphoton microscopy that will likely improve signal strength, depth penetration and/or the experimental usefulness of this approach.

1 Introduction

Direct imaging of the individual cell types of the immune system in their native context undeniably provides the most accurate spatiotemporal assessment of the system-wide properties of the immune response. Over the past 6 years, this technique has increasingly been facilitated by imaging methods utilizing two-photon laser-scanning microscopy (TPLSM) (Cahalan and Parker 2008). Unlike other methods, the behavior of individual cells are observed via this technology in an largely intact microenvironment containing, by definition, physiological concentrations of soluble mediators, growth factors, as well as cell–cell contacts with other components of the system.

Central to this is the ability of multiphoton excitation to provide improved depth penetration and reduced phototoxicity over longer observation periods (Cahalan et al. 2002; Williams et al. 2001), which is a key aspect to observing biology over time within healthy tissues. To achieve such observational accuracy, however, is not without challenges. These include proper experimental design to highlight specific cells without perturbing the overall biology, optimal sample preparation to minimize artifacts due to whole animal surgery and, of course, the best possible instrumentation for detecting optical signals from the deepest possible location within complex organs. This latter component is an area of intense development and an area whose improvement simplifies experimental and sample-preparation considerations. In this review, we will highlight the optical and instrumentation approaches that are improving this technology, and compare a variety of TPLSM implementations used in immune-imaging. As an example of a custom system that has been highly successful in imaging immune responses, we will elaborate the details of a custom-instrument that we have implemented, and which is increasingly being adopted for its relative ease of implementation, flexibility, cost, and importantly, imaging quality. Finally, we will describe emerging technologies and how they are likely to improve spatial and temporal aspects of this approach.

1.1 The Power of Imaging for Addressing Issues/Answering Questions in the Immune System

What are the benefits of live-cell imaging in the immune response, generally? Studies undertaken by Wülfing and Davis (Wülfing et al. 1997), Delon and Trautmann (Delon et al. 1998), Negulescu and Cahalan (Negulescu et al. 1996) and Dustin and Unanue (Dustin et al. 1997) in the mid-1990s highlighted the distinct power of observing single cell dynamics with reference to the calcium response. Notably, all these groups were able to take advantage of the fact that imaging single-cell dynamics permits the direct observation of an activation 'timeline'. In particular, they were to observe each cell from the start of its interaction with an antigen-presenting cell, followed by the full course of activation and calcium influx dynamics as it related to cell shape change and motility arrest (Negulescu et al. 1996; Dustin et al. 1997; Delon et al. 1998), and subsequently how it was influenced by the nature of antigen-presenting cells (APC) (Delon et al. 1998) and peptide–MHC complexes (pMHC) (Wülfing et al. 1997).

The imaging of single events in their entirety provides clear benefits over other 'bulk' methods such as flow cytometry for observing kinetic relationships since it avoids the 'temporal smear' generated by variations in the population behavior. Take, for example, the onset, magnitude, and duration of calcium signaling in a population of cells. A variation within the population in any of these parameters can be both observed and normalized when all cells are measured in their entirety during the period of interest. In contrast, a bulk measurement of the population over time simply measures the average behavior. If there is great variation in any of the kinetic parameters, the maximum magnitude of the others can be misrepresented due to contributions from cells that are at different stages of the response. Analysis of single cells allows all parameters to be viewed in their direct relationship to each other (i.e., in the recording of each single cell). These can later be pooled by common feature for statistical analysis (e.g., always starting at the time of onset of the responses), thus providing dramatically improved details of downstream kinetic relationships within the population.

The direct observation of cells undergoing biological activity using real-time microscopy also provides important spatial information. This information can help describe subcellular events and/or characterize the local microenvironment. In the first case, direct imaging of immune cells *in vitro* permits subcellular analysis of responses, such as the very early imaging of antibody-driven calcium influx within cytotoxic T lymphocytes (CTLs). Such observations have provided evidence that much of the influx was polarized, emanating from a single side of the cell (Poenic et al. 1987). This is now known to represent release of intracellular calcium stores, often from polarized endoplasmic reticulum (ER). Similarly, the movement of subcellular signaling molecules within cells, tagged with GFP, permits the assembly of a wide variety of signaling complexes to be analyzed with reference to calcium signaling, morphology, and/or motility (Krummel et al. 2000; Schaefer et al. 1999; Bunnell et al. 2001; Varma et al. 2006; Yokosuka et al. 2005). In the second case, cellular behavior in complex tissues is not constant but rather varies depending on local factors such as the presence of chemokines. For example, naive T cells arrest more on dendritic cells (DC) that are already involved in activation via other T cells (Hugues et al. 2007; Castellino et al. 2006). This type of problem, in particular, has benefited from multiphoton imaging of immune responses *in vivo*.

Within this area, multiphoton microscopy has, or is poised, to address the following types of questions with respect to cellular activation:

- What are the single-cell dynamics of activation? For example, Miller and Cahalan (Miller et al. 2002, 2004), and Mempel and von Andrian (Mempel et al. 2004) have shown that T cells undergo activation in multiple phases, corresponding to initial scanning of antigen-bearing surfaces and culminating in firm arrest of T cells on activated DCs.
- Where does activation occur spatially within a given tissue? For example, it was found that B cells can be activated on DCs, following entry into the lymph node, and as they traverse the T cell zone rather than solely in B cell zone (Qi et al. 2006).
- What is the effect of the spatial milieu (different microenvironments or different tissues)? For example, activating B cells near the B cell zone/T cell zone junction are sensitive to a chemokine gradient in that zone which attracts them toward the T cell zone. In contrast, more distal cells show no evidence of motion toward this potential source of T cell 'help' (Okada et al. 2005).
- Which cell types are present during an immune response *in vivo* and how do they contribute? For example, DCs have been highlighted to be the partners for T cell activation using chemical dye-labeling *in vivo* (Miller et al. 2004), dye-labeling prior to their adoption (Mempel et al. 2004; Bousso and Robey 2003), antibody-labeling *in vivo* (Hugues et al. 2004), and genetic marking (Lindquist et al. 2004; Shakhar et al. 2005).

Multiphoton microscopy has the potential to produce multiple types of readouts. Currently, a majority is limited to determinations of cell–cell interactions and positional detail although improving instrumentation is now permitting greater use of subcellular markers. Such resolution and sensitivity will begin to allow direct measurements of signaling protein aggregation, endocytosis, exocytosis, and polarity, in addition to the creation of custom biosensors to report on specific signaling cascades. Some of the most frequently applied include:

- Cell arrest. For example, the arrest of T cells following exposure to antigen (Miller et al. 2002; Bousso and Robey 2003).
- Cell morphology. For example, T cells undergoing motility tend to do so in a hand-mirror morphology, similar to amoeba (Miller et al. 2002).
- Calcium influx. For example, B cells encountering antigen-bearing DCs upon entry into the lymph node can be observed to flux calcium as measured by Fluo4 (Qi et al. 2006). This method is limited at present as the fluorescent dyes for these types of studies are rapidly vesicularized.
- Cell–cell association (or persistence). For example, T cells activating in a lymph node often do so in 'clusters' (Bousso and Robey 2003; Tang et al. 2006; Beuneu et al. 2006; Sabatos et al. 2008) and such clusters can permit long-lived cell–cell contacts between adjacent cells leading to directed cytokine secretion (Sabatos et al. 2008).
- Subcellular analysis of particle distribution. For example, antigens taken up by macrophages in the cortical sinus are distributed within subcellular compartments there (Phan et al. 2007).

1.2 Functional Requirements for Imaging Immune Function In Vivo

Leukocytes have defined characteristics that define the requirements for imaging events *in vivo*. On the one hand, the immune system is highly motile and thus requires sufficiently fast sampling to quantify this motion. In addition, this is a system containing multiple cell types and, thus, requires multiple labeling and detection strategies to differentiate cell types. It is also not limited in its activity to the surface of tissues or organs – in fact the cells of the immune system can lie deep within tissues, surveying for or defending against foreign organisms.

Functional requirements that need be considered for imaging are speed of sampling, sensitivity of detection, particularly with reference to depth penetration, multicolor acquisition, and the tradeoff between all of these and microscopic spatial resolution and macroscopic field of view (FOV).

1.2.1 Frame-Rate and Speed of Acquisition Considerations

Lymphocytes, key players in the immune responses, are intrinsically motile, with center-of-mass displacements that reach >25 μm min^{-1} (Miller et al. 2002). In addition, projections from the lymphocyte surface can appear and significantly change within seconds. As the path taken by T cells is rarely linear in tissues such as lymph nodes but more closely resembles a 'random walk' (Miller et al. 2002) (though it is not likely to be truly 'random'; Bajenoff et al. 2006), higher sampling frequencies are essential to provide an accurate readout of instantaneous velocities. Regardless of the underlying guiding force for movement, it is clear that sampling frequency plays a large part in determining the accuracy of a measurement of velocity, particularly when cells are persistent for only a finite period of time. A brief practical measure is that the sampling interval t should be considerably less than the persistence time P.

As an illustrative example, consider theoretical displacements based on motion described by a typical equation for a Persistent Random Walk (Othmer et al. 1988):

$$< d(t)^2 > = nS^2 \, [Pt - P^2 \, (1 - e^{-t/P}),$$

where $d(t)$ is the observed displacement, n is the number of dimensions, S is the speed, and P is the persistence (period of time without a turn). When the observed time period is less than the cellular directional persistence time ($t < P$), the equation reduces to:

$$d\,(t) = St$$

which is to say that the observed displacement is closely approximated by measures of speed multiplied by time. This is the ideal measurement scenario – that the measured displacement is in fact an accurate measure of the true instantaneous velocities.

For a theoretical cell with persistence of 2 min whose behavior followed this type of description (S set to 10), measurements at 30-s intervals ($P > t$) would yield displacement values of 5.8 μm/30 s and therefore velocities of 11.6 μm min^{-1}. Intervals of collection of 2 min ($P = t$) would on average obtain measurements of "instantaneous velocities" as ~21 μm 2 min^{-1} or 10.5 μm min^{-1}. Further reduced sampling rates of 4 and 10 min intervals ($P < t$) would on average obtain measurements of "instantaneous velocities" as 8.7 and 6.9 μm min^{-1}, respectively. The true magnitude of the errors obviously depends upon the underlying biology and doubtless adheres to a significantly more complex description. Indeed, there are a plethora of variations of models for modeling cell motility (Codling et al. 2008). However, this example serves to indicate the value of fast sampling in theory, but may also indicate a practical explanation for modest differences in reported motility parameters for naive T lymphocytes depending upon the frame-rate of data acquisition.

In practice, lymphocytes can cover considerable distances during a short time course of analysis. This in turn creates the important consideration of total X–Y–Z FOV since one cannot track a cell that has left the observation volume. Practically, this will create conflicts with requirements for spatial resolution, since complete sampling of a larger volume can either be done by capturing more pixels (longer collection times) or with the same number of larger pixels. Practically, this often entails variations in the number of Z-slices that are collected since varying this parameter can quickly multiply the time taken to acquire data. For example, presuming 30 z-stacks are required to capture a sufficient volume of data for analysis, 3-s acquisition times result in a full 90 s-between frames, thus potentially under-representing the very fastest movements as described above. Lowering spatial resolution in the Z-axis by taking larger Z-steps may initially appear an attractive way to minimize these bottlenecks. However, the resulting decrease in ability to accurately assess the cell center of mass in the Z-axis results in inaccurate high and low velocities being reported and, more generally, a broadening of the velocity distribution. A more detailed analysis of the effect of both temporal and spatial sampling frequency on the accuracy of velocity measurements is presented in Codling and Hill (2005).

1.2.2 Number of Detection Channels

As the examples above illustrate, measurements of cells and their behaviors *in vivo* most frequently rely upon the relationships amongst cell types and their environment. Consequently, the number of detection channels should match the number of fluorescent tags. To this extent, a two-cell interaction requires at least two distinct channels in which to collect the fluorescence emissions from two distinct dyes. For example, interactions between T cells and DCs can be achieved using a red fluorophore (e.g., the vital dye CMTMR) to label T cells and the distinct green fluorophore (e.g., CFSE) labeling method for the DCs (Miller et al. 2004).

However, as the compartmentalization of the immune response is studied further, it becomes apparent that even a 'simple' organ like a lymph node comprises distinct zones and behaviors which vary according to those zones. This requires the use of additional fluorophores or combinations of fluorophores to highlight critical regions;

we term this fiduciary labeling as the fluorophore which thus serves to highlight key positions within organs and tissues. So, for example, activating B cells near the T–B border move toward the border as a result of gradients of CCR7 ligands whereas those that are a distance away fail to do so – despite both being within the same B cell follicle (Okada et al. 2005). Additional fluorophores that highlight this border (in this case those that highlight the T cell zone) prove useful in determining the spatially distinct behavior. As another example, T cells in ectopic EL-4 tumors tend to migrate along paths parallel to blood vessels. Markers that highlight the blood vessels are necessary to reveal this – otherwise the T cells show guided migration without any mechanistic insight behind the nature of this confinement.

Thus, the use of fiduciary labeling leads to a frequent need to provide at least a third channel. More complex biology (i.e., those involving more than two cell types plus obligate fiduciary labels) can easily require 4 or more channels and this should be considered for choosing or adapting a microscope. In practice, the number of fluorophores and thus channels does not need to equal the number of distinct species to be labeled due to the possibility for multiplexing. In this way, three cell types can be labeled using combinations of just two dyes (green, red and green–red together create three distinct species in just two channels). As a general rule, the number of populations that can be distinctly labeled can be defined as:

$$C = 2^n - 1$$

where C is the number of distinct species that can be separately distinguished based on binary determinations of the presence of a dye and n is the number of channels of distinct detection. In practice, the true number that can be distinguished depends a bit on the application. For example, if labeled species (e.g., cells) never get very close, then different levels of each dye can be introduced into collections of cells leading to much larger variation in species and thus discrimination. As an example, Lichtman and colleagues generated populations of neurons *in vivo* using differing levels of CFP, YFP, and dsRed (three fluorophores) which permitted discrimination of approximately 90 different ‘colors’ based on combinations of just these three (Livet et al. 2007). On the other hand, the ability to discriminate based on quantitative measures of each component (component analysis) is made more difficult by fluorophores with wide-emissions that ‘bleed’ light into each other's emission channels and thus somewhat resemble ‘dual-colored species’. Component analysis can often still distinguish these, except when cells or structures bearing fluorophores get very close to one another. When species overlap in the same measurement space (i.e., voxel), the quantitative contributions also blur and it is often difficult to tease apart the borders of the two. Under such circumstances, more distinct fluorophores may be required.

1.2.3 Detection Sensitivity

A critical requirement for effective deep-tissue imaging of immune cells is efficient detection of emission light. There are two aspects of detection sensitivity: detector sensitivity and detection path efficiency. In general, there are two main detector options:

photomultiplier tubes (PMTs) and charge-coupled device (CCD) cameras. CCD cameras typically have better quantum efficiency, the ability to convert light into electrical signals suitable to digitize. However, while both detector types can have high gain capabilities, PMTs often excel in multichannel and scanning applications because they combine good sensitivity with a large measurement bandwidth. PMT-based systems are also less affected by scattering in the emission path because fluorescence is assigned to a particular point in the sample based on its instantaneous excitation time rather than by its position at the detector. Additionally, PMT number scales more cheaply than increases in camera number. This is particularly important for systems employing three or four emission channels. While it is possible to combine a CCD camera with an emission filter wheel to get spectral separation this results in considerably higher total collection times since each volume of data needs to be illuminated multiple times. This in turn can affect the overall collection speed and increase the likelihood of photobleaching and phototoxicity at the sample.

The nature of the detection path also determines the collection efficiency of the optical system and is therefore an important parameter. The detection path extends from the back aperture of the objective lens through to the detector itself. From an optical design perspective, short detection paths generally provide the greatest collection efficiency. Furthermore, the correct choice of dichroic mirrors and barrier filters along this path are critical to maximizing total detection efficiency, minimizing the detection of autofluorescent wavelengths and minimizing bleed-over or crosstalk between emission channels.

1.2.4 Number of Lasers

Another way to distinguish between different fluorescent species (and thereby discriminate between more cell types or fiduciary species) is to employ different excitation wavelengths. Switching between different excitation wavelengths is not easily achieved with current Ti-sapphire lasers. However, we and a number of other groups have begun to integrate multiple lasers into a single system. While not a cheap option, this approach benefits from the ability to excite more fluorophores and distinguish between fluorophores with overlapping emission spectra but resolvable excitation spectra. However, simultaneously illuminating a sample with two lasers increases the likelihood of phototoxicity and complicates the discrimination of fluorophores with overlapping excitation spectra. For this reason, dual or multiple laser systems typically require recording from interlaced frames, which increases the amount of time required to scan the sample and thus decreases the overall collection rate.

1.2.5 Depth Penetration

Much of the interesting biology of the immune response occurs below the surface of organs. Unfortunately, most mammalian tissues scatter visible light significantly and do so in a wavelength-dependent fashion. Measured scattering constants (μs) are typically of the order of 50 cm^{-1} (Collier et al. 2003). The probability of transmission T of the photon without redirection by scattering after a path length L (cm) is given by the equation:

$$T = e^{-\mu s L}.$$

Under these circumstances, at depths of just 50 μm, approximately 25% of the incident beam is typically scattered (and lost) during excitation and a similar percentage is redirected during fluorescence emission. This rises to 40% at 100 μm and 72% at 250 μm making such depths practically inaccessible.

Additionally, tissue is not just scattering but also absorptive of both incident light and emitted fluorescence, an effect that is linearly related to the tissue thickness (depth) subtended during excitation. Absorption is also highly dependent on the wavelength of incident light (or fluorescence) and is somewhat tissue specific. Typically, visible light is more likely to be absorbed. In fact, it has long been recognized that there exists an "optical window" (i.e., 600–700 nm) where major cell and tissue absorbers, such as melanin and hemoglobin, exhibit the least amount of absorbance (König 2000). This optical window has been exploited by many *in vivo* microscopic techniques (Frangioni 2003).

Scanning of the sample (vs full field illumination), coupled with collection at a PMT, permits spatial assignment of all of the measured emitted light intensity. This occurs regardless of the scattered path this light takes en route to the PMT and thus effectively eliminates a large portion of emission scatter. However, the most critical feature in deep tissue is absorption and this is where many confocal approaches often fail. In practice, imaged depths of up to 100 μm have been reported with spinning disk (Egeblad et al. 2008) or even scanning-based confocal microscopy (Stoll et al. 2002). The former, when operated up to these depths, can sometimes be significantly better than multiphoton imaging as a result of the larger quantum efficiencies of current-generation cascade-based CCD cameras (as compared to PMTs). There can also be some overall benefit in resolution at shallow depths due to the combined effects of lower wavelengths used in single-photon and confocal pinhole on the overall point-spread function. However, starting around 50 μm in many tissues, the scatter into adjacent pixels (blurring) combined with the loss of incident excitation light due to single-photon absorption becomes unacceptably high.

1.3 Advantages of Multiphoton Imaging

While many forms of optical imaging can be used to study the immune system, multiphoton imaging has many clear advantages. These advantages include superior tissue penetration, less image degradation by light scattering leading to better resolution, and high image quality deep inside tissues. Furthermore, tissue autofluorescence elicited with infra-red (IR) excitation is significantly reduced over equivalent visible

Table 1 Current multiphoton instrument options

	Source	Scanning mechanism		Detector	Colors	Sensitivity	Imaging rate (fps)	Spatial resolution (max. lines per image)	Pulse conditioning	Ease of implementing	Cost
		X and Y	Z (range)								
Custom	Parker–Sanderson Krummel–Bullen	Resonant + galvo	Pifoc (350 μm)	PMT	2 4	High	Up to 60	400	No Yes	Difficult Moderate	$$$
Custom	Svoboda	Dual galvo	???	PMT	2	High	2	???	No	Difficult	$$$
Custom	Kleinfield	Dual galvo	Focus drive automation	PMT	2	High	3	512	No	Difficult	$$$
Mixed[1]	Yuste[2]	Dual galvo	Focus drive automation	PMT	1 + 1[3]	High	2	512	No	Moderate	$$$$
Mixed[1]	Ridsdale[4]	Dual galvo	Focus drive automation	PMT	3	High	2	512	No	Moderate	$$$$
Commercial	Zeiss LSM 710 NLO	Dual galvo	Piezo (??μm)	PMT	2 or 3	High	5	Up to 6,144 per dimension	Nonstandard	Easy	$$$$$
Commercial	Olympus FV1000 MPE	Dual galvo	Focus drive automation	PMT	2	High	4	64–4,096 per dimension	Yes	Easy	$$$$$
Commercial	Lecia TCS SP5 MP	Tandem systems[5]	Focus drive automation	PMT	4	High	1[5] or 25[7]	8,196[8] or 512[9]	Nonstandard	Easy	$$$$$
Commercial	Prairie Ultima	Tandem systems[10]	Focus drive automation	PMT	2 or 3	High	1[11] or 30[12]	512	Nonstandard	Easy	$$$$
Commercial	LaVision TriM Scope	Multibeam	Piezo[13]	CCD camera	1[14]	Low	$30[15]	1,000[16]	Yes	Easy	$$$$$

[1] Adaptation of existing commercial system
[2] Modification of Olympus Fluoview microscope
[3] One detector for fluorescence and one for SHG signal
[4] Modification of Nikon C1 confocal scope
[5] Tandem scanning system includes dual galvos scanner and resonant scanner operating in bidirectional scan mode
[6] Using dual galvo scanner
[7] Using fast resonant scanner
[8] Using dual galvo scanner
[9] Using fast resonant scanner
[10] Tandem scanning system includes dual galvos scanner and acoustic optic scanner.
[11] Using dual galvo scanner
[12] Using acousto optic scanner
[13] Signal limited to first 150 μm
[14] More possible with extra camera and/or emission filter wheel
[15,16] Depends on CCD camera

wavelengths, which improves signal specificity and image brightness against background. When compared to confocal systems, multiphoton systems typically exhibit better optical efficiency with no signal loss arising from an emission pinhole. So, while absolute signal magnitude is typically less when compared to confocal applications, the overall signal-to-noise ratio is typically improved. Multiphoton excitation also produces highly resolved z excitation planes that leads to extremely good z-registration and consequently better three dimensional reconstructions. This localized excitation also produces lower levels of photodamage (i.e., toxicity) and thereby allows increased recording episodes. In short, the tissue penetrating power of infrared light makes multiphoton excitation especially suitable for *ex vivo* and in situ imaging.

Multiphoton imaging systems are available in many forms. These options include commercial systems, user adaptations of existing confocal microscopes, and custom systems. Many of these current options and the relative merits of each type are documented in Table 1. Current commercial two-photon imaging systems are commonly extensions of existing confocal microscopes and therefore enjoy the advantages of being commercial grade instruments with the flexibility to perform different kinds of imaging.

In contrast, custom-built systems can be constructed at a considerably reduced cost and offer greater potential for customization. In particular, they offer the ability to quickly add extra features and improved capabilities. There are also several free software packages [e.g., ScanImage (Pologruto et al. 2003) and MPScope (Nguyen et al. 2006)] available that facilitate the construction of these custom-built systems. Existing confocal microscopes can also be retrofitted by users to become two-photon scopes, and there are several reports documenting how this can be done (e.g., Fan et al 1999; Majewska et al. 2000; Nikolenko et al. 2003; Ridsdale et al. 2004).

2 Description of Custom Two-Photon Instrumentation

We have assembled a scanning two-photon system based on a resonant scanner as first used by Tsien and colleagues (Fan et al. 1999). Our design was modeled after second generation resonant scanner designs by Ian Parker and Mike Sanderson (Sanderson and Parker 2003; Callamaras and Parker 1999). Their early designs functioned as confocal systems but more recent incarnations extend the technology (and actually simplify the design) for two-photon excitation (Nguyen et al. 2001). In addition to minor modifications in the overall design of the Parker/Sanderson prototypes, our instruments have been expanded to include four PMTs and also place those PMTs within the infinity space of the objective. Four-channel collection (and higher channel numbers) are now possible using high-speed acquisition cards. Furthermore, as we are essentially biologists first, our system is facilitated by a collection of scan-head parts available from Sutter Instruments. This availability dramatically reduces the need for novices to construct parts.

2.1 A Custom Design Composed of Off-the-Shelf Parts Simplifies System Construction

The important elements in this design are shown schematically in Fig. 1. Generally, this system can be assembled from readily available parts. Each of these parts and their important properties are discussed below. Additional and more detailed information regarding this design is available on our website (http://pathology.ucsf.edu/krummel/2PhotonHome.html). Included on this site is a comprehensive parts list.

2.1.1 Laser

Highly specialized lasers are required for multiphoton excitation. In particular, this type of excitation requires the concentration of photons in space and time. Spatial concentration of photons is normally achieved by focusing a laser beam to a small spot with a high numerical aperture objective lens as in traditional single-photon microscopy. In contrast, the temporal concentration of photons is accomplished by compressing photons from a continuous source into ultra-short (i.e., femtosecond) pulses. Near-infrared pulses of this type are commonly produced by a Ti-sapphire oscillator driven by a continuous-wave pump laser and typically exhibit high peak intensities but low average power.

Early Ti-sapphire lasers were notable for their limited wavelength range and poor user friendliness. In particular, these early systems were intolerant to variations

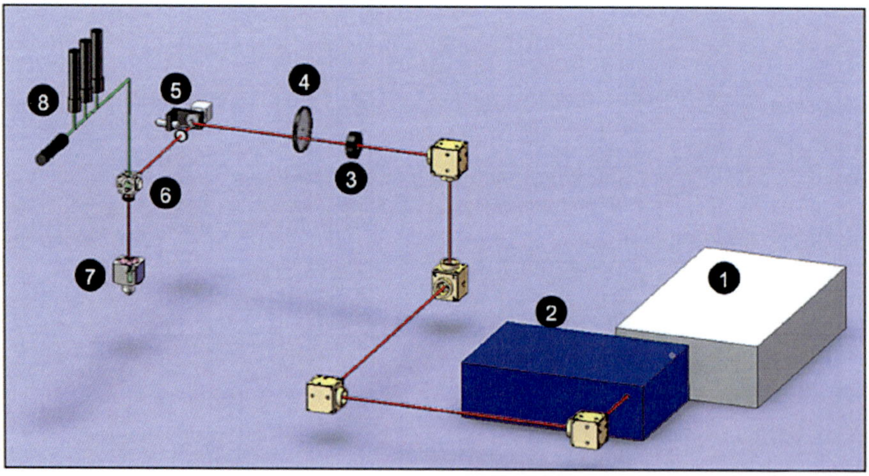

Fig. 1 Instrument design scheme including: (1) Ti–sapphire laser; (2) pulse conditioner; (3) mechanical shutter; (4) neutral density filter wheel; (5) scanning mirrors; (6) primary dichroic mirror; (7) z-focus drive including objective; and (8) PMTs

in room temperature and required constant adjustment. Now, however, there are modern one-box solutions that encompass both the pump laser and the regenerative amplifier which perform more robustly. These systems have simplified laser control and operation to the point where everyday laser operation no longer requires a specialized laser technician.

Likewise, extended range Ti:Sapphire lasers are now available that give access, and sufficient power, to a wider spectral range (700–1,040 nm). This is important for those interested imaging with red fluorescent protein (RFP) and other indicators that require longer excitation wavelengths. This improved spectral bandwidth also enables the use of more fluorophores within the same experiments.

Our multiphoton imaging systems all employ some version of the Newport/ Spectraphysics Maitai product, but equivalent lasers from other suppliers are available that perform similarly. This laser has a pulse length of approximately of ~100 fs and is clocked at 80 MHz. Lasers are available that possess shorter pulse lengths and in theory these may improve multiphoton excitation provided that these pulses can be propagated all the way to the sample. These Ti–sapphire systems also come in a range of sizes. Typically, 6 W systems produce sufficient power to support most kinds of imaging. Because of the nature of this laser, and the downstream imaging hardware, it is firmly mounted to a large vibration isolation table.

2.1.2 Pulse Conditioning Unit

Immediately downstream of the laser is a pulse-conditioning unit that prechirps the femtosecond pulses in a way that maximizes the efficiency of multiphoton excitation. The role of this device and its utility is described in a later section. We have employed two different version of this technology: firstly, a freestanding device from APE (Berlin, Germany) called FemtoControl, and secondly, on a separate system, we have installed an add-on called DeepSee, from Newport/Spectraphysics, to an existing Ti–sapphire laser. Both systems are functionally equivalent.

2.1.3 Translation Optics

The beam steering optics used in this design translate the laser beam between devices laid out on a large vibration isolation table. Typically we have chosen to use optical components that are optimized for ultra-fast laser applications (mostly from Newport Corporation). These broadband optics are designed to operate over the spectral range, and at the power levels, of the Ti–sapphire laser while exhibiting maximizing reflectivity and minimizing pulse dispersion.

2.1.4 Scanning Mirrors

The vast majority of laser scanning microscopes use galvanometer mirrors. They have excellent optical properties and allow zooming and image rotation. Their major drawback is their relatively slow speed (>1 ms per line). Resonant mirrors are an alternative often used for high frame rate imaging. We have chosen to use a dual configuration including a fast (8 kHz) resonant scanner for fast line scans (CRS) and a slower closed-loop galvanometric scanner for vertical scanning (M3S). Bi-directional horizontal scanning enables line scans of approximately ~60 μs in duration which corresponds to video rate imaging with 480 lines per image. We obtained both these scanning mirrors from General Scanning Inc. (GSI; Billerica, MA). Each scanner is under the control of custom electronics also supplied by GSI. These electronics, power supplies and related hardware required to synchronize this system with the video acquisition board are contained within a mirror control box (Sutter Instruments, Novato, CA). These scanners are mounted on an optical breadboard enclosed within a scanning enclosure. This enclosure was also manufactured by Sutter Instruments and is mounted adjacent to Olympus BX51 microscope base.

One problem with these resonant mirrors is that they introduce some image distortion that requires digital correction. This distortion arises because the mirror velocity is not linear but rather has a sinusoidal profile. The pixel manipulation and field selection required to overcome this image distortion is conducted automatically behind the scenes and is transparent to the user. Details of the mathematical procedure underlying this correction have been described extensively elsewhere (Sanderson 2004; Leybaert et al. 2005).

2.1.5 Objectives, Field Size and Stage Movement

The objective lens(es) used with this system determines the level of spatial resolution and the efficiency of signal capture. Any lens considered must also be able to work in media or physiological fluids and possess sufficient working distance for use with tissues and animals. Moreover, it must exhibit high transmittance for both the pulsed IR excitation light and a wide range of emitted fluorescence. We predominantly use an XLUMP FL20XW from Olympus. This objective combines intermediate magnification (20×) with relatively high NA (0.95) and is particularly well suited for imaging in scattering tissue.

This system also possesses electronic control of the field size. The user is able to choose between two pixel sizes (i.e., 0.4 or 0.7 μm pixel^{-1}). Correspondingly, the field size scanned is either 192 × 160 μm or 336 × 280 μm.

While most of the experimental work conducted on this system is focused on a microscopic level of detail, there are instances where users want to combine both microscopic and macroscopic levels of resolution. This is achieved by stage scanning with a motorized stage (Prior 101A, Boston, MA). This process is automated and controlled in software (described below).

2.1.6 PMT Selection

Photomultiplier tubes are hand-made devices with a surprising amount of batch-to-batch variability in their signal- and noise-amplification characteristics. While each supplier typically provides average values for sensitivity and noise performance, these can be misleading. In particular, root–mean–square (rms) measurements of noise performance are essentially time-averages that disguise significant peak-to-peak signal fluctuations. In high bandwidth applications, such as video rate scanning, short but significant bursts of noise, which are undetectable in rms measurements, become significant problems. For example, in a system collecting pixels at MHz frequency, we have observed PMTs with transient bursts of noise lasting just a microsecond or two that cause isolated pixels to become completely saturated. Depending on the frequency of these bursts, such PMTs may prove unsuitable for imaging. Some suppliers allow batches of 10–20 PMTs to be individually tested as part of a purchase. In general, there continues to be advances in the development of lower-noise, high gain PMTs. For instance, some latest generation GaAs-based detectors have shown exceptional sensitivity and signal-to-noise performance. However, such PMTs can also become easily saturated and transiently insensitive. This places an additional burden on the user to carefully manage their light levels to avoid such damage.

2.1.7 Emission Split Allows Four-Color Imaging

The four emission channels included in this design provide coverage of most of the visual spectrum normally used for fluorescence imaging. A scheme showing the specifics of these individual channels is documented in Fig. 2. The spectral placement of these channels allows one to collect signal from four independent fluorophores. Moreover, the relative placement of these filters was chosen to (1) match the spectra from currently available fluorescent proteins, (2) provide the best spectral separation between fluorophores, and (3) to maximize the signal capture in each channel. With our current hardware, there is some flexibility to switch emission filters and thereby fine-tune these channels even further, but changing dichroic mirrors is more difficult. In cases where the spectra of two fluorophores overlap in adjacent channels, it is possible, via image math, to separate out the underlying contribution of each indicator. Likewise, linear unmixing can be used to remove the contribution of autofluorescence or overlapping signals. Examples of these procedures are shown in below Fig. 6

2.1.8 Software

Our system employs several different software applications from multiple vendors. These applications are run on two separate computers and in turn connect to several devices or control boxes. A scheme documenting the relationships between these computers, applications and devices is shown in Fig. 3. Coordination of these disparate elements during image acquisition is achieved by a single master application (Confocal; IO Industries, London, Ontario, Canada). The Confocal application controls the scheduling and relative timing of all the hardware devices used in a typical experiment. It also interacts with a commercial video recording software suite (i.e., VideoSavant) that is used to reconstruct and record images.

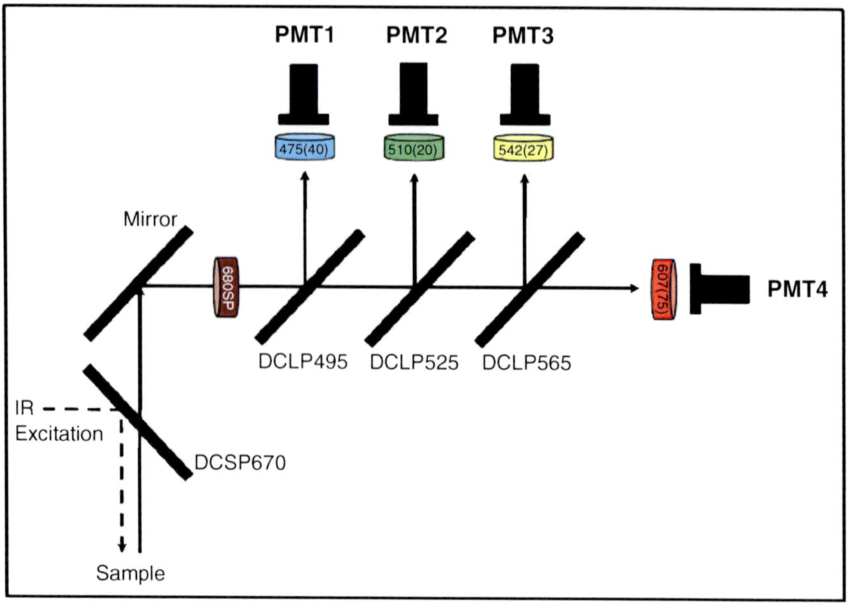

Fig. 2 Emission split. This optical configuration is optimized to capture the maximum fluorescence signal from commonly used fluorescence proteins (especially CFP, GFP, YFP and dsRed or similar). This setup includes a strong excitation blocking filter that excludes any backscattered excitation light without impacting the emitted fluorescence. *DCSP* Dichroic short-pass mirror, *DCLP* dichroic long-pass mirror. Barrier filters described by their center wavelength (and full width at half maximum)

The graphical user interfaces of these programs are shown in Fig. 4. The confocal application is composed of a Main tab and three derivative tabs. Data can be acquired as a single time point, a z-series, a z-series over time, or any of the above using multiple stage positions. Files are exported in multiimage TIFF format that is accessible to Image J, Metamorph, Imaris, and the simpler Windows image viewers.

2.2 *Optimizing Two-Photon Instrumentation*

It is relatively common nowadays to be able to buy, or retrofit, an existing confocal microscope with a laser suitable for multiphoton excitation. However, such instruments provide relatively poor imaging capabilities and many are unable to achieve the necessary frames rates described earlier. Furthermore, the relative efficiency of IR delivery and fluorescence capture in these systems is often suboptimal. Based on these factors and others we have learned from experience, the following section describes several critical considerations in achieving high quality multiphoton images and physiologically relevant results.

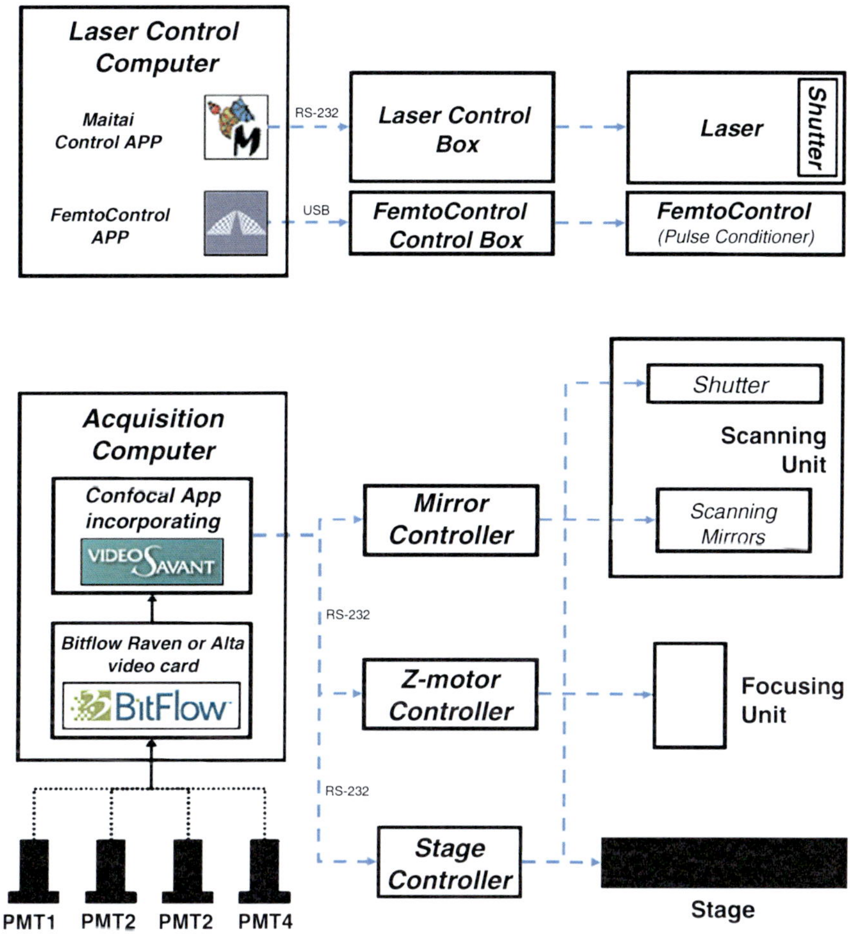

Fig. 3 Software control scheme showing the relationship between the acquisition and laser control computers, device control boxes and each physical device in the larger system. *Dotted blue lines* represent outbound control signals. *Black lines* represent acquired or processed signals

2.2.1 A Simplified Light Path Minimizes the Effect of Dispersion

The strength of multiphoton excitation is inversely proportional to pulse duration. Although very short pulses (<100 fs) give the strongest signal, they are no longer purely monochromatic and, as they propagate through different optical elements (i.e., optical fibers and objective lens), some dispersion, or pulse broadening, occurs. This chromatic dispersion arises because separate spectral components are retarded differentially depending on their wavelength. Pulse dispersion reduces the overall efficiency of multiphoton excitation in general but is most

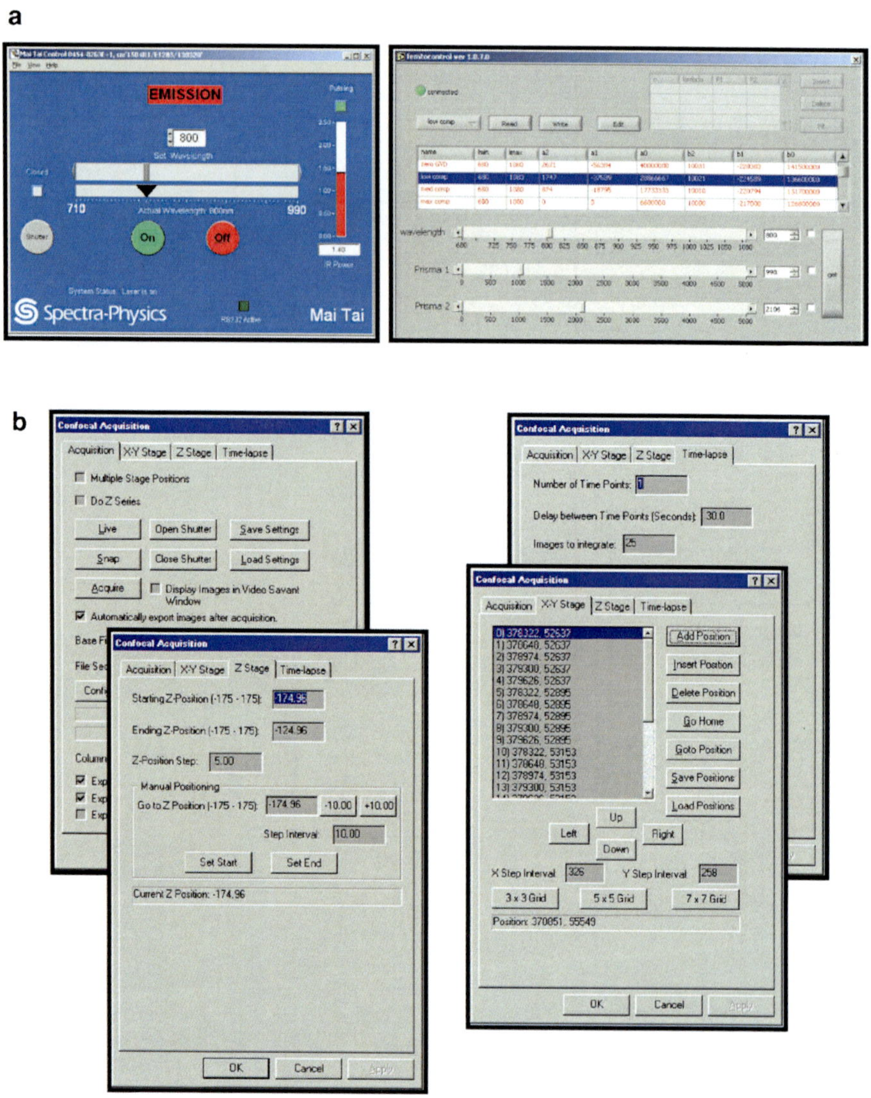

Fig. 4 Software user interface. (**a**) Applications running on the laser control computer. (**b**) Different control tabs of the Confocal application running on the acquisition computer

problematic in applications that use complicated optical configurations such as intravital microscopy (IVM). In theory, pulses of greater starting intensity could be used to overcome dispersion, but these can cause photodamage. Similarly, pulse power can be amplified at the expense of repetition rate (Theer et al. 2003) but again, only under conditions in which photodamage or background fluorescence are not limiting.

It should be noted that simpler optical systems limit dispersion because they have fewer optical elements (lenses, mirrors, etc.) which are the primary site where dispersion arises. Since a dedicated two-photon instrument can be easily assembled with only a few optical elements (the required light path consists of approximately 3 cm of glass), dispersion can be kept low and excitation maximal. Thus, our strategy to overcome the effects of pulse dispersion is to minimize amount of glass in the system, but obviously some glass is always required. In contrast, many recent commercial systems have suffered from the use of highly dispersive elements, notably large numbers of relay lenses as well as dispersive elements such as acousto–optic deflectors (AODs). The resultant loss in peak power made many of these extremely inefficient at producing the high peak powers that is necessary for producing the two-photon effect.

2.2.2 Pulse Conditioning can be Used to Overcome the Effect of Dispersion

New methods of pulse shaping are now available that completely counteract pulse dispersion and thereby improve the efficiency of multiphoton imaging. Pulse-shaping methods (also called prechirping) impart a 'negative dispersion' (also known as negative group velocity dispersion or GVD) to longer wavelength components in each pulse and thereby allow pulses to travel further through a scattering tissue before they become too weak to be effective (Fig. 5a). This manipulation improves the effective depth penetration by up to three times (McConnell 2006) relative to unconditioned pulses without causing any additional photodamage. In general, pulse compression produces better images by optimizing multiphoton excitation. Some tissues may also add additional dispersion effects such as bone when imaging through the skull or various connective tissues capsules around other organs. Unexpectedly, there are even reports that relatively simple biological fluids can also add small amounts of pulse dispersion (Coello et al. 2007). Additional dispersion compensation can be used to overcome these tissue and fluids effects. Commercial devices are now available that precondition pulses for dispersion compensation and many of these are now being built into one-box laser solutions (e.g., Maitai DeepSee).

2.2.3 Resonant Scanning is Required to Achieve High Frame Rates

A majority of commercial multiphoton instruments typically scan the sample on-demand. For most commercial systems, the slow rate of scanning mirror movement, and other factors, conspire to limit the image acquisition to approximately 1–4 s per image. In contrast, a resonant scanner is always oscillating at a higher rate and much higher frame rates are easily achieved. If this higher frame is not required then multiple frames can be averaged to increase the signal-to-noise ratio. The version that we utilize typically produces an image of 400 × 480 pixels 30 times per second. When these scanners are coupled with fast piezoelectric z-drives (with settling times of approximately 10 ms) the result is effectively 3–6 z-sections per second,

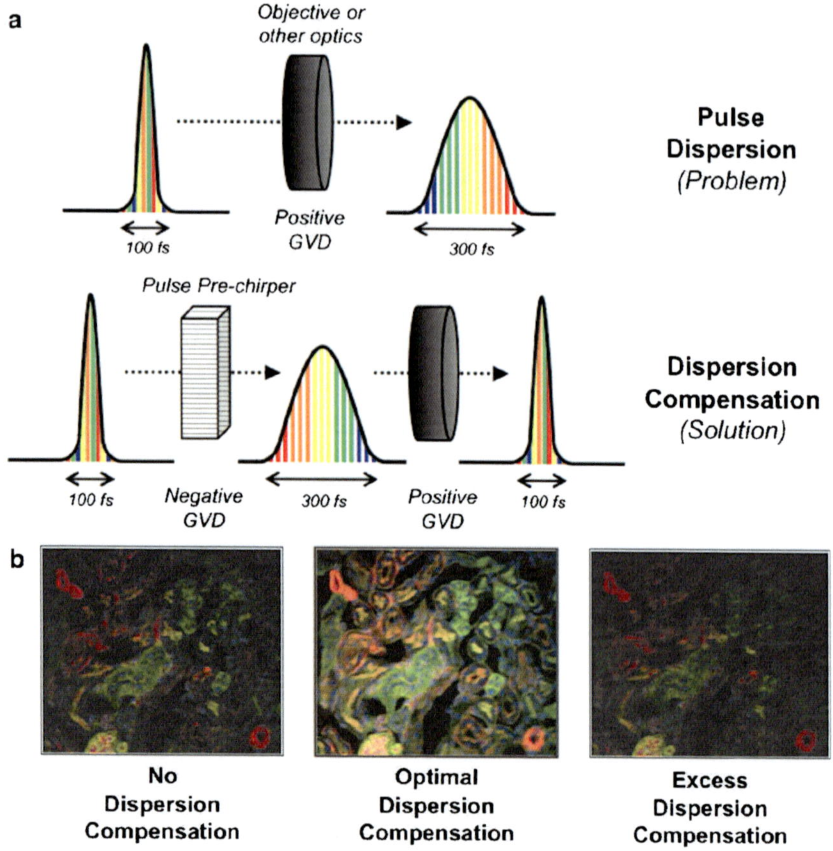

Fig. 5 Pulse compression for improved image brightness. (**a**) Illustrates the problem of pulse dispersion whereby system optics smear and attenuate laser pulses. Pulse chirping preconditions these pulses so that when they reach the sample they are in an optimal form for two-photon excitation. (**b**) Representatives with and without dispersion compensation

which is a dramatic improvement over conventional scanning modes. One downside of this technology is that the scanning pattern is relatively fixed. There is little scope for zooming or scanning smaller regions of interest. This rules out FRAP-type bleaching or faster scanning of smaller regions. However, in practice, this limitation is not significant for many current *in vivo* applications where FRAP has not yet seen great utility and where large fields are desirable.

2.2.4 Frame Averaging can Improve Image Quality

Video-rate image capture sometimes suffers from diminished image quality, especially when the underlying signal is weak. This occurs because pixel dwell times are typically quite short in comparison to nonresonant scanning technologies. Frame averaging can be used to overcome this problem. The extent of signal improvement is governed by sampling theory and therefore significant image improvements can be achieved by averaging 5–30 frames. Further averaging typically produces only small improvements in image quality and this occurs at the expense of increase exposure times. In practice, most applications require some amount of frame-averaging (typically 5–20 frames averaged) resulting in effective frame rates of 3–12 frames per second.

2.2.5 Maintaining Tissue Viability and Animal Health

We (Friedman et al. 2005) and others (Miller et al. 2002) have shown that various aspects of immune physiology are exquisitely sensitive to variations in temperature and oxygen concentration. For this reason, it is important to carefully monitor and control the temperature and oxygenation state of any tissue being examined with two-photon microscopy. Numerous commercial systems are readily available for this purpose. Likewise, if mice or other whole animals are being studied, it is critical to control their temperature, respiration, and state of anesthesia.

3 Representative Data

Our four-channel TPLSM system has recently allowed us (Gardner et al. 2008) to image three unique cell populations and the lymph node capsule, as well as to isolate our fluorescent signals from the autofluorescence present in the tissue (See Fig. 6). In this example, the collagen capsule is visualized by using collagen's second harmonic generation, leaving three channels available to detect our labeled cell populations.

In the examples shown in Fig. 6, transgenic Adig mice express IGRP–GFP under the Aire promoter in mTECs (medullary thymic epithelial cells) and in a population of cells in secondary lymphoid organs termed eTACs (extra-thymic Aire-expressing cells). GFP expression levels in eTACs are low, especially when compared to the high levels of green autofluorescence in the region of the lymph node where these cells are present. While the GFP fluorescence is primarily present in the green channel, the tissue autofluorescence is more broadly detectable. This allowed us to use image math to eliminate the autofluorescent signal from the green channel that obscured the GFP signal (Fig. 6a).

The goal of these studies was to analyze the interactions between T cells and eTACs (Gardner et al.2008). Since T cell behavior varies depending on microenvironment, often the best control is to simultaneously image control cell populations in the same microenvironment as experimental cell populations. To image control and experimental T cell populations together with the GFP labeled eTACs and the second harmonic, we labeled one population of T cells with CMTMR and the second with both CMTMR and CFSE. By separating out the dually labeled cells based

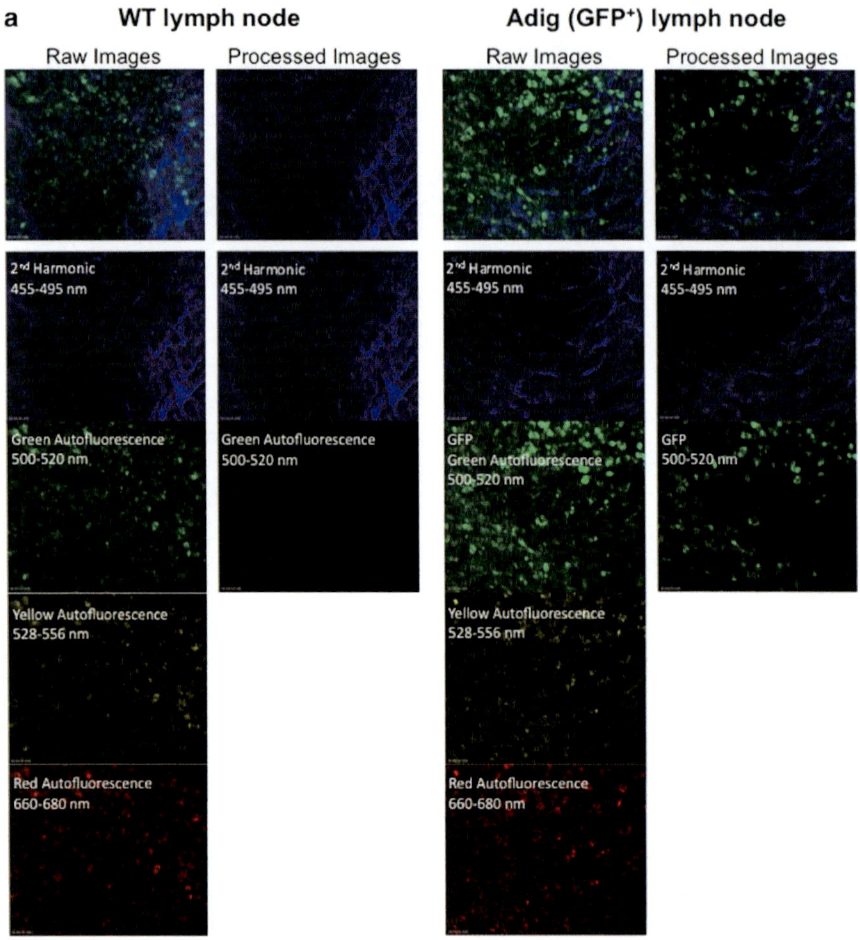

Fig. 6 Representative data. (**a**) Using autofluorescent signals in unused channels to eliminate autofluorescence in the channel of interest. Images show maximum intensity *Z* projections of florescence levels in the blue and green channels overlayed (*top*), and in individual channels (*bottom*) before and after image processing. The green (500–520 nm) channel contained a large amount of autofluorescent signal that obscured the dim GFP signal. The autofluorescence was also present in the yellow (528–556 nm) and red (660–680 nm) channels, while the GFP signal was primarily present in the green channel. Using a sample without GFP fluorescence (*left*) we established an equation to subtract out the autofluorescent signal where "*a*" was determined based on the overlap between autofluorescence in the green and yellow channels and "*b*" was determined based on the overlap between the green and red channels.

$$\text{GFP Fluorescence} = \text{Green Fluorescence} - a \times \text{Yellow Fluorescence} - b \times \text{Red Fluorescence}$$

When applied to a sample without GFP fluorescence this equation eliminated the green autofluorescence (*left*), and when applied to a sample containing GFP labeled cells (*right*), the equation eliminated autofluorescence without eliminating GFP fluorescence, making the GFP signal clearly identifiable. Following image arithmetic to eliminate autofluorescence, the images were processed with a Gaussian filter to reduce noise. Image processing was done using Imaris and Matlab.

on the colocalization of emission in the red and green channels, and subsequently applying image arithmetic, we were able to isolate the signals from each of the cell populations and the second harmonic into four unique channels (Fig. 6b).

4 Future Perspectives

4.1 Pulse Manipulations

In addition to the femtosecond pulse manipulations described earlier, there have been a number recent advances that further enhance signal intensity and/or reduce phototoxicity. Some these advances are described below. Adoption of these are likely to further expand the utility of TPLSM in the future.

Fig. 6 (continued) (b) Using image arithmetic to separate populations with different fluorescence profiles. As described in Fig. 6a, the dim GFP signal is obscured by strong autofluorescence in the green channel. Additionally, two populations of T cells were added to the sample. One population was labeled with the dye CMTMR which is detected in the red channel, and the other population was labeled with both CMTMR which is detected in the red channel and CFSE which is detected in the green and yellow channels. Our goal was to establish clearly separated channels for each of the three cell populations and to eliminate autofluorescence. To do this, we first established a new channel that contained the regions of bright colocalized green and red fluorescence, where "c" and "d" were determined based on the fluorescent levels of the CMTMR/CFSE labeled cells in the red and green channels respectively.

Colocalized CMTMR/CFSE Fluorescence = Pixels with Red > c and Green > d

This channel specifically contained the T cell population that was labeled with both CMTMR and CFSE dyes. The equation previously established to eliminate autofluorescence in the green channel was then applied.

GFP Fluorescence = Green Fluorescence − a × Yellow Fluorescence − b × Red Fluorescence

In this case, the equation served to both eliminate green autofluorescence and to eliminate fluorescence in the green channel from the CMTMR/CFSE labeled cell population, leaving only the GFP signal in the green channel. We then applied an equation to eliminate autofluorescence in the red channel, as well as to eliminate fluorescence in the red channel from the CMTMR/CFSE labeled cell population where "e" was determined based on the overlap between autofluorescence and CMTMR/CFSE fluorescence in the red and yellow channels and "f" was determined based on the overlap between the red and green channels.

CMTMR Fluorescence = Red Fluorescence − e × Yellow Fluorescence − f × Green Fluorescence

When applied to the sample this equation eliminated autofluorescence and fluorescence from the CMTMR/CFSE labeled cell population without eliminating fluorescence from the cell population labeled with only CMTMR. Following image arithmetic to separate the cell populations and eliminate autofluorescence, the images were processed with a Gaussian filter to reduce noise. Image processing was done using Imaris and Matlab

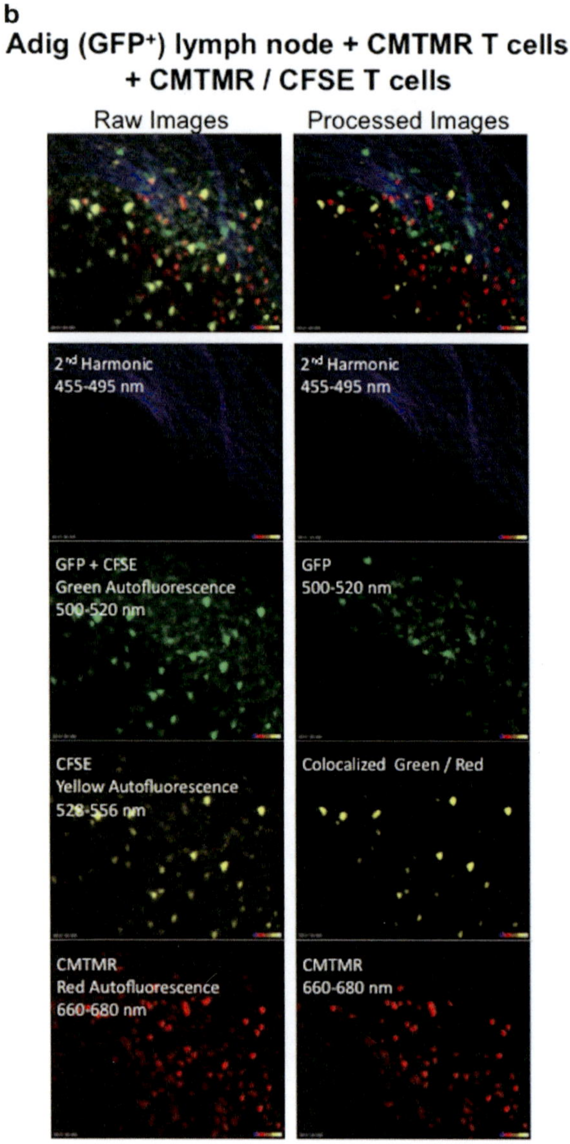

Fig. 6 (continued)

4.1.1 Increasing the Efficiency of Multiphoton Excitation by Manipulating the Spacing Between Pulses

Donnert and colleagues (Donnert et al. 2007) have recently reported that a substantial improvement in multiphoton efficiency (that is, stronger fluorescence and less photobleaching) could be achieved by increasing the time between successive pulses enough to allow complete dark-state relaxation. Normally, the time between successive pulses in two-photon microscopy is 10–25 ns. By lengthening this time to >1 μs these authors were able to show a substantial augmentation (that is, 20–25×) in fluorescence intensity and significantly less photobleaching. Although this lengthened interpulse interval will probably translate into fewer pulses per pixel dwell time and/or slower frame rates, the magnitude of this signal augmentation is highly significant and will certainly improve image quality in many applications.

4.1.2 Passive Pulse Splitting can also be Used to Enhance Signal Intensity and Reduce Phototoxicity

Magee and Betzig have recently documented a passive pulse splitting strategy whereby each femtosecond pulse is split into many mini-pulses (Ji et al. 2008). These streams of mini-pulses prove to be substantially more efficient for multiphoton excitation (up to 100-fold). Concurrently, these authors described a corresponding 6- to 20-fold decrease in photodamage.

4.1.3 Higher Order, or Pulse Phase, Manipulations can also Help Improve the Effectiveness and Repeatability of Ultra-Short Pulses

These coherence, or phase adjustments, minimize constructive/destructive interference normally occurring within each pulse and thereby promote more homogeneous excitation. Apparently, these phase corrections can increase signal intensity and improve contrast compared with uncompensated pulses. Conveniently, these coherence manipulations are maintained following passage through scattering tissues (Dela Cruz et al. 2004) and there is evidence to suggest that photodamage is also lessened. Interestingly, this method can also be used for selective excitation (Schelhas et al. 2006) of closely spaced fluorophores or ratiometric imaging of different excitation bands.

4.2 Alternative Scanning Methods

Adaptive illumination is a relatively new technique that varies laser intensity with depth and thereby counteracts the effect of decreasing excitation strength with increasing depth. It has been used in other types of imaging, especially confocal microscopy. Under this scheme, multiphoton excitation is enhanced at the points beyond the normal depth penetration (Chu et al. 2007). Because multiphoton excitation only occurs in a small focal volume the extra energy does not cause any deleterious effects. The profile of increased laser intensity used by this technique can typically be determined empirically, although most often a simple linear ramp is sufficient to produce consistent signal quality improvements at greater depths than usually possible.

Another particularly exciting development in ultra-short pulse-shaping technologies is that some variations may enable axial focusing without having to move the terminal focusing element (that is, without moving an objective lens) (Zhu et al. 2005). In fact, it has been shown that under specific optical conditions insertion of a variable GVD can be used to adjust z-focus. Such a remote axial focusing capability would be a major improvement to current methods in three-dimensional IVM. It may be possible to use this method in combination with lower numerical aperture lenses, which possess correspondingly longer working distances, to reach previously inaccessible tissue depths. Other configurations (Helmchen and Denk 2005) that reduce the effective numerical aperture of an objective (that is, by incompletely illuminating its back focal plane) are also candidates to combine with this temporal focusing effect.

Finally, recent reports (Duemani Reddy et al. 2008; Reddy and Saggau 2005; Saggau 2006) have highlighted the possibility that advanced three-dimensional acousto–optic scanning methods may be applied in multiphoton microscopy. The utility of these methods is that they supposedly allow very fast beam positioning in three-dimensional space without any mechanical movement of scanning mirrors or objectives lenses.

References

Bajenoff M, Egen JG, Koo LY, Laugier JP, Brau F, Glaichenhaus N, Germain RN (2006) Stromal cell networks regulate lymphocyte entry, migration, and territoriality in lymph nodes. Immunity 25:989–1001

Beuneu H, Garcia Z, Bousso P (2006) Cutting edge: cognate CD4 help promotes recruitment of antigen-specific CD8 T cells around dendritic cells. J Immunol 177:1406–1410

Bousso P, Robey E (2003) Dynamics of CD8(+) T cell priming by dendritic cells in intact lymph nodes. Nat Immunol 4:579–585

Bunnell SC, Kapoor V, Trible RP, Zhang W, Samelson LE (2001) Dynamic actin polymerization drives T cell receptor-induced spreading: a role for the signal transduction adaptor LAT. Immunity 14:315–329

Cahalan M, Parker I (2008) Choreography of cell motility and interaction dynamics imaged by two-photon microscopy in lymphoid organs. Annu Rev Immunol 26:585–626

Cahalan MD, Parker I, Wei SH, Miller MJ (2002) Two-photon tissue imaging: seeing the immune system in a fresh light. Nat Rev Immunol 2:872–880

Callamaras N, Parker I (1999) Construction of a confocal microscope for real-time x-y and x-z imaging. Cell Calc 26:271–279

Castellino F, Huang AY, Altan-Bonnet G, Stoll S, Scheinecker C, Germain RN (2006) Chemokines enhance immunity by guiding naive CD8 + T cells to sites of CD4 + T cell-dendritic cell interaction. Nature 440:890–895
Chu K, Lim D, Mertz J (2007) Enhanced weak-signal sensitivity in two-photon microscopy by adaptive illumination. Opt Lett 32:2846–2848
Codling E, Hill N (2005) Sampling rate effects on measurements of correlated and biased random walks. J Theor Biol 233:573–588
Codling E, Plank M, Benhamou S (2008) Random walk models in biology. J R Soc Interf 5:813–834
Coello Y, Xu B, Miller T, Lozovoy V, Dantus M (2007) Group-velocity dispersion measurements of water, seawater, and ocular components using multiphoton intrapulse interference phase scan. Appl Opt 46:8394–8401
Collier T, Arifler D, Malpica A, Follen M, Richards-Kortum R (2003) Determination of Epithelial Tissue Scattering Coefficient Using Confocal Microscopy. IEEE J Sel Top Quant Electr 9:307–313
Dela Cruz J, Pastirk I, Comstock M, Lozovoy V, Dantus M (2004) Use of coherent control methods through scattering biological tissue to achieve functional imaging. Proc Natl Acad Sci USA 101:16996–17001
Delon J, Bercovici N, Liblau R, Trautmann A (1998) Imaging antigen recognition by naive CD4 + T cells: compulsory cytoskeletal alterations for the triggering of an intracellular calcium response. Eur J Immunol 28:716–729
Delon J, Bercovici N, Raposo G, Liblau R, Trautmann A (1998) Antigen-dependent and -independent Ca2 + responses triggered in T cells by dendritic cells compared with B cells. J Exp Med 188:1473–1484
Delon J, Gregoire C, Malissen B, Darche S, Lemaltre F, Kourilsky P, Abastado JP, Trautmann A (1998) CD8 expression allows T cell signaling by monomeric peptide-MHC complexes. Immunity 9:467–473
Donnert G, Eggeling C, Hell S (2007) Major signal increase in fluorescence microscopy through dark-state relaxation. Nat Methods 4:81–86
Duemani Reddy G, Kelleher K, Fink R, Saggau P (2008) Three-dimensional random access multiphoton microscopy for functional imaging of neuronal activity. Nat Neurosci 11:713–720
Dustin ML, Bromley SK, Kan Z, Peterson DA, Unanue ER (1997) Antigen receptor engagement delivers a stop signal to migrating T lymphocytes. Proc Natl Acad Sci USA 94:3909–3913
Egeblad M, Ewald A, Askautrud H, Truitt M, Welm B, Bainbridge E, Peeters G, Krummel M, Werb Z (2008) Visualizing stromal cell dynamics in different tumor microenvironments by spinning disk confocal microscopy. Dis Models Mech 1:155–167
Fan G, Fujisaki H, Miyawaki A, Tsay R, Tsien R, Ellisman M (1999) Video-rate scanning two-photon excitation fluorescence microscopy and ratio imaging with cameleons. Biophys J 76:2412–2420
Frangioni J (2003) In vivo near-infrared fluorescence imaging. Curr Opin Chem Biol 7:626–634
Friedman RS, Jacobelli J, Krummel MF (2005) Mechanisms of T cell motility and arrest: deciphering the relationship between intra- and extracellular determinants. Semin Immunol 17:387–399
Gardner J, Devoss J, Friedman R, Wong D, Tan Y, Zhou X, Johannes K, Su M, Chang H, Krummel M, et al (2008) Deletional tolerance mediated by extrathymic Aire-expressing cells. Science 321:843–847
Helmchen F, Denk W (2005) Deep tissue two-photon microscopy. Nat Methods 2:932–940
Hugues S, Fetler L, Bonifaz L, Helft J, Amblard F, Amigorena S (2004) Distinct T cell dynamics in lymph nodes during the induction of tolerance and immunity. Nat Immunol 5:1235–1242
Hugues S, Scholer A, Boissonnas A, Nussbaum A, Combadiere C, Amigorena S, Fetler L (2007) Dynamic imaging of chemokine-dependent CD8 + T cell help for CD8 + T cell responses. Nat Immunol 8:921–930
Ji N, Magee J, Betzig E (2008) High-speed, low-photodamage nonlinear imaging using passive pulse splitters. Nat Methods 5:197–202
König K (2000) Multiphoton microscopy in life sciences. J Microsc 200:83–104
Krummel MF, Sjaastad MD, Wülfing C, Davis MM (2000) Differential Assembly of CD3z and CD4 During T cell Activation. Science 289:1349–1352

Leybaert L, de Meyer A, Mabilde C, Sanderson M (2005) A simple and practical method to acquire geometrically correct images with resonant scanning-based line scanning in a custom-built video-rate laser scanning microscope. J Microsc 219:133–140

Lindquist RL, Shakhar G, Dudziak D, Wardemann H, Eisenreich T, Dustin ML, Nussenzweig MC (2004) Visualizing dendritic cell networks in vivo. Nat Immunol 5:1243–1250

Livet J, Weissman T, Kang H, Draft R, Lu J, Bennis R, Sanes J, Lichtman J (2007) Transgenic strategies for combinatorial expression of fluorescent proteins in the nervous system. Nature 450:56–62

Majewska A, Yiu G, Yuste R (2000) A custom-made two-photon microscope and deconvolution system. Pflugers Arch 441:398–408

McConnell G (2006) Improving the penetration depth in multiphoton excitation laser scanning microscopy. J Biomed Opt 11:054020

Mempel TR, Henrickson SE, von Andrian UH (2004) T-cell priming by dendritic cells in lymph nodes occurs in three distinct phases. Nature 427:154–159

Miller MJ, Safrina O, Parker I, Cahalan MD (2004) Imaging the single cell dynamics of CD4 + T cell activation by dendritic cells in lymph nodes. J Exp Med 200:847–856

Miller MJ, Wei SH, Parker I, Cahalan MD (2002) Two-photon imaging of lymphocyte motility and antigen response in intact lymph node. Science 296:1869–1873

Negulescu PA, Krasieva TB, Khan A, Kerschbaum HH, Cahalan MD (1996) Polarity of T cell shape, motility, and sensitivity to antigen. Immunity 4:421–430

Nguyen Q, Tsai P, Kleinfeld D (2006) MPScope: a versatile software suite for multiphoton microscopy. J Neurosci Methods 156:351–359

Nguyen QT, Callamaras N, Hsieh C, Parker I (2001) Construction of a two-photon microscope for video-rate Ca(2+) imaging. Cell Calc 30:383–393

Nikolenko V, Nemet B, Yuste R (2003) A two-photon and second-harmonic microscope. Methods 30:3–15

Okada T, Miller MJ, Parker I, Krummel MF, Neighbors M, Hartley SB, O'Garra A, Cahalan MD, Cyster JG (2005) Antigen-engaged B cells undergo chemotaxis toward the T zone and form motile conjugates with helper T cells. PLoS Biol 3:e150

Othmer H, Dunbar S, Alt W (1988) Models of dispersal in biological systems. J Math Biol 26:263–298

Phan TG, Grigorova I, Okada T, Cyster JG (2007) Subcapsular encounter and complement-dependent transport of immune complexes by lymph node B cells. Nat Immunol 8:992–1000

Poenie M, Tsien RY, Schmitt-Verhulst AM (1987) Sequential activation and lethal hit measured by [Ca2+]i in individual cytolytic T cells and targets. EMBO J 6:2223–2232

Pologruto T, Sabatini B, Svoboda K (2003) ScanImage: flexible software for operating laser scanning microscopes. Biomed Eng Online 2:13

Qi H, Egen J, Huang A, Germain R (2006) Extrafollicular activation of lymph node B cells by antigen-bearing dendritic cells. Science 312:1672–1676

Reddy G, Saggau P (2005) Fast three-dimensional laser scanning scheme using acousto-optic deflectors. J Biomed Opt 10:064038

Ridsdale A, Micu I, Stys P (2004) Conversion of the Nikon C1 confocal laser-scanning head for multiphoton excitation on an upright microscope. Appl Opt 43:1669–1675

Sabatos CA, Doh J, Chakravarti S, Friedman RS, Prandurangi PG, Tooley AJ, Krummel MF (2008) A synaptic basis for paracrine interleukin-2 signalling during homotypic T cell interaction. Immunity 29:238–248.

Saggau P (2006) New methods and uses for fast optical scanning. Curr Opin Neurobiol 16:543–550

Sanderson MJ, Parker I (2003) Video-rate confocal microscopy. Methods Enzymol 360:447–481

Sanderson MJ (2004) Acquisition and correction of multiple real-time images for laser-scanning microscopy. Microsc Anal 18:17–23

Schaefer BC, Marrack P, Fanger GR, Kappler JW, Johnson GL, Monks CRF (1999) Live cell fluorescence imaging of T cell MEKK2: Redistribution and activation in response to antigen stimulation of the T cell receptor. Immunity 11:411–421

Schelhas L, Shane J, Dantus M (2006) Advantages of ultrashort phase-shaped pulses for selective two-photon activation and biomedical imaging. Nanomedicine 2:177–181

Shakhar G, Lindquist RL, Skokos D, Dudziak D, Huang JH, Nussenzweig MC, Dustin ML (2005) Stable T cell-dendritic cell interactions precede the development of both tolerance and immunity in vivo. Nat Immunol 6:707–714

Stoll S, Delon J, Brotz TM, Germain RN (2002) Dynamic imaging of T cell-dendritic cell interactions in lymph nodes. Science 296:1873–1876

Tang Q, Adams JY, Tooley AJ, Bi M, Fife BT, Serra P, Santamaria P, Locksley RM, Krummel MF, Bluestone JA (2006) Visualizing regulatory T cell control of autoimmune responses in nonobese diabetic mice. Nat Immunol 7:83–92

Theer P, Hasan M, Denk W (2003) Two-photon imaging to a depth of 1000 microm in living brains by use of a Ti:Al2O3 regenerative amplifier. Opt Lett 28:1022–1024

Varma R, Campi G, Yokosuka T, Saito T, Dustin ML (2006) T cell receptor-proximal signals are sustained in peripheral microclusters and terminated in the central supramolecular activation cluster. Immunity 25:117–127

Williams R, Zipfel W, Webb W (2001) Multiphoton microscopy in biological research. Curr Opin Chem Biol 5:603–608

Wülfing C, Rabinowitz JD, Beeson C, Sjaastad MD, McConnell HM, Davis MM (1997) Kinetics and extent of T cell activation as measured with the calcium signal. J Exp Med 185:1815–1825

Yokosuka T, Sakata-Sogawa K, Kobayashi W, Hiroshima M, Hashimoto-Tane A, Tokunaga M, Dustin ML, Saito T (2005) Newly generated T cell receptor microclusters initiate and sustain T cell activation by recruitment of Zap70 and SLP-76. Nat Immunol 6:117–127

Zhu G, van Howe J, Durst M, Zipfel W, Xu C (2005) Simultaneous spatial and temporal focusing of femtosecond pulses. Opt Exp 13:2153–2159

Visualizing Intermolecular Interactions in T Cells

Nicholas R.J. Gascoigne, Jeanette Ampudia, Jean-Pierre Clamme, Guo Fu, Carina Lotz, Michel Mallaun, Nathalie Niederberger, Ed Palmer, Vasily Rybakin, Pia P. Yachi, and Tomasz Zal

Contents

N.R.J. Gascoigne(✉), J. Ampudia, J.-P. Clamme, G. Fu, and V. Rybakin
Department of Immunology and Microbial Science, The Scripps Research Institute, 10550 North Torrey Pines Rd, La Jolla, CA, 92037, USA
e-mail: Gascoigne@scripps.edu

M. Mallaun and E. Palmer
Department of Biomedicine, University Hospital Basel, Switzerland

C. Lotz
Department of Hematology, University Medical Center Utrecht, Lundlaan 6, 3584EA, Utrecht, The Netherlands

N. Niederberger
Serono Pharmaceutical Research Institute, Geneva, Switzerland

P.P. Yachi
Applied Molecular Evolution,3520, Dunhill Street, San Diego, CA 92121, USA

T. Zal
Department of Immunology, University of Texas MD Anderson Cancer Center, 7455 Fannin, Houston, TX 77030, USA

D. McGavern and M. Dustin (eds.), *Visualizing Immunity,* Current Topics in Microbiology and Immunology 334,
DOI: 10.1007/978-3-540-93864-4_2,

Abstract The use of appropriate fluorescent proteins has allowed the use of FRET microscopy for investigation of intermolecular interactions in living cells. This method has the advantage of both being dynamic and of working at the subcellular level, so that the time and place where proteins interact can be visualized. We have used FRET microscopy to analyze the interactions between the T cell antigen receptor and the coreceptors CD4 and CD8. This chapter reviews data on how these coreceptors are recruited to the immunological synapse, and how they interact when the T cell is stimulated by different ligands.

1 Introduction

Fluorescent proteins based on the *Aequoria victoria* green fluorescent protein (GFP) and improvements in microscopy have made it possible both to locate cells in living tissues and to locate proteins within living cells. Within immunology, this has made great contributions to studies of subcellular localization and translocation during, for example, T cell activation. Use of different fluorescent proteins or complementary fragments of fluorescent proteins has allowed analysis of intermolecular interactions using Förster resonance energy transfer (FRET) or bimolecular fluorescence complementation (BiFC). Our work has used GFP variants primarily to investigate the movements and interactions of T cell receptor (TCR) components and the coreceptors CD4 and CD8.

Early immunofluorescence studies of T cells interacting with antigen-presenting cells (APC) showed that the CD4 or CD8 coreceptors became concentrated at the interface between the cells (Kupfer and Singer 1989). Later improvements in imaging technology made it possible to identify features of this interface in some detail. These revealed that the TCR and signaling components such as protein kinase Cθ became concentrated in a central region of the interface – the central supramolecular activation complex (cSMAC) surrounded by a region where intracellular adhesion molecules were concentrated, the peripheral (p)-SMAC, with other proteins such as the phosphatase CD45 pushed out to a distal (d)-SMAC (Grakoui et al. 1999; Johnson et al. 2000; Monks et al. 1997, 1998). The whole structure has become known as the immunological synapse (IS). However, time-lapse studies using cell–cell interactions or experimental systems where T cells interact with purified proteins associated with a supported lipid bilayer have demonstrated that the immunological synapse as originally defined, described above, is a mature version of a structure that in fact is extremely dynamic (Grakoui et al. 1999; Hailman et al. 2002; Revy et al. 2001; Ritchie et al. 2002; Trautmann and Valitutti 2003; Zal et al. 2002). Indeed, recent data cast doubt on the idea that the immunological synapse is itself a signaling structure. These experiments indicate that signaling occurs in small accumulations of TCRs (microclusters) at the distal area of the synapse, and that these are brought to the cSMAC to be internalized and degraded (Campi et al. 2005; Varma et al. 2006). Other experiments showed that initial phosphorylation of TCR by the kinase Lck occurs at the distal region of the immunological synapse (Lee et al. 2002), although signaling is within the immunological synapse at later time points (Cemerski et al. 2008).

Even relatively early experiments showed that the immunological synapse structure of cSMAC and pSMAC forms dynamically over time, with cell adhesion molecules starting centrally and then switching places with the TCR to end in the mature structure (Grakoui et al. 1999). Experiments with dendritic cells rather than B cells or lipid bilayers as the antigen-presenting surface, revealed that coreceptors are recruited to the interface between T cell and dendritic cell even when antigen is not present (Zal et al. 2002), and that the interaction between T cell and nonstimulatory dendritic cell is sufficient to cause a small Ca^{2+} response in the T cell (Revy et al. 2001). It has become clear that, even in the absence of overt activation of the T cell, some information is transferred, such that interactions with endogenous peptides allow T cells to get survival signals, or thymocytes to get signals for positive selection (Goldrath and Bevan 1999; Stefanova et al. 2002). The original definition of the immunological synapse is rather too restrictive to cover all of the different interfaces between T cells and APCs, or their form at different stages. To quote US President Bill Clinton, *it depends on what IS is*. We prefer a broad definition, that the immunological synapse is the interface between a T cell and APC, whether or not antigen is involved, and whether or not the T cell becomes activated.

2 Molecular Recruitment to the Immunological Synapse

2.1 T Cell Receptors and Coreceptors

Mutagenenesis of the *A. victoria* green fluorescent protein and discovery of fluorescent proteins in other taxa resulted in different colored versions which could report localization of two or more proteins inside cells. A very useful resource identifying different FPs and their uses is available (Shaner et al. 2005). Our experiments have predominantly used the enhanced cyan and yellow derivatives of GFP: CFP and YFP. These were chosen because, having distinguishable but overlapping spectra, they were well suited as a FRET donor and acceptor pair to investigate intermolecular interactions as well as for conventional tracking of receptor movement to the immunological synapse. We were particularly interested in the dynamics and localization of interactions between the TCR and CD4 duo of MHC Class II-binding receptor and coreceptor. We prepared gene constructs encoding CFP fused to the CD3ζ subunit of the TCR–CD3 complex and YFP fused to CD4, which were expressed in a T cell hybridoma with defined TCR that was deficient for CD4 expression (Zal et al. 2002). Thus, the transfected cells expressed only the fluorescent CD4. Although the cells expressed endogenous, nonfluorescent CD3ζ, this was at a relatively low level, such that expression of TCR on the cell surface was weak, as CD3ζ is limiting for TCR expression (Klausner et al. 1990). CD3ζ–CFP expression therefore raised cell surface TCR expression to a level comparable to normal T cells (Zal et al. 2002). We demonstrated that the chimeric proteins were functional and that they allowed the transfected cells to respond to antigen similarly to cells expressing nonfluorescent versions of CD4 and CD3ζ.

Time-lapse movies of fluorescent T cells interacting with APCs showed that recruitment of CD4 to the immunological synapse is almost instantaneous, with concentrations of CD4 discernible within 10 s of contact between the T cell and the APC. This concentration of CD4 built up over the next minute and, over a period of several minutes, the contact between T cell and APC was seen to be very dynamic. Although TCR–CD3 also became concentrated at the synapse, this was to a much smaller degree than for CD4, 1.3- to 3-fold higher than on the rest of the cell surface compared to 4- to 8-fold for CD4. Recruitment of CD4, but not CD3ζ, was found when the T cell interacted with APCs that were not loaded with antigen. We found that this did not occur with all the different APCs, but that it occurred with mature dendritic cells and with a subset of B cell tumors. It did not occur with immature dendritic cells nor with macrophages, certain B cell tumors, or MHC class II-transfected fibroblasts. This nonantigen-specific recruitment of coreceptor was inhibited with antibodies against various cell surface molecules, including anti-MHC class II and CD4, as well as the B7.1, B7.2-blocker CTLA4-Ig, and antibodies against the cell adhesion molecules LFA1 and ICAM1. In contrast, anti-CD45 enhanced the CD4 recruitment. We found that CD4 recruitment to the immunological synapse in the absence of antigenic stimulation was slower than that seen when antigen was present, showing that recognition of antigen increased the rate of CD4 translocation to the immunological synapse (Zal et al. 2002).

Similar experiments were performed with MHC class I-restricted cells expressing CD3ζ–CFP and CD8α–YFP, or normal CD8α plus CD8β–YFP (Yachi et al. 2005, and unpublished). As with CD4$^+$ cells, the coreceptor was recruited rapidly to the immunological synapse, although here we did not notice a significant difference between the presence or absence of antigen, except that the conjugates formed more slowly and were less frequent in the absence of antigen (Fig. 1). Again, coreceptor concentration was much more evident than TCR concentration.

Experiments with MHC class I-restricted T cells have a technical advantage over those using class II-restricted T cells in that we can use the cell line RMA-S as the APC, which can be loaded with defined peptides in a well-controlled manner. RMA-S is a thymoma cell line that is deficient in transporter of antigen processing (Tap)-2 expression. Because of this, it is unable to load most peptides onto class I molecules, and its cell surface MHC class I expression is thus much lower than that of its parental cell line, RMA. Early flow cytometry data showed class I-expression on RMA-S to be 5–10% of the amount of class I on RMA (Ljunggren and Karre 1985; Townsend et al. 1989), and we estimate that there are ~550 molecules of H2-K^b on the cell surface in the absence of added peptides (Yachi et al. 2007).

In fact, the class I molecules associated with β2-microglobulin are transported to the cell surface, but the vast majority are unstable because they lack bound peptide. At below-physiological temperatures, the molecules remain stable, so it is possible to culture the cells at 32°C in the presence of suitable peptides, which then become

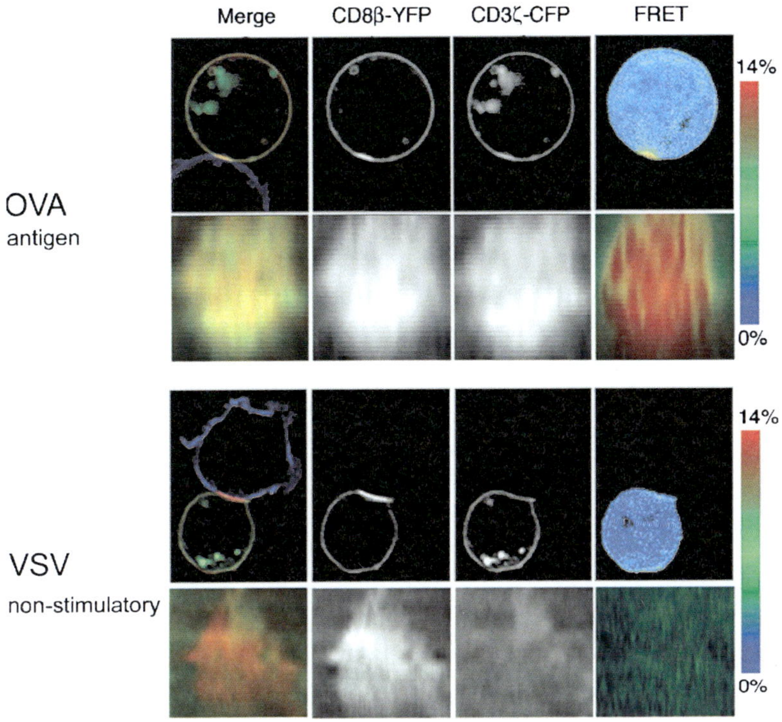

Fig. 1 OT-I T cell recognizing antigen (OVA, *top panels*) or a nonstimulatory peptide (VSV, *lower panels*). The *top set of images of each panel* shows a single z-slice through a T cell:APC couple, showing, from left to right, a merged color image (CD8β–YFP in *red*, CD3ζ–CFP in *green,* the APC labeled with Cy5, shown in *blue*), CD8β–YFP in monochrome, CD3ζ–CFP in monochrome, and the FRET image, heat-colored to represent the FRET efficiency (using the *scale* shown on the *far right*). Reproduced from (Yachi et al. 2005)

integrated into the MHC molecule. When the temperature is raised to 37°C, class I molecules into which peptide has been folded remain stable and can therefore be recognized by T cells, while the other, "empty," class I molecules are degraded. Thus, it is possible to control both the specific peptides that are presented by class I molecules on the cell surface, and the total amount of MHC class I, which we used to dissect the forces that drive TCR and CD8 recruitment to the immunological synapse. We found that either antigenic peptide or a peptide that is not recognized by the T cell's TCR allowed CD8 to be recruited to the immunological synapse and concentrated there (Fig. 2). The amount of CD8 recruited to the immunological synapse correlated with the quantity of these peptides loaded into the MHC class I molecules, indicating that the CD8 recruitment is caused by the noncognate interaction between CD8 and class I (Yachi et al. 2005). Recent experiments performed with a

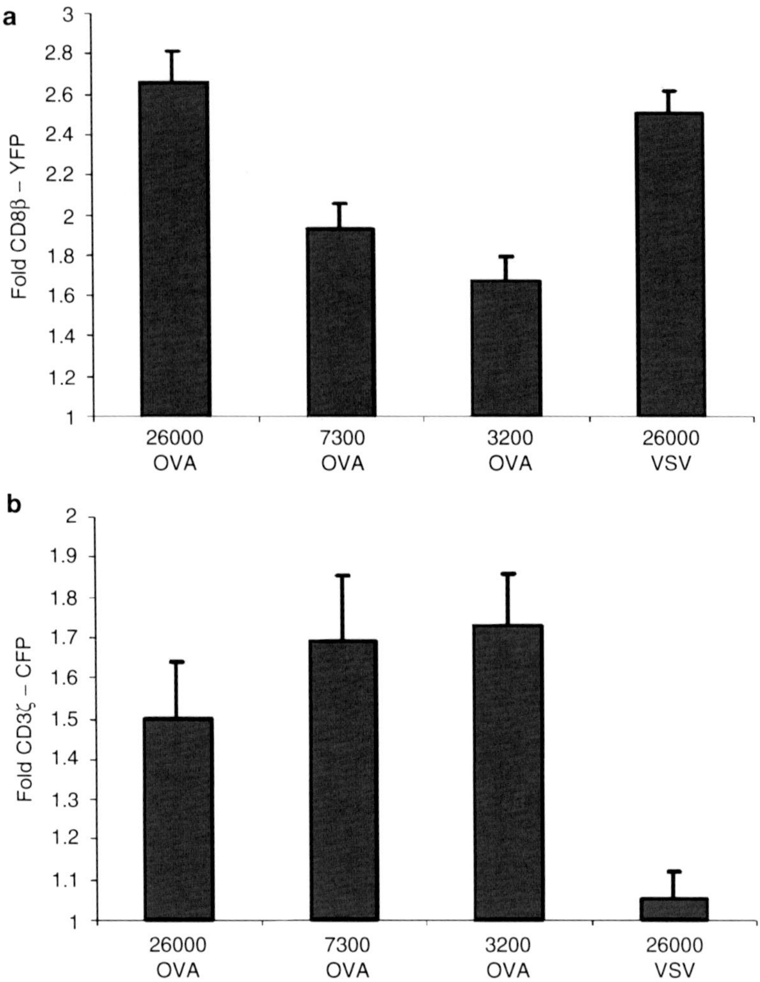

Fig. 2 Recruitment of CD8β–YFP and CD3ζ–CFP in response to different quantities of peptide (OVA or VSV) presented on RMA-S cells. The *numbers* refer to the numbers of $H2K^b$ molecules on the cell surface or the RMA-S cells. Reproduced from (Yachi et al. 2005)

T hybridoma that does not express TCRα or β-chains, but which we transfected with CD8α and CD8β–YFP, demonstrated that the CD8 recruitment to the synapse was independent of TCR signaling (Yachi, unpublished).

2.2 *Signaling Proteins*

We are currently extending our studies to analyze various aspects of signal transduction in the immunological synapse. Analysis of protein kinases C originally indicated that only PKCθ was associated with the immunological synapse (Monks et al. 1997, 1998). The recruitment of PKCθ required signaling through the kinase Lck and costimulation (Huang et al. 2002). We recently identified PKCη as another PKC that is strongly and specifically recruited to the synapse (G. Fu and N. Niederberger, in preparation). Interestingly, PKCη is upregulated during thymocyte positive selection (Mick et al. 2004; Niederberger et al. 2005), and is seen in synapses formed by mature thymocytes, whereas PKCθ is expressed from earlier stages of development and is found in the IS of both immature $CD4^+$ $CD8^+$ and mature thymocytes (Fu and Niederberger, in preparation). We are currently testing the hypothesis that PKCθ and PKCη have some redundant functions in thymocyte development.

3 Molecular Interactions in the Immunological Synapse

3.1 FRET Analysis

By choosing enhanced CFP and YFP as our fluorescent reporters, we could employ FRET to detect molecular range proximity between these fluorophores. FRET is the nonradiative transfer of energy between an excited molecule and a nonexcited molecule. For this to happen, the donor must have an emission spectrum that overlaps with the acceptor's excitation spectrum, and the fluorescent cores of the molecules must be within about 10 nm of each other. Thus, this is a useful technique to analyze whether molecules are in close proximity with each other. However, there are certain difficulties in using the method in live cells. When the two fluorescent moieties are on separate molecules, it is important that sufficient acceptors are available (typically 1:1 to 3:1 acceptor to donor ratio) and that the dependence of FRET on acceptor concentration is determined to differentiate between FRET due to specific interactions and crowding effects.

The simplest method for analysis is the "donor recovery" method, where an image of the donor, for example, CFP, is taken before and after photobleaching of the acceptor (YFP). If FRET occurs, then the donor image taken after the photobleaching will be brighter by the amount of energy that was transferred to the acceptor before photobleaching. This method is unsuitable for living cells where the relevant molecules are likely to move between taking the two images. For live imaging, we therefore used an extended ratiometric method where images for emission at both CFP and YFP wavelengths were taken (Zal and Gascoigne 2004; Zal et al. 2002). This is best performed with simultaneous imaging of CFP and YFP emissions to minimize movement artifacts (Zal and Gascoigne 2004). If FRET occurs, then the YFP emission increases while the CFP decreases during CFP excitation. In addition, the YFP-only image and correction factors for bleed-through between the channels are necessary to deduce FRET between independently expressed donors and acceptors. This method is suitable for FRET analysis of live cells, but corrections must also be made for the photobleaching that occurs when taking z-stacks and during time-lapse imaging (Zal and Gascoigne 2004).

3.2 TCR–Coreceptor Interactions

We used FRET imaging to investigate the interaction between the TCR–CD3 complex and CD4 or CD8 during antigen recognition. Many studies have addressed whether TCR and coreceptors are associated, but the data are highly equivocal, with some showing a constitutive interaction (Anderson et al. 1988; Arcaro et al. 2001; Beyers et al. 1992; Doucey et al. 2003; Gallagher et al. 1989; Kwan Lim et al. 1998; Rojo et al. 1989; Suzuki et al. 1992; Takada and Engleman 1987), and others showing an antigen-induced interaction (Anel et al. 1997; Block et al. 2001; Mittler et al. 1989; Osono et al. 1997). Some of these studies used cocapping as a measure of interaction (Anderson et al. 1988; Kwan Lim et al. 1998; Rojo et al. 1989; Takada and Engleman 1987), which may be flawed by the molecules residing on the same membrane microdomains, and almost all used activation by antibodies rather than by MHC–peptide complexes on an APC. Thus, our studies allowed us to investigate the induction of the interaction during recognition of antigens presented on APC. We found that, for both CD4 and CD8, the FRET signal between TCR and CD4 or CD8β increased significantly during antigen recognition, indicating that the TCR–coreceptor interaction is induced during TCR recognition of antigen (Yachi et al. 2005; Zal et al. 2002). Figure 1 shows the FRET response between TCR and CD8 by antigen (OVA), but not by a nonstimulatory peptide (VSV). A certain low level constitutive association between the molecules is indicated by a low background FRET signal (~2%), but the interaction is strongly enhanced by antigen recognition.

Interestingly, the time course of the FRET signal was different with different systems. In the case of the CD4–TCR interaction, we found that FRET was induced clearly within 3 min, and that it remained strong for up to 30 min (Zal et al. 2002). In the CD8–TCR experiments, we found that FRET was clearly above background levels (averaged over the whole of the IS) within 5 min, and peaked at around 10 min, then declined (Yachi et al. 2005, 2006). These experiments were performed using RMA-S cells as APCs, but in other experiments using APC such as EL4, where processing of peptides and MHC class I expression is normal, we did not see such a steep decline in the FRET signal (unpublished).

3.3 TCR:CD8 Interactions Induced by Positive and Negative Selecting Ligands

We used FRET microscopy to analyze the recruitment of TCR and CD8, and their interaction in the immunological synapse, during recognition of ligands of different types by the OT-I TCR for which the affinities for a variety of K^b–peptide complexes have been determined (Alam et al. 1996, 1999 ; Gascoigne et al. 2001; Rosette et al. 2001). These different ligands have defined biological activities, as agonists, weak agonists, and antagonists of mature cells, and as negative and positive selectors of thymocytes in fetal thymus organ culture (FTOC) (Hogquist et al. 1994; Jameson et al. 1994). We found that, as expected, CD8 recruitment was not affected by the different ligands, although TCR recruitment was affected – weaker ligands caused less CD3ζ to be concentrated at the immunological synapse (Yachi et al. 2006). The induction of CD3ζ–CD8β FRET followed different time courses for negative selectors (strong agonists) than for positive selectors (weak agonists and antagonists). Negative selecting ligands showed peak FRET at 10 min, whereas the positive selectors peaked at 20 min. The negative selectors showed a faster induction of FRET that then faded out, whereas the positive selectors gave a slower induction of FRET. The magnitude of the FRET response was not as clearly correlated with the bioactivity, as a weak agonist that is a positive selecting ligand gave a strong FRET signal but with the slower time course found with other positive selectors (Yachi et al. 2006).

More recently, we have expanded the analysis of ligands for the OT-I TCR to include those that straddle the rather sharp boundary between positive and negative selection. The weakest ligand that induces only negative selection, and the strongest that causes only positive selection (Daniels et al. 2006), induced similar levels of CD3–CD8 FRET, yet were separated in their time courses (Mallaun et al. 2008): the positive selector showed a slower induction of FRET than the negative selector. Despite considerable variability in the strength of FRET induced, the differences in the timing of peak FRET were maintained between the positive and negative selecting ligands.

Most interestingly, there was one ligand that could induce either positive or negative selection in FTOC – at higher concentration it induced negative selection, but at lower concentration it induced positive selection. This switch occurred abruptly within a 10-fold concentration range (Daniels et al. 2006). In this case, the induction of FRET between TCR and CD8 was induced with kinetics similar to the negative selecting ligands, but FRET remained at a high level for longer, into the time-range when the positive selectors induce FRET (Yachi, unpublished).

From these results, we infer that the time course of the induction of the intimate CD8–TCR interaction is related to the signal eventually transduced by the TCR, or vice versa. When we analyzed the induction of phosphorylation of Erk in the cells responding to the positive and negative selecting ligands, we found that phospho-Erk was induced more rapidly and strongly at the immunological synapse by the negative selectors than by the positive selectors (Yachi et al. 2006). In another series of experiments, phospho-Erk and other members of the Ras/MAP kinase signaling pathway were found predominantly at the cell surface after stimulation with negative selectors, but were found in an intracellular compartment (apparently the Golgi apparatus) with the positive selectors (Daniels et al. 2006). Phosphorylated JNK was only found intracellularly. This separation of ERK and JNK signals in negative but not positive selection signaling is believed to underlie the induction of differentiation or apoptosis by the two types of signal (Daniels et al. 2006).

3.4 *Antagonism of T Cell Activation*

T cell activation by antigen can be antagonized by other MHC–peptide complexes, usually, but not necessarily, closely related to the antigen's sequence. The mechanisms by which this works have been a subject of much debate. The original definition of a TCR antagonist came from studies on altered peptide ligands for an MHC class II-restricted T cell (Evavold and Allen 1991), and many examples are now known. For example, in the MHC class I-restricted T cell OT-I, changing one of the potential TCR-contacting residues (residues 1, 4, 6, or 7 of the 8-amino acid peptide antigen) was often sufficient to change the peptide from antigen (agonist) to antagonist (Hogquist et al. 1994; Jameson et al. 1993, 1994; Rosette et al. 2001). In other cases, the change was to a weak agonist or to a nonstimulatory peptide. As mentioned earlier, the antagonists were also found to induce positive selection (Hogquist et al. 1994; Jameson et al. 1994; Rosette et al. 2001). They showed weaker affinity and usually faster dissociation kinetics than did the strong agonists (Alam et al. 1996, 1999; Gascoigne et al. 2001; Rosette et al. 2001).

We made use of FRET microscopy to investigate how antagonism of T cell activation affects the interaction of coreceptor with TCR. We stimulated an MHC class II-restricted T cell hybridoma that responds to a self-antigen derived from a complement protein (Zal et al. 1994) with the antigenic C5 peptide. Two antagonist peptides have been identified for this TCR, each differing from the antigen by a single amino acid (Volkmann et al. 1998; Zal et al. 2002). Addition of either antagonist peptide, or a peptide that does not stimulate this TCR (but which binds to the same MHC molecule), did not affect the recruitment of the TCR (measured by CD3ζ–CFP) to the synapse (Fig. 3). The CD4–YFP recruitment to the synapse was also similar to when the cells were stimulated with antigen alone. We had earlier found that stimulation with the antagonist peptide alone led to a very weak FRET signal, much weaker than that obtained with agonist (Zal et al. 2002), and that this is also the case for antagonists in the OT-I system (Yachi et al. 2006). When both agonist and antagonist ligands were present, the FRET signal induced by agonist was abrogated (Fig. 3), but this did not occur when the nonstimulatory peptide was used in place of the antagonists. This demonstrated that the antagonists somehow inhibited the close range interaction between the TCR and the coreceptor, despite not affecting the ability of either of these molecules to be recruited to the immunological synapse (Zal et al. 2002). Since the efficiency of FRET is linearly dependent on the proportion of donor (TCR) that is in complex with the coreceptor (CD4 or CD8), we have interpreted this finding as supporting a model where the stability of the coreceptor–TCR interaction is dependent on the quality of TCR binding to the MHC–peptide. Moreover, our data indicate that MHC molecules are bound first by the coreceptor, irrespective of the peptide, and then by TCR. The half-life of TCR binding to agonists is usually longer than that of antagonists (Alam et al. 1996; Gascoigne et al. 2001; Kersh et al. 1998; Lyons et al. 1996). We envisage that the coreceptor–MHC complexes in the immunological synapse have even more dramatic differences in the kinetics of their interaction with TCR depending on the peptide than it is expected based on the measurements of MHC–TCR interaction without coreceptor in solution. In other words, we suggest a threshold-setting function for the coreceptor that helps discriminate between MHC-peptides that otherwise interact with TCR with only slightly different kinetics. Since the antagonists were present in the experiment at a higher concentration than

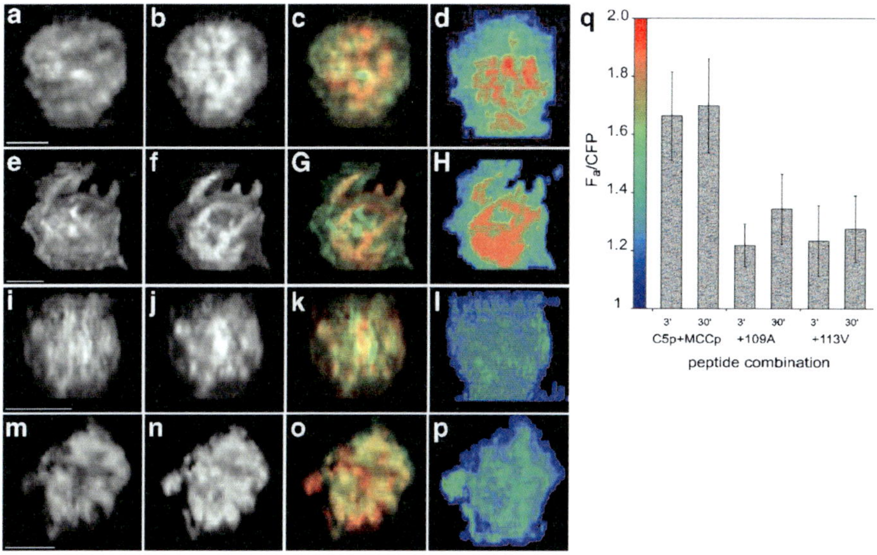

Fig. 3 Antagonists of the TCR inhibit the close interaction between TCR and CD4. These images all show en face projections of the immunological synapse. (**a–d**) 3-min time point, (**e–h**) 30-min time point for antigen plus nonstimulatory peptide. (**i–j**) 3-min time point for antigen plus antagonist 109A, (**m–p**), the same at 30 min. *Left panel* (**a**) and below show CD3ζ–CFP images. (**b**) and images below show CD4–YFP. (**c**) and pictures below show merged false color images showing CD3ζ–CFP in *green* and CD4–YFP in *red*. *Right panel* (**d**) and below are FRET images, heat-colored as in **Fig. 1**. (**q**) shows FRET signals obtained from multiple measurements in the IS with antigen plus the nonstimulatory peptide MCC, or the C5 antagonists 109A and 113V. Reproduced from (Zal et al. 2002)

the agonist, they would bind TCR more frequently than the agonists (due to higher frequency of the ligands) but TCR would fall off more rapidly. The overall signal for T cell activation would depend on the accumulation of sufficient TCR–coreceptor complexes in a short time.

4 Variability in the Form of the T Cell:APC Interface

4.1 *Nanotubes*

The ability to image cell surface molecules during interactions between cells led to many surprises, for example, the structure of the immunological synapse itself, and the choreography of different proteins entering and exiting the synapse at different times. The form of the interaction site between T cell and APC alters over time, with the earliest interaction showing very active membrane spreading, then a period of contraction when the mature immunological synapse is formed. At later time points, we noticed very fine processes joining T cells to APCs, often as the T cells detached and started to move away from the APC (Zal et al., unpublished) (Fig. 4). However, they could also form early in T:APC interactions or could be intermediates between periods of full immunological synapse-like contact. The point of contact of these membrane tethers to the APC showed concentrations of TCR and coreceptor. Similar structures were identified in *Drosophila* wing buds, providing cell–cell interactions between cells in different regions of the imaginal disk, and in mouse limb buds. These were termed cytonemes (Ramirez-Weber and Kornberg 1999). Similar "nanotubes" formed between mammalian cells in culture were shown to allow transport of organelles and membrane vesicles from cell to cell (Rustom et al. 2004). In immune system cells, nanotubes have been shown to allow intercellular transfer of cell membrane proteins (Onfelt et al. 2004). More recently, nanotube connections between T cells have been found to be routes for HIV transfer as well as information transfer (Sowinski et al. 2008). The emerging data on nanotubes have recently been reviewed (Davis and Sowinski 2008; Gerdes and Carvalho 2008). The resolution and sensitivity of two-photon microscopy is not yet sufficient to be able to identify these kind of structures in the immune system *in vivo*. However, it is very tempting to speculate that they could be involved in the

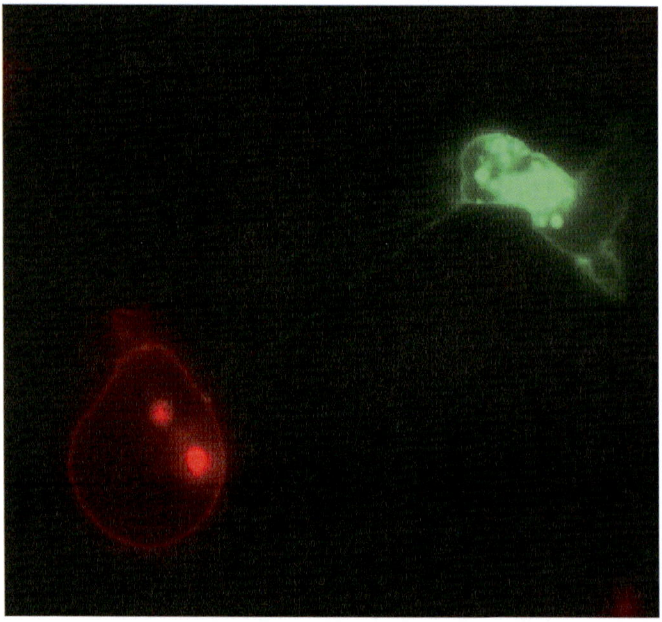

Fig. 4 A nanotube extending from a CD4$^+$ T cell expressing CD3ζ–CFP (*green*) interacting with an antigen-expressing APC (*red*). This is from a movie lasting about 30 min. The T cell started in synapse with the APC, then moved away. It remained tethered for the duration of the movie

scanning activity of T cells migrating through the lymph node, for instance, in the early phase of T cell stimulation by antigen where T cells do not make long-term contacts with APC (Mempel et al. 2004). They may conceivably retain these fine contacts with numerous APCs that they have previously contacted, perhaps adding signals until they are sufficiently stimulated to make a full stop response and make a stable synapse.

Nanotubes are still very poorly understood; much remains to be learned about their functional importance, particularly *in vivo*. This section serves to illustrate the idea that imaging of the immune response gives an opportunity to discover surprising and perhaps delightful facts about the cells that could not have been predicted by other methods.

5 Concluding Remarks

Imaging of molecules in living cells has led to many exciting discoveries about how cells interact. These could not have been predicted by any knowledge that was available before the advent of high resolution microscopic techniques. We are confident that improvements in resolution, especially to the single molecule level, and newer techniques, such as total internal reflection microscopy, will lead to more surprises about cellular interactions and signaling. Use of multiphoton methods for *in vivo* studies will also be more useful when we can image cell surface molecules rather than simply labeled cells. FRET as a method for investigating intermolecular interactions is a useful technique but currently suffers from relatively low dynamic range. This is expected to improve as newer fluorescent protein FRET partners are developed. More importantly, because of the extensive "image math" that must be performed for each subcellular region, such as the IS, FRET is extremely low throughput. Significant improvements in image analysis software are needed to overcome this problem.

Acknowledgements Work from this lab was funded by the NIH (R01GM065230 and AI074074 to N.R.J.G., T32HL07195 to P.P.Y., T32AI07290 and K22AI065688 to T.Z.). C.L. was supported by Deutsche Krebshilfe. J-P.C. was supported by a Cancer Research Institute Postdoctoral Fellowship. This is TSRI manuscript number 19758.

References

Alam SM, Davies GM, Lin CM, Zal T, Nasholds W, Jameson SC, Hogquist KA, Gascoigne NRJ, Travers PJ (1999) Qualitative and quantitative differences in T cell receptor binding of agonist and antagonist ligands. Immunity 10:227–237

Alam SM, Travers PJ, Wung JL, Nasholds W, Redpath S, Jameson SC, Gascoigne NRJ (1996) T cell receptor affinity and thymocyte positive selection. Nature 381:616–620

Anderson P, Blue M-L, Schlossman SF (1988) Comodulation of CD3 and CD4. Evidence for a specific association between CD4 and approximately 5% of the CD3:T cell receptor complexes on helper T lymphocytes. J Immunol 140:1732–1737

Anel A, Martinez-Lorenzo MJ, Schmitt-Verhulst AM, Boyer C (1997) Influence on CD8 of TCR/CD3-generated signals in CTL clones and CTL precursor cells. J Immunol 158:19–28

Arcaro A, Gregoire C, Bakker TR, Baldi L, Jordan M, Goffin L, Boucheron N, Wurm F, van der Merwe PA, Malissen B, et al (2001) CD8β endows CD8 with efficient coreceptor function by coupling T cell receptor/CD3 to raft-associated CD8/p56(lck) complexes. J Exp Med 194:1485–1495

Beyers AD, Spruyt LL, Williams AF (1992) Molecular associations between the T-lymphocyte antigen receptor complex and the surface antigens CD2, CD4, or CD8 and CD5. Proc Natl Acad Sci USA 89:2945–2949

Block MS, Johnson AJ, Mendez-Fernandez Y, Pease LR (2001) Monomeric class I molecules mediate TCR/CD3ε/CD8 interaction on the surface of T cells. J Immunol 167:821–826

Campi G, Varma R, Dustin ML (2005) Actin and agonist MHC-peptide complex-dependent T cell receptor microclusters as scaffolds for signaling. J Exp Med 202:1031–1036

Cemerski S, Das J, Giurisato E, Markiewicz MA, Allen PM, Chakraborty AK, Shaw AS (2008) The balance between T cell receptor signaling and degradation at the center of the immunological synapse is determined by antigen quality. Immunity 29(3):414–422

Daniels MA, Teixeiro E, Gill J, Hausmann B, Roubaty D, Holmberg K, Werlen G, Hollander GA, Gascoigne NRJ, Palmer E (2006) Thymic selection threshold defined by compartmentalization of Ras/MAPK signalling. Nature 444:724–729

Davis DM, Sowinski S (2008) Membrane nanotubes: dynamic long-distance connections between animal cells. Nat Rev Mol Cell Biol 9:431–436

Doucey MA, Goffin L, Naeher D, Michielin O, Baumgartner P, Guillaume P, Palmer E, Luescher IF (2003) CD3δ establishes a functional link between the T cell receptor and CD8. J Biol Chem 278:3257–3264

Evavold BD, Allen PM (1991) Separation of IL-4 production from Th cell proliferation by an altered T cell receptor ligand. Science 252:1308–1310

Gallagher PF, Fazekas de St Groth B, Miller JFAP (1989) CD4 and CD8 molecules can physically associate with the same T-cell receptor. Proc Natl Acad Sci U S A 86:10044–10048

Gascoigne NRJ, Zal T, Alam SM (2001) T-cell receptor binding kinetics in T-cell development and activation. Exp Rev Mol Med:1–17 (http://www.expertreviews.org/01002502h.htm)

Gerdes HH, Carvalho RN (2008) Intercellular transfer mediated by tunneling nanotubes. Curr Opin Cell Biol 20:470–475

Goldrath AW, Bevan MJ (1999) Selecting and maintaining a diverse T-cell repertoire. Nature 402:255–262

Grakoui A, Bromley SK, Sumen C, Davis MM, Shaw AS, Allen PM, Dustin ML (1999) The immunological synapse: a molecular machine controlling T cell activation. Science 285:221–227

Hailman E, Burack WR, Shaw AS, Dustin ML, Allen PM (2002) Immature CD4(+)CD8(+) thymocytes form a multifocal immunological synapse with sustained tyrosine phosphorylation. Immunity 16:839–848

Hogquist KA, Jameson SC, Heath WR, Howard JL, Bevan MJ, Carbone FR (1994) T cell receptor antagonist peptides induce positive selection. Cell 76:17–27

Huang J, Lo PF, Zal T, Gascoigne NRJ, Smith BA, Levin SD, Grey HM (2002) CD28 plays a critical role in the segregation of PKCθ within the immunologic synapse. Proc Natl Acad Sci USA 99:9369–9373

Jameson SC, Carbone FR, Bevan MJ (1993) Clone-specific T cell receptor antagonists of major histocompatibility complex class I-restricted cytotoxic T cells. J Exp Med 177:1541–1550

Jameson SC, Hogquist KA, Bevan MJ (1994) Specificity and flexibility in thymic selection. Nature 369:750–752

Johnson KG, Bromley SK, Dustin ML, Thomas ML (2000) A supramolecular basis for CD45 tyrosine phosphatase regulation in sustained T cell activation. Proc Natl Acad Sci USA 97:10138–10143

Kersh GJ, Kersh EN, Fremont DH, Allen PM (1998) High- and low-potency ligands with similar affinities for the TCR: the importance of kinetics in TCR signaling. Immunity 9:817–826

Klausner RD, Lippincott-Schwartz J, Bonifacino JS (1990) The T cell antigen receptor: insights into organelle biology. Annu Rev Cell Biol 6:403–431

Kupfer A, Singer SJ (1989) Cell biology of cytotoxic and helper T cell functions: immunofluorescence microscopic studies of single cells and cell couples. Annu Rev Immunol 7:309–337

Kwan Lim GE, McNeill L, Whitley K, Becker DL, Zamoyska R (1998) Co-capping studies reveal CD8/TCR interactions after capping CD8β polypeptides and intracellular associations of CD8 with p56lck. Eur J Immunol 28:745–754

Lee KH, Holdorf AD, Dustin ML, Chan AC, Allen PM, Shaw AS (2002) T cell receptor signaling precedes immunological synapse formation. Science 295:1539–1542

Ljunggren HG, Karre K (1985) Host resistance directed selectively against H-2-deficient lymphoma variants. Analysis of the mechanism. J Exp Med 162:1745–1759

Lyons DS, Lieberman SA, Hampl J, Boniface JJ, Chien Y, Berg LJ, Davis MM (1996) A TCR binds to antagonist ligands with lower affinities and faster dissociation rates than to agonists. Immunity 5:53–61

Mallaun M, Naeher D, Daniels MA, Yachi PP, Hausmann B, Luescher IF, Gascoigne NRJ, Palmer E (2008) The T cell receptor's α-chain connecting peptide motif promotes close approximation of the CD8 coreceptor allowing efficient signal initiation. J Immunol 180:8211–8221

Mempel TR, Henrickson SE, Von Andrian UH (2004) T-cell priming by dendritic cells in lymph nodes occurs in three distinct phases. Nature 427:154–159

Mick VE, Starr TK, McCaughtry TM, McNeil LK, Hogquist KA (2004) The regulated expression of a diverse set of genes during thymocyte positive selection in vivo. J Immunol 173:5434–5444

Mittler RS, Goldman SJ, Spitalny GL, Burakoff SJ (1989) T-cell receptor-CD4 physical association in a murine T-cell hybridoma: induction by physical antigen receptor ligation. Proc Natl Acad Sci U S A 86:8531–8535

Monks CRF., Freiberg BA, Kupfer H, Sciaky N, Kupfer A (1998) Three-dimensional segregation of supramolecular activation clusters in T cells. Nature 395:82–86

Monks CRF., Kupfer H, Tamir I, Barlow A, Kupfer A (1997) Selective modulation of protein kinase C-θ during T-cell activation. Nature 385:83–86

Niederberger N, Buehler LK, Ampudia J, Gascoigne NRJ (2005) Thymocyte stimulation by anti-TCR-β, but not by anti-TCR-α, leads to induction of developmental transcription program. J Leukoc Biol 77:830–841

Onfelt B, Nedvetzki S, Yanagi K, Davis DM (2004) Cutting edge: membrane nanotubes connect immune cells. J Immunol 173:1511–1513

Osono E, Sato N, Yokomuro K, Saizawa MK (1997) Changes in arrangement and in conformation of molecular components of peripheral T-cell antigen receptor complex after ligand binding: analyses by co-precipitation profiles. Scand J Immunol 45:487–493

Ramirez-Weber FA, Kornberg TB (1999) Cytonemes: cellular processes that project to the principal signaling center in Drosophila imaginal discs. Cell 97:599–607

Revy P, Sospedra M, Barbour B, Trautmann A (2001) Functional antigen-independent synapses formed between T cells and dendritic cells. Nat Immunol 2:925–931

Ritchie LI, Ebert PJR., Wu LC, Krummel MF, Owen JJT., Davis MM (2002) Imaging synapse formation during thymocyte selection: inability of CD3ζ to form a stable central accumulation during negative selection. Immunity 16:595–606

Rojo JM, Saizawa K, Janeway CA, Jr. (1989) Physical association of CD4 and the T-cell receptor can be induced by anti-T-cell receptor antibodies. Proc Natl Acad Sci U S A 86:3311–3315

Rosette C, Werlen G, Daniels MA, Holman PO, Alam SM, Travers PJ, Gascoigne NRJ, Palmer E, Jameson SC (2001) The impact of duration versus extent of TCR occupancy on T cell activation: a revision of the kinetic proofreading model. Immunity 15:59–70

Rustom A, Saffrich R, Markovic I, Walther P, Gerdes HH (2004) Nanotubular highways for intercellular organelle transport. Science 303:1007–1010

Shaner NC, Steinbach PA, Tsien RY (2005) A guide to choosing fluorescent proteins. Nat Methods 2:905–909

Sowinski S, Jolly C, Berninghausen O, Purbhoo MA, Chauveau A, Kohler K, Oddos S, Eissmann P, Brodsky FM, Hopkins C, et al (2008) Membrane nanotubes physically connect T cells over long distances presenting a novel route for HIV-1 transmission. Nat Cell Biol 10:211–219

Stefanova I, Dorfman JR, Germain RN (2002) Self-recognition promotes the foreign antigen sensitivity of naive T lymphocytes. Nature 420:429–434

Suzuki S, Kupsch J, Eichmann K, Saizawa MK (1992) Biochemical evidence of the physical association of the majority of CD3 δ chains with the accessory/co-receptor molecules CD4 and CD8 on nonactivated T lymphocytes. Eur J Immunol 22:2475–2479

Takada S, Engleman EG (1987) Evidence for an association between CD8 molecules and the T cell receptor complex on cytotoxic T cells. J Immunol 139:3231–3235

Townsend ARM, Öhlén C, Bastin J, Ljunggren H-G, Foster L, Karre K (1989) Association of class I major histocompatibility heavy and light chains induced by viral peptides. Nature 340:443–448

Trautmann A, Valitutti S (2003) The diversity of immunological synapses. Curr Opin Immunol 15:249–254

Varma R, Campi G, Yokosuka T, Saito T, Dustin ML (2006) T cell receptor-proximal signals are sustained in peripheral microclusters and terminated in the central supramolecular activation cluster. Immunity 25:117–127

Volkmann A, Barthlott T, Weiss S, Frank R, Stockinger B (1998) Antagonist peptide selects thymocytes expressing a class II major histocompatibility complex-restricted T cell receptor into the CD8 lineage. J Exp Med 188:1083–1089

Yachi PP, Ampudia J, Gascoigne NRJ, Zal T (2005) Nonstimulatory peptides contribute to antigen-induced CD8-T cell receptor interaction at the immunological synapse. Nat Immunol 6:785–792

Yachi PP, Ampudia J, Zal T, Gascoigne NRJ (2006) Altered peptide ligands induce delayed and reduced CD8-TCR interaction – a role for CD8 in distinguishing antigen quality. Immunity 25:203–211

Yachi PP, Lotz C, Ampudia J, Gascoigne NRJ (2007) T cell activation enhancement by endogenous pMHC acts for both weak and strong agonists but varies with differentiation state. J Exp Med 204:1747–2757

Zal T, Gascoigne NRJ (2004) Photobleaching-corrected FRET efficiency imaging of live cells. Biophys J 86:3923–3939

Zal T, Volkmann A, Stockinger B (1994) Mechanisms of tolerance induction in major histocompatibility complex class II-restricted T cells specific for a blood-borne self-antigen. J Exp Med 180:2089–2099

Zal T, Zal MA, Gascoigne NRJ (2002) Inhibition of T-cell receptor-coreceptor interactions by antagonist ligands visualized by live FRET imaging of the T-hybridoma immunological synapse. Immunity 16:521–534

Multiscale Analysis of T Cell Activation: Correlating *In Vitro* and *In Vivo* analysis of the Immunological Synapse

Michael L. Dustin

Contents

Abstract Recently implemented fluorescence imaging techniques, such as total internal reflection fluorescence microscopy and two-photon laser scanning microscopy, have made possible multiscale analysis of the immune response from single molecules in an interface to cells moving in lymphoid tissues and tumors. In this review, we consider components of T cell sensitivity: the immunological synapse, the coordination of migration, and antigen recognition *in vivo*. Potency, dose, and detection threshold for peptide-MHC determine T cell sensitivity. The immunological synapse incorporates T cell receptor microclusters that initiate and sustain signaling, and it also determines the positional stability of the T cells through symmetry and symmetry breaking. *In vivo* decisions by T cells on stopping or migration are

M.L. Dustin (✉)
Department of Pathology, Program of Molecular Pathogenesis,
Skirball Institute of BioMolecular Medicine, NYU School of Medicine, 540 First Avenue,
2nd Floor, New York, 10016, NY, USA

D. McGavern and M. Dustin (eds.), *Visualizing Immunity*, Current Topics in Microbiology and Immunology 334,
DOI: 10.1007/978-3-540-93864-4_3, © Springer-Verlag Berlin Heidelberg 2009

based of antigen stop signals and environmental go signals that can sometimes prevent arrest of T cells altogether, and thus can change the outcome of antigen encounters.

1 Introduction

A fundamental challenge for the adaptive immune system is to bring rare antigen-specific T cells together with antigen-presenting cells early in an infection (Zinkernagel and Doherty 1975). This challenge appears to be met by sensitive T cell receptor signaling at the cellular level, T cell movement within secondary lymphoid tissues to make contact with antigen-presenting cells at the tissue level, and attraction of T cells and APC to sites of infection at the whole animal level. Each of these processes is studied on a different scale. Quantitative *in vitro* studies have led to our current understanding of T cell sensitivity to antigen. Total internal reflection fluorescence microscopy (TIRFM) and high speed confocal microscopy are important technologies in analysis of sensitivity at the single molecule (Fig. 1). TIRFM is based on reflecting light off the interface between a high refractive index substrate and lower refractive index aqueous media, including cells (Axelrod 1989; Mattheyses and Axelrod 2006). A recent advance in high speed confocal imaging of cell–cell interfaces is based on using laser tweezers to orient the conjugate so the entire interface can be captured in one confocal plane (Oddos et al. 2008). At the tissue level, anatomical analysis of lymph nodes yielded some clues regarding searching strategies (Kaldjian et al. 2001), but dynamic imaging in intact tissues was essential to fully develop current concepts (Miller et al. 2002). Two-photon laser scanning microscopy (TPLSM) enables imaging deep in explanted or *in vivo*

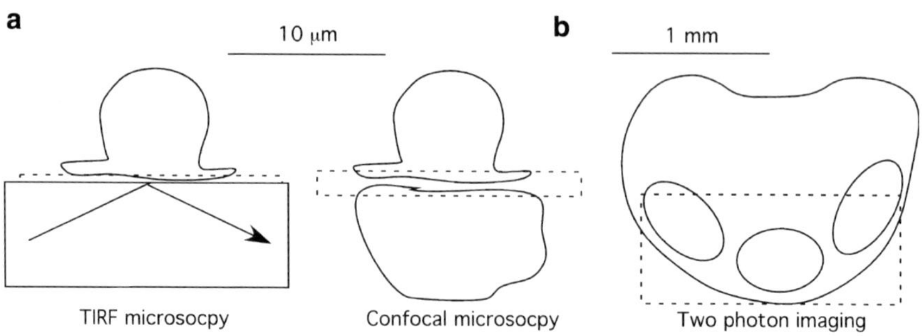

Fig. 1 Microscopy modes used for studying the immune system. (**a**) TIRFM has a depth of 100–200 nm and lateral resolution limited by the diffraction limit and crowding of fluorescent groups. (**b**) High speed confocal of optimally oriented conjugates requires a combination of micromanipulation and imaging. One potential combination is the use a laser tweezer combined with a spinning disc confocal imaging system. (**c**) TPLSM is based on using femtosecond pulses of red light to excite fluorophores deep in tissues (typically up to 0.5 mm, but up to 1 mm)

tissues (Fig. 1). TPLSM works by exciting fluorescence using two photons of infrared light delivered in femtosecond pulses that can penetrate up to 1 mm in tissues with the collection of all possible emitted visible photons (Zipfel et al. 2003). At the whole animal level, the multistep paradigm for lymphocyte migration by adhesion molecules and chemokines, and new information about control of lymphocyte egress from tissues based on pharmacological and genetic experiments, have provided insights into how responses are regulated and optimized (Springer 1995). An important concept that emerged from early *in vitro* studies was that antigen delivers a stop signal for T cell movement and induces formation of a stable immunological synapse, which is an important element in models for control of sensitivity, migration, differentiation, and specificity. Comprehending the breadth of this problem requires analysis at multiple scales, from single molecule to whole animal. Making microscopic observations *in vivo* is particularly important because carefully performed experiments inform us of the cellular dynamics of bona fide immune responses, and this has been vital in establishing the appropriate conditions for these more readily controlled *in vitro* experiments. In this review, I will discuss recent progress in integration of *in vitro* fluorescence microscopy and *in vivo* two-photon microscopy. I will focus on T cells by first introducing parameters that control T cell responses and sensitivity

2 Parameters in T Cell Activation

2.1 Three Dimensions of T Cell Sensitivity

T cells are specialized to detect a variety of chemical structures that are presented on the surface of specialized cells in the context of different types of immune responses. The majority of T cells have diverse repertoires of T cell receptors (TCR) composed of α and β chains that are specialized to recognize 8–9 amino acids of peptides bounds to major histocompatibility complex proteins (peptide–MHC). The three dimensions of sensitivity that we will consider are coreceptor type, TCR–peptide–MHC interaction quality, and peptide–MHC detection threshold.

There are two types of MHC molecules that present peptides generated in different intracellular compartments of the antigen-presenting cell to T cells with different coreceptors (Trombetta and Mellman 2005) (Fig. 2). MHC class I antigens bind peptides that are generated in the cell cytosol by the proteosome, pumped into the endoplasmic reticulum where newly synthesized MHC class I molecules are associated with chaperones in a peptide receptive state. Once the peptides of 8–9 amino acids total length are bound, the folding of the MHC molecules is completed and this peptide–MHCI complex is exported to the plasma membrane via the Golgi apparatus and constitute secretion pathways. While MHC class I can accommodate C-terminal extensions of peptides, efficient trimming operations almost always result in 8–9 amino acid peptides in exported complexes. MHC class II molecules are associated with a transmembrane chaperone referred to as invariant chain (CD74), which blocks the peptide binding groove in the endoplasmic reticulum and

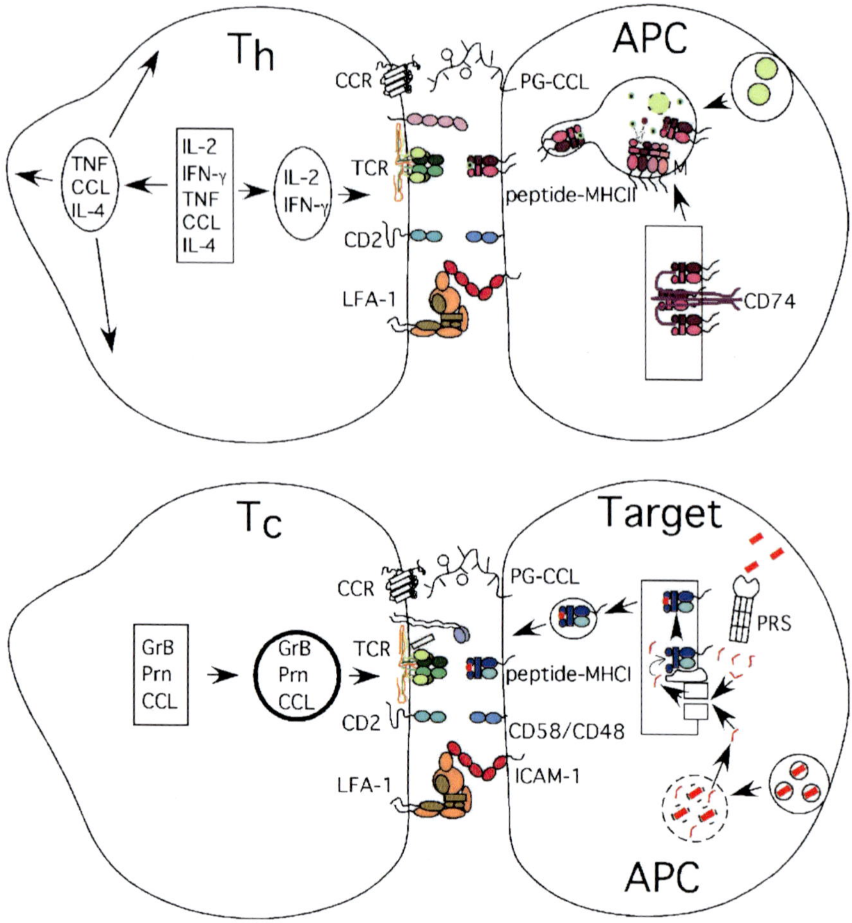

Fig. 2 Schematic of molecules in involved in helper T cell (**a**) and cytotoxic T cell (**b**)

direct trafficking of the molecule to the endosomal compartments via trans Golgi apparatus and plasma membrane clathrin-coated vesicles. Once in late endosomes, cystein proteases degrade invariant chain and leave the invariant chain-derived blocking peptide "CLIP" in the peptide-binding groove. Exchange of CLIP with other peptides generated in the endosomal–lysosomal system is facilitated by lateral interaction with of MHC class II with H–2M (mouse) or HLA–DM (human). The M protein facilitates "editing" of peptides by accelerating dissociation of poorly bound peptides such that MHC class II generally is exported to the surface as a highly stable complex. Peptides bound to MHC class II have the same 8–9 amino acid core in the binding groove, but trimming reactions are less efficient so peptides with the same core sequence can have different length extensions at the

N and C terminus, such that sequence analysis reveals families of peptides with the same core sequence and different flanks. The flanking residues can determine not only processing but also the conformation of the core residues seen by the TCR (Lovitch et al. 2006). With regard to sensitivity, the most important part of MHC class I and class II molecules is the nonpolymorphic binding site for coreceptors. MHC class I binds to the coreceptor CD8 and MHC class II binds to the coreceptor CD4. CD8 and CD4 are both Ig-superfamily members, but CD8 can exist as a disulfide-linked αβ heterodimer or a αα homodimer, whereas CD4 is a monomer. Both CD8 and CD4 interact with Src family kinase Lck, and this interaction is functionally significant for T cell signaling (Weiss and Littman 1994). CD8 binds MHC class I with greater affinity then CD4 binding MHC class II and this difference is likely of significance in determining sensitivity of CD8 versus CD4 expressing T cells. CD4 does not appear to contribute to the binding of peptide–MHC in a meaningful manner, but the precise mechanism by which CD4 facilitates signaling is still largely unknown and will be discussed below in the context of different triggering models. CD8 binds strongly enough to MHC class I to be readily detected in both adhesion experiments and through the binding of multivalent particles such as tetramers and MHC class I coated quantum dots (Anikeeva et al. 2006). This binding activity appears to be regulated, and thus it has been difficult to arrive at a single number of the affinity of CD8 or CD4 to the respective MHC molecules. Nonetheless, it is important to pay attention to whether studies are performed with CD8 or CD4 T cells since most evidence suggests that generalization is difficult between these two types of T cells that result from different developmental programs.

Each TCR expressed by a mature T cell can be considered to have a spectrum of interactions with the syngeneic MHC molecules based on alterations in the peptide. This dynamic range could be further expanded if the MHC molecule itself is altered as in allorecognition or when the residues flanking the peptide-binding groove are otherwise mutated for engineering or analytical reasons. A two-step binding model has been proposed in which the TCR docks with MHC based on contacts with the peptide flanking parts of the MHC molecule to define the k_{on} and then undergoes a conformational change to optimize binding to the peptide, which then sets the k_{off} (Wu et al. 2002). Biologically, each TCR needs to recognize two types of ligands to function: a weakly interacting peptide–MHC is needed for *positive selection* in the thymus, and a more strongly interacting peptide for *T cell activation* in the periphery. When a strongly interacting peptide–MHC complex is recognized in the thymus this leads to *negative selection* to avoid self-reactivity. This continuum can be explored experimentally using altered peptide ligands (APLs), in which the peptide is systematically mutated and functional parameters related to activation as assessed (Evavold et al. 1993; Sloan-Lancaster et al. 1993). The mutated peptide–MHC may be weaker or stronger in inducing biological functions than the parent peptide–MHC.

APLs are ranked based on activity in T cell growth assays (IL-2 or proliferation) with strong and partial agonists, antagonists, coagonist, and null peptide–MHCs (Rabinowitz et al. 1996). Agonists induce proliferation and IL-2 production. Judging whether an agonist is strong or weak is primarily based on the dose needed to stimulate the T cell with APC-like splenocytes or B cell tumors in relation to the starting agonist peptide. Weak or partial agonists cover a large continuum, but may be divided into two fairly discrete and quite functionally distinct types of ligands. We propose that strong agonists are those that can stimulate TCR downregulation and activate early TCR signals in a CD28 independent fashion, whereas partial agonists do not induce TCR downregulation and are CD28-dependent for TCR signal generation (Bachmann et al. 1997a, b), but these ligands can induced strong activation when sufficient costimulation is available (Cemerski et al. 2007). This is likely to be a biologically important distinction since TCR recognition processes that are CD28-dependent may display defective peripheral tolerance since the ligands would be poorly recognized in the steady state conditions where CD80 and CD86 are low (Steinman et al. 2003). Thus, partial agonists, defined by CD28 requirements for recognition, will probably be important in autoimmunity. It is important to distinguish the requirement for CD28 in TCR–peptide–MHC binding from the requirement for CD28 in priming of naive T cells, which also applies to strong agonists. Since T cells can signal in response to strong agonist peptide–MHC complexes, the CD28 signal can be used to gauge the level of innate immune stimulation and can contribute to peripheral tolerance versus immunity decisions. In contrast, because T cells are not effectively activated by weak agonist peptide–MHC in the absence of CD28 engagement, there is no mechanism to induce peripheral tolerance with these ligands since tolerance induction requires activation (Huang et al. 2003; Redmond et al. 2003). Antagonists are peptide–MHC that inhibit responses to agonists (De Magistris et al. 1992). These complexes generally do not induce proliferation at all in physiological dose ranges, but can induce peripheral tolerance through a calcineurin-independent, farnesylation-dependent signaling pathway that induces partial TCR phosphorylation (Skokos et al. 2007; Sloan-Lancaster et al. 1993, 1994). Coagonists are ligands that do not induce proliferation on their own, but synergize with an agonist to extend sensitivity (Lee et al. 2002; Wulfing et al. 2002). The signaling mechanisms involved in coagonism are still a matter of debate. Davis and colleagues have suggested that the TCR interacting with a rare agonist peptide–MHC can promote the phosphorylation of many TCR that interact transiently with abundant TCR that are interacting with coagonist self-peptide–MHC in a CD4-dependent fashion (Irvine et al. 2002; Krogsgaard et al. 2005). It is likely that positive selection operates in the same kinetic range as coagonists, so in many cases the process of positive selection sets up a situation in which the sensitivity to foreign antigens is boosted by the availability of many self-coagonist peptide–MHC. In rare situations, the positively selecting ligands are antagonists for mature T cells, and in these cases the sensitivity of the mature T cell is likely impaired (Santori et al. 2001). An alternative model for self-MHC function has emerged from studies with CD8 T cells that are not peptide-specific. Early studies from Mescher and recent work from Gascoigne's laboratory have shown that CD8 has sufficient avidity to bind any MHC class I molecules to promote T cell sensitivity (O'Rourke and Mescher 1992; O'Rourke et al. 1990; Yachi et al. 2007). This pathway is probably CD8-specific since CD4 interacts with MHC class II very weakly and is unlikely to contribute to adhesion at all. CD8 clearly has sufficient avidity to bind quantum dots clustering 10–12 MHC class I molecules in a peptide-independent fashion, and synergy of agonist and irrelevant peptides was also observed in this system (Anikeeva et al. 2006). A third way that self-peptide–MHC contributes to T cells sensitivity, but maintaining polarity in naive T cells, will be discussed later in the context of signaling.

The definition of weak and strong or low or high potency in this case is eventually based on interactions in the contact area between the T cell and APC, which are only partially understood. Surface plasmon resonance (SPR) and differential scanning calorimetry (DSC) are methods which can be used to determine the equilibrium binding constants, kinetic rates, and thermodynamics of interactions between the TCR and peptide–MHC in solution. These measurements neglect the impact of the orientation and mechanical forces that are a natural part of the interaction between molecules that are tethered to cells. Therefore, it is not surprising that different studies have revealed that different 3D interaction parameters correlate with T cell activation for different MHC–peptide–MHC in CD8 and CD4 systems. The intensively studied I-E^k moth cytochrome C (MCC) CD4 system appears to be dominated by a composite parameter that includes off-rate and heat capacity (Qi et al. 2006), whereas the extensively studied K^b chicken ovalbumin (Ova) CD8 system appears to be well correlated with affinity (Alam et al. 1996). There are a number of methods for evaluating 2D interaction parameters, but these are all more difficult than the 3D measurements and typically have phenomenological components that cannot be uniquely traced to isolated physical parameters (Dustin et al. 1997b, 2007). For example, we have made use of 2D Kd measurements based on fluorescence in cell-supported planar bilayer interfaces, and comparison of these measurements to 3D Kd allows estimation of a "confinement length" parameter which combines all the differences between the 2D and 3D measurements (Bell 1978; Dustin et al. 1997b). These differences include the loss of up to 2 degrees of rotational freedom, concentration of the binding sites in a small volume due to membrane alignment, the effect of cellular forces on the interactions, and the impact of rebinding of dissociated receptors and ligands in the confines of the contact area. Surface force microscopy determines the effect of forces on populations of interactions, but it can be difficult to determine exactly how many bonds are generating the observed force profiles (Bayas et al. 2007). Single molecule analysis with a "biomembrane force probe," which can examine the response of single interactions to force at different rates of force application, provides one method to determine the effects of force on interactions at nm resolution (Chen et al. 2008). Application of an ensemble of 2D/surface interactions methods to the TCR–peptide–MHC interaction should advance our understanding of peptide–MHC quality to a more fundamental level in the next 5 years.

The third parameter related to T cell activation is the dose of a peptide–MHC needed to stimulate a particular response. Due to positive feedback characteristics in downstream signaling, biological responses like IL-2 production and cytokine production can show "switch"-like responses with sharp dose and quality thresholds making it meaningful to discuss the number of peptide–MHC complexes needed to activate a T cell (Daniels et al. 2006). This threshold defines important functional characteristics, such as how effectively processes like negative selection can be expected to operate as a function of self-antigen expression in the thymus and how early T cells can respond after the initial encounter with a pathogen. Naturally, APLs show different dose potencies for a single activation parameter. An excellent parameter for ranking the potency of different ligands is the upregulation of CD69, which depends upon Ras/MAP kinase activation, is a relatively fast transcriptional dependent process, and does not depend upon a complete activation of early TCR signaling parameters such as cytoplasmic Ca^{2+} elevation (Daniels et al. 2006). It has been demonstrated that antigen dose can switch a positively selecting ligand to a negatively selecting ligand in CD8 systems.

3 Immunological Synapses

3.1 Cooperation with Adhesion Molecules

Sensitive antigen recognition by T cells requires the cooperation of multiple classes of adhesion molecules with the TCR (Springer et al. 1987) (Fig. 1). For example, when the integrin family adhesion molecule LFA-1 is knocked out, the sensitivity of T cells to antigen is decreased 100-fold (Bachmann et al. 1997a). Springer was the first to emphasize that the molecular architecture of TCR–peptide–MHC complex interaction appeared to differ from that of the LFA-1–ICAM-1 interaction (Springer 1990). Modeling these interactions as rigid structural elements like columns in a building, it can be schematized that the TCR–peptide–MHC interaction spans about 13 nm, whereas the LFA-1–ICAM-1 interaction could span up to 40 nm. Springer speculated that LFA-1–ICAM-1 interactions might segregate laterally from TCR–peptide–MHC interactions, that some adhesion molecules like CD2–CD58, which are also predicted to span 13 nm, would colocalize with the TCR–peptide–MHC interactions, and that large, nonligated molecules like the tyrosine phosphatase could be excluded from the 13-nm contact regions. He also proposed that CD45 exclusion could promote tyrosine kinase-mediated signaling in the TCR–peptide–MHC rich domains (Springer 1990). van der Merwe and Davis and Shaw and M.L.D. further elaborated on this concept in the context of adhesion and costimulation, respectively (Davis and van der Merwe 1996; Shaw and Dustin 1997; van der Merwe et al. 1995). A series of experiments by van der Merwe and colleagues have supported the validity of this model with regard to the importance of the size of the TCR and peptide–MHC complexes, and small adhesion molecules and like CD2 and CD58, both in relative and absolute terms (Wild et al. 1999). Extending CD48, the rodent ligand for T cell adhesion molecule CD2, by 2 or 3 immunobulin-like (Ig-like) domains results in a decrease in T cell sensitivity to peptide–MHC ligands in a dominant fashion by inhibiting TCR–peptide–MHC engagement (Wild et al. 1999). Increasing the size of the peptide–MHC complex itself by 2–3 Ig-like domains further decreased TCR triggering and compromised CD45 exclusion (Choudhuri et al. 2005). A series of experiments with supported planar bilayers (Fig. 2) has also provided insight into the importance of molecular size.

3.2 Self-Assembly Versus Active Adhesion

Self-assembly aspects of cell adhesion mediated by small adhesion systems have been well illustrated by studies with supported planar bilayers. The planar bilayer system, originally developed by McConnell and colleagues, and extended by my group in studies with Springer, Golan and independently, is a reconstitution method for testing the activity of purified ligands and visualizing interactions at membrane interfaces (Chan et al. 1991; Dustin 1997; Dustin et al. 1996, 1997b) (Fig. 2). The CD2–CD58 interaction was reconstituted by incorporating purified glycophosphatidylinositol (GPI) labeled CD58 in a supported planar bilayer and documenting the highly efficient and specific adhesion (Dustin et al. 1987). Fluorescent labeling of CD58 enabled visualization of the CD2–CD58 interaction in the contact area and measurement of the two-dimensional (2D) dissociation constant (Kd) for this interaction (Dustin 1997; Dustin et al. 1996, 1997b). The interaction of CD2 and CD58 resulted in self-assembly of large contact areas with uniform intermembrane spacing generating an effective interaction volume of only 3–5 nm. The 2D Kd was of the order of 1.1–1.6 molecules μm^{-2} (Dustin et al. 1997b, 2007). This process is efficient at low temperature suggesting that it is largely energy independent (Dustin et al. 1987). The interaction of KIRs with MHC class I ligands appears to drive a similar, energy-independent assembly of adhesion domains in NK cell interactions with cells that are protected from killing by self-recognition (Oddos et al. 2008). A similar 2D Kd was observed for the CD28–CD80 interaction, but at physiological densities of CD28 this interaction appears to be unfavorable and requires TCR–peptide–MHC interactions to be observed (Bromley et al. 2001). The LFA-1–ICAM-1 interaction, in contrast, is associated with irregular spacing at the interface and a strong temperature, cell activation, and F-actin dependence. Small adhesion systems like CD2–CD58 appear to use self-assembly to form adhesive domains and are regulated by expression level and cooperation with like sized adhesion systems. Larger adhesion systems like integrins and cadherins appear to be poor at self-assembly and engage in cooperative mechanisms that involved active cytoskeletal mechanisms, the physical basis of which is poorly understood. When the ICAM-1 and CD58 are included together in the same planar bilayer and cells are activated to enable LFA-1–ICAM-1 interactions, the two interactions are concentrated in distinct microscale domains that are perfectly segregated from each other with submicron transitions between distinct domains (Dustin et al. 1998). Thus, there is evidence for distinct self-assembly modes for different adhesion mechanisms, and the systems with distinct membrane spacing requirements segregate laterally to induce different domains in the same contact area.

3.3 Supramolecular Activation Clusters

Visualization of T–B interfaces *in vitro* revealed micron scale segregation of engaged LFA-1 and TCR in the stable junction between a T cell and antigen-presenting cell (Monks et al. 1998). Kupfer reported that these interactions were organized into two supramolecular activation clusters (SMACs): the central SMAC (cSMAC) enriched in TCR, peptide–MHC complexes, and PKC–θ; and the peripheral SMAC (pSMAC) enriched in LFA-1, ICAM-1 and talin (Monks et al. 1998). The distal SMAC (dSMAC) was later proposed based on a peripheral compartment enriched in CD45 (Freiberg et al. 2002). SMACs were initially defined structurally with no data supporting function except the spatial–temporal correlation of receptors with biochemically relevant signaling and cytoskeletal proteins.

3.4 Immunological Synapse Reconstitution

We proposed the use of the term immunological synapse (IS) to describe these stable, polarized cell–cell junctions in conjunction with our work on organization of adhesion domains (Dustin et al. 1998). This term was utilized previously in two commentaries in 1984 and 1994 that hypothesized that the TCR might be an ion channel like the acetylcholine receptor or that IL-4 would be secreted in a directed fashion from T cells to B cells (Norcross 1984; Paul and Seder 1994). While the specifics of both of these speculations have been disproved, the idea is compelling as a description of a provisionally stable cell–cell communication interface for the immune system. Planar bilayers played a key early role of working out the dynamics of these systems. Mark Davis' laboratory had generated GPI-anchored MHC class II molecule I-E^k for crystallization studies, and we used this as a starting point for reconstitution of functional MHC–peptide complexes based on extensive exchange of the undefined material bound in the groove of the I-E^k-GPI expressed in CHO cells for strongly bound MCC 88–103 or Hemoblobin 64–76 peptides at low pH (Grakoui et al. 1999). In order to maintain this high degree of selection for stable MHC–peptide complexes, the agonist peptides were diluted with mutated peptides with conserved MHC binding residues, but no T cell activating potential, in order to control the density of the agonist ligands while maintaining a uniform population of stably folded I-E^k-GPI molecules (Grakoui et al. 1999). Purified ligands can also be attached to the bilayers using poly-His tagged soluble proteins, which greatly simplifies the process of preparing the reagents and bilayers (Fig. 3). When these stable I-E^k–peptide complexes were coreconstituted in the planar bilayers with ICAM-1, they mediated efficient T cell activation at low agonist MHC–peptide densities. The dynamics of IS formation involve dominantly peripheral engagement of the TCR in the first moments and then the inversion of this pattern to generate the cSMAC and pSMAC. It was quite remarkable that this entire process could be triggered in the T cell by two highly purified molecules in an artificial bilayer. This maturation process was F-actin-dependent and thus appeared to represent an active process. We now appreciate that this inversion is mediated at the level of submicron clusters, called microclusters (MCs), which move from the periphery to the center of the synapse and are also observed at early time in synapse formation in cell–cell systems (Krummel et al. 2000; Oddos et al. 2008). In the standard

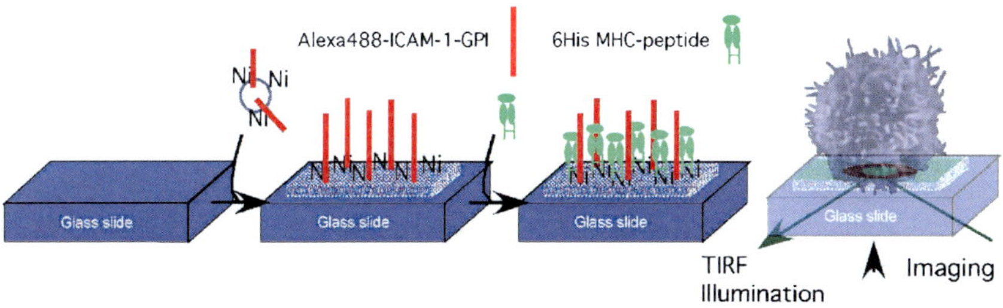

Fig. 3 Formation of supported planar bilayers. Stepwise process of forming supported bilayer for T cell stimulation. Glasss supported planar bilayers are formed with liposomes containing Cy5-ICAM-1-GPI and Ni^{2+} chelating lipids. Poly-His tagged MHC class II molecules are then added to the bilayer. Live T cells are then applied to this surface for imaging and activation studies

planar bilayer preparations described above, the TCR accumulated in the cSMAC displayed highly stable interactions with the agonist MHC–peptide ligands as indicated by a failure of the MHC–peptide complexes in the bilayers to recover following photobleaching (Grakoui et al. 1999). Subsequently, we found that these stable TCR–MHC–peptide interactions were surprisingly inactive in signaling. Signaling was instead sustained by peripheral MCs, which had a life span of less than 2 min (Varma et al. 2006).

This pattern of signaling, with activation signaling in peripheral microclusters and no signaling activity associated with the stable components of the cSMAC, can be fully reproduced in two additional systems based on linking defined soluble peptide–MHC complexes to bilayers via 6His tags and Ni^{2+} chelating lipids. The OT-II TCR transgenic T cells can be stimulated with I-A^b loaded with Ovalbumin peptide 323–339 with an identical pattern of TCR engagement and signaling (Tseng et al. 2008). This system is less potent, probably because the Ova323–339 peptides binds to I-A^b in 3-registers and only one of the minor binding registers activates OT-II cells (Robertson et al. 2000). AND TCR transgenic T cells are similarly activated in a similar pattern by a 6His tagged form of I-E^k loaded with MCC 88–103 peptide and then purified by gel filtration (Varma and Vardhana, unpublished observations).

Recently, Cemerski et al. have modified the planar bilayer procedure to use lower concentrations of agonist peptides, resulting in a large population of I-E^k-GPI complexes of undefined composition along with lower densities of agonits and with distinct T cell activating properties in which perpheral microclusters are replaced by an actively signaling cSMAC (Cemerski et al. 2008). A major advantage of the planar bilayer system is the ability to have defined components. It remains to be determined what additional forms of peptide–MHC are generated by this new loading procedure. It will also be important to reproduce this result with more defined components generated through an independent procedure, such as highly purified soluble peptide–MHC attached to bilayers through 6His tags as described above.

3.5 Synaptic Perspective on T Cell Sensitivity

T lymphocyte recognize antigen as a complex of major histocompatibility complex (MHC) protein and a short peptide. Dendritic cells (DC) are thought to be primary antigen-presenting cells for naive T cells and different DCs appear to be specialized for direct presentation CD4 T cells versus cross-presentation to $CD8^+$ T cells (Dudziak et al. 2007). The efficiency with which APC convert intact proteins into MHC class II has been measured at 1% for a dominant epitope; thus 500 molecules of intact hen egg lysozyme (HEL) need to be endocytosed by a B cell line to generate one I-A^k-HEL48-–62 complex, whereas DC are about 10-fold more efficient (Latek and Unanue 1999). DCs appear to have unique features in their endosomal system to slow the degradation process and prolong/enhance peptide generation (Delamarre et al. 2005; Trombetta and Mellman 2005). Given the intrinsic challenges of making peptide–MHC complexes from ingested antigenic structures, it is important that T cells have a high sensitivity to complexes that are generated early in infection. T cell sensitivity has been estimated in two ways. The first approach is based on populations of APC and T cells and provides an average value for that population. This generated estimates of the order of 300 peptide–MHC complexes needed to activate a naive T cell (Demotz et al. 1990; Harding and Unanue 1990; Latek and Unanue 1999; Peterson et al. 1999). In contrast, effector responses such as $CD8^+$ T cell-mediated killing were found to be much more sensitive with robust reactions triggered with an average of 1 peptide–MHC complex per target cell (Sykulev et al. 1996). Thymic negative selection was also found to be more sensitive that naive T cell priming by ~100-fold such that there appeared to be a safety margin between the negative selection and peripheral T cell activation (Latek and Unanue 1999; Peterson et al. 1999). The second approach is based on single cell analysis.

Mark Davis' laboratory has developed a single cell/single molecule imaging approach to determining the number of peptide–MHC required to induce signaling and full activation in T cells (Irvine et al. 2002; Krogsgaard et al. 2005). This is based on loading APC with small numbers of biotinylated peptides and then detecting these with phycoerythrin–streptavidin. This is a very challenging assay and thus far other laboratories have not adopted this approach. The challenging part of the method is ensuring that the streptavidin reagent detects all peptide–MHC complexes in a monovalent fashion. Consistent with earlier results from Sykulev, they found that T cells can respond to a single peptide–MHC complex in the interface and can be fully restimulated by 10-peptide–MHC complexes in the interface with an APC. This analysis may accurately reflect the number of peptides that are needed in a snapshot of the interface, but it is not clear that this fixed number of exogenously added peptides would sustain signaling over hours or days. In most of these experiments, synthetic peptides were added to APC, resulting in exchange reactions at the cell surface that are not well understood. The peptide–MHC complexes formed by addition of exogenous peptides are structurally different from complexes formed by natural processing of peptides that involve the H-2M or HLA-DM proteins to facilitate peptide editing and formation of a stable complex (Pu et al. 2002, 2004). The more common type of MHC class II restricted T cell clone (Type A) can recognize peptide–MHC complexes whether they are loaded without HLA-DM or on MHC class II. TCR can distinguish these structural differences. There are distinct T cell clones (Type B) that only recognize peptide–MHC complexes that are generated by exogenous loading of preprocessed peptides and cannot detect peptides processed from intact antigens intracellularly with the aid of H-2M/HLA-DM proteins, but appear to recognize residues in more loosely bound peptides. It is likely that these structural differences cut across different classes of TCR–peptide–MHC functional interactions from strong agonists to coagonists. Most experiments have been done with Type A T cells. DCs appear to have antigen-processing strategies to sustain delivery of new complexes to the surface (Delamarre et al. 2005). While a single DC can maintain antigen expression over many hours, it is likely that T cells will interact with more than one DC during priming and differentiation (Celli et al. 2005). Therefore, even though T cells are highly sensitive to peptide–MHC, the requirement for presentation over a period of days (at least for helper T cell responses) and the interaction of the T cell with more than one DC will likely require hundreds of peptide–MHC complexes in foci within the secondary lymphoid tissue to initiate a full T cell response.

It has been proposed in multiple studies that T cells are sensitive to single MHC–peptide complexes, but the requirements for triggering TCR signaling are still not well established. It had been assumed based on precedents set by hormone receptors that signaling by the TCR would require at least dimerization, if not higher order cluster formation (Germain 1997). TCR–peptide–MHC complexes spontaneously oligomerize in solution, but the stoichiometry of the oligomers was not determined and this line of experimentation was not extended (Reich et al. 1997). Another possibility is that TCR signaling can be induced by monovalent ligand in the presence of specific accessory molecules like LFA-1–ICAM-1 or CD2–CD58. The LFA-1–ICAM-1 interaction increases T cell sensitivity by 100-fold *in vivo* and thus is important for T cells to achieve the ultrahigh sensitivity described above (Bachmann et al. 1997a). This 100-fold increase in sensitivity is also consistent with relaxing or eliminating requirements for TCR dimerization. Titration experiments with peptide–MHC in the planar bilayer system suggest that a single peptide–MHC ligand can trigger the formation of an active MC containing multiple TCR (Varma et al. 2006). This is an F-actin-dependent process. It is possible that ligation stabilizes existing transient clusters that are form in response to LFA-1–ICAM-1 interaction (Varma et al. 2006). TCR preorganization has been detected by other static methods (Lillemeier et al. 2006; Schamel et al. 2005), but TCR appears to diffuse independently by fluorescence correlation spectroscopy on naive T cells (James et al. 2007).

4 Coordination of Antigen Recognition and Migration

4.1 Stop Signals and Go Signals

Activated lymphocytes are intrinsically motile *in vitro* and will move rapidly on 2D surfaces coated with integrin ligands, particularly ICAM-1 (Dustin et al. 1992). When agonist MHC–peptide complexes are presented by antigen-presenting cells

(Negulescu et al. 1996) or along with the ICAM-1 on a surface (Dustin et al. 1997a), it stops activated T lymphocyte motility for several hours. Arrest of hybridoma's and naive T cells is Ca^{2+}-dependent (Negulescu et al. 1996; Skokos et al. 2007), whereas arrest of T cell blasts seems to be at least in part Ca^{2+} independent (Dustin et al. 1997a). The arrest of T cells by antigen is not absolute, but can be overcome by other environmental signals. This has been demonstrated *in vivo* by using a Boyden chamber assay in which two compartments are separated by a filter coated with ICAM-1 with or without agonist MHC–peptide complexes. Activated T cells spontaneously transmigrate through filters coated with ICAM-1 only, but cocoating the filters with small amounts of agonist MHC–peptide complexes arrested transmigration (Bromley et al. 2000). This system could then be subjected to chemokine gradients that would urge the T cell to continue to transmigrate rather than arrest in response to the TCR signal. Chemokines binding to CXCR4 and CCR5 had no effect on TCR-mediated T cell arrest, but chemokines binding to both CXCR3 and CCR7 induced T cells to move despite the antigen stop signal (Bromley et al. 2000). Thus, the competition between TCR stop signals and chemokine go signals is determined by distinct signaling pathways. Recently, it has been suggested that the ability of CXCR4 and CCR5 to be coopted into the IS prevents them from inducing go signals in the presence of TCR stop signals, whereas the failure of CCR7 to enter the IS may account for its ability to compete with TCR stop signals (Molon et al. 2005).

Naive T cells are not intrinsically motile like activated T cells. The relative immobility of naive and resting memory T cells in most ex vivo settings contrasted with their highly dynamic behavior in collagen gels containing DC (Gunzer et al. 2000). Two-photon laser scanning microscopy in explanted lymph nodes and later lymph nodes of living mice revealed that resting T cells are highly motile *in vivo* with maximum speeds of 30 $\mu m\ min^{-1}$ and average speeds of ~12 $\mu m\ min^{-1}$ (Miller et al. 2003, 2002). This migration processes follows the stromal network of the T cell zones (Bajenoff et al. 2006), which scaffolds an overlying DC network (Kaldjian et al. 2001; Lindquist et al. 2004). Since the stromal network is highly branched, the migration of the cells appears to follow a "random walk." The random walk executed by rapidly moving lymphocytes over a relatively stationary DC network is an excellent search strategy since each DC can make contact with 5,000 T cells per hour (Miller et al. 2004), which allows 10 antigen bearing DC to sort through 1.2 million T cells per day, which would allow efficient T cell priming even with only a few hundred precursors in each mouse. It also appears that activated DC can attract $CD8^+$ T cells under conditions where innate immune stimulation is present (Castellino et al. 2006). This effect is mediated by chemokines and is not antigen-specific, but it may further accelerate the search of antigen bearing DC that have become activated by innate signals or T cells and release attractive chemokines like CCL3 and CCL4. The CCR5 receptor that recognizes these chemokines is upregulated upon innate stimulation of T cells.

There appear to be a number of signals that stimulate rapid T cell motility in secondary lymphoid tissues. Pertussis toxin treatment of T cells dramatically reduces T cell speed (Huang et al. 2007; Okada and Cyster 2007). This indicates the $G\alpha i$ coupled receptors play an important role in motility, but there are many such receptors expressed on lymphocytes. A prime candidate to mediate T cell motility was CCR7, but the CCR7 knockout produces a relatively small decrease on T cell motility in a lymph node (Huang et al. 2007; Okada and Cyster 2007; Worbs et al. 2007), supporting a role for two or more factors in inducing rapid T cell migration through Gαi coupled receptors. While the CCR7 signal is not required for rapid T cell motility in lymph nodes, the ligands for CCR7, CCL19 or CCL21, when adsorbed to a glass surface are sufficient to mediate rapid T cell migration on planar surfaces (Woolf et al. 2007). This was a surprising finding in that the T cell polarity and motility did not require any other adhesion system and including ICAM-1 with the CCR7 ligand did not alter the migration velocity. This was consistent with data that T cell migration speed is relatively normal in lymph nodes of ICAM-1 or LFA-1 β–subunit-deficient mice (Scholer et al. 2008; Woolf et al. 2007). The concept that chemokines were capable of acting as adhesion molecules had been demonstrated earlier for CX3CR1 and its transmembrane ligand CX3CL1 (Imai et al. 1997). The result that T cells can use a chemokine receptors for induction of polarity and traction maybe relevant to recent findings that DC lacking any integrins can still migrate *in vivo* (Lammermann et al. 2008). We have reproduced the rapid CCR7-dependent motility reported by Woolf et al (2007). We have extended these results by applying interference reflection microscopy (IRM) to the preparations, which allows visualization of the contact areas. Activated T cells migrating on ICAM-1 have a large contact area of 50–100 μm^2 (Dustin et al. 1997a). Naive T cells forming IS on substrates with ICAM-1 and peptide–MHC complexes have slightly smaller, but still substantial, contacts with surfaces bearing ICAM-1 (Bromley et al. 2001). In contrast, the contact areas formed by naive T cells on a surface coated with CCL21 are too small to detect. The T cell appears to contact the surface only through processes that are too small to resolve by IRM (<0.5 μm diameter) (Huang and MLD, unpublished observations). The naive T cells also formed no contacts with surfaces coated with ICAM-1 alone and did not migrate, as reported by Woolf et al. and consistent with prior data that LFA-1 is inactive on resting lymphocytes (Dustin and Springer 1989). However, when CCL21 and ICAM-1 were combined the cells migrated rapidly, but also formed readily detectable contact areas by IRM. The formation of these contact areas is evidence that CCR7 signaling activates LFA-1 in the absence of shear flow. We would argue that this chemokine-mediated activation of LFA-1 will be essential for the LFA-1-dependent increase in T cell sensitivity to antigen which has been documented in LFA-1-deficient mice (Bachmann et al. 1997a).

The transition from this Gαi coupled receptor driven motility and arrest on antigen bearing cells can be almost immediate or take several hours depending upon the antigen dose (Henrickson et al. 2008). The *in vitro* studies by Gunzer et al. (2000) and recent work with antagonist type APLs, which induce tolerance *in vivo* without inducing any change in migration speed (Skokos et al. 2007), demonstrate that T cells can integrate signals without forming a stable synapse. One of us has proposed the term kinapse to describe transient T–APC junctions that transduce signals (Dustin 2007, 2008a, b). So the structure described above, in which T cells migrate rapidly on substrates coated with CCL21 or CCL21 and ICAM-1, are both types of kinapses-moving junctions that

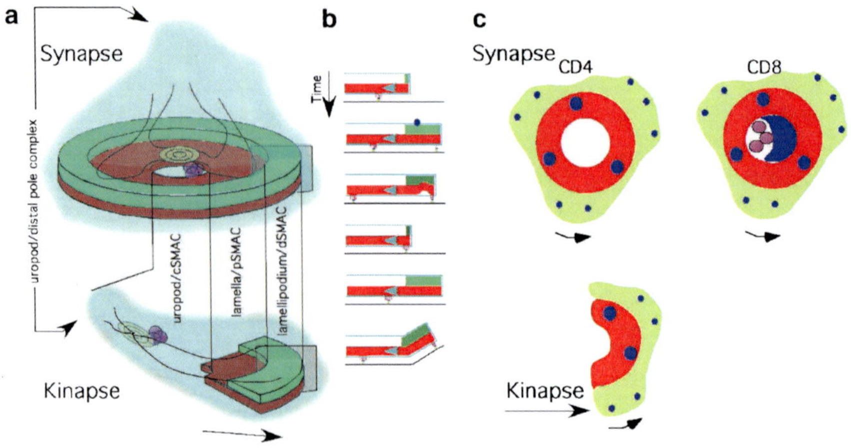

Fig. 4 Synapses versus kinases. (**a**) Schematic of multilayer actin cytoskeleton in the periphery of the immunological synapse. Centripetal action flow is involved in antigen gathering. When this system is symmetric the cell forms a stable synapse. When its asymmetric the cell gains net traction and crawls using the same machinery. (**b**) Model for contractile oscillations; after Sheetz (Giannone et al. 2007). The oscillations may be important for force sensing via that TCR, which is a key function of this process in fibroblasts. (**c**) Schematic of immunological synapses in $CD4^+$ and $CD8^+$ T cells. In $CD4^+$ T cells signaling is largely turned off in the cSMAC, whereas in $CD8^+$ CTL they retain significant cSMAC signaling

integrate the chemokine signal (Fig. 4). The signal that is integrated is the polarization signal delivered by the chemokine and perhaps other signals.

The nature of signaling during these encounters of mature T cells with self-ligands appears to be based on partial phosphorylation of the T cell receptor, but the downstream signaling from this is not known (Stefanova et al. 2002). It does not appear that Ras/mitogen activated kinase pathway signaling is induced since CD69 is not upregulated. Maintenance of T cells polarity appears to be an important contribution of self-peptide–MHC signaling since blocking of these signals reduces TCR polarity rapidly and, over a period of days, the cells also lose the ability to migrate in secondary lymphoid tissues (Fischer et al. 2007). The downstream signals induced by the antagonist peptide–MHC that do not induce a decrease in migration speed compared to self-peptide–MHC are beginning to be elucidated, but there are still many questions open. Early in the analysis of APLs, Paul Allen's laboratory described partial phosphorylation of TCR ITAMs as a characteristic of signaling by antagonists to induce tolerance (Sloan-Lancaster et al. 1993). While the antagonist peptide–MHC do not induce elevation of Ca^{2+}, they do rapidly upregulate CD69. We concluded from this that Ras is activated (Skokos et al. 2007). Consistent with this, tolerance induced by antagonist peptide–MHC complexes *in vivo* was significantly reduced by farnesylation inhibitors. Current models for Ras activation require two stages to generate a switch-like response (Roose et al. 2007). First, RAS-GRP is activated by diacylglycerol to produce a small amount of active RAS. This active RAS can then bind to the allosteric site of SOS, which needs to be independently recruited to the appropriate membrane. Since that antagonist peptide–MHC does not increase Ca^{2+} it is reasonable to postulate that phospholipase C–γ is not activated and little or no DAG is likely to come from this pathway to activate RAS-GRP. Mark Philips' laboratory has identified an alternative pathway for DAG production to activate RAS at the plasma membrane in T cells. This is based upon the ability of LFA-1 and the TCR to collaborate to activate phospholipase D1 (Mor et al. 2007). It is not known if antagonist peptide–MHC can induce this LFA-1-dependent signal; an important question for the future.

4.2 Mechanisms of T Cells Arrest

Arrested T cells often display reorientation of the microtubule-organizing center (MTOC) toward the interface (Dustin et al. 1997a). Early on, this was interpreted as evidence for directed secretion of constitutively secreted elements into the interface (Kupfer et al. 1991), but this is not always the case since a number of cytokines are randomly secreted even in cells that display directed secretion of other components (Huse et al. 2006). MTOC polarization does reflect a symmetric arrangement of microtubule–membrane anchoring sites in the pSMAC (Kuhn and Poenie 2002). The critical determinant for T cells arrest appears to be the symmetrization of the integrins to form the pSMAC (Sims et al. 2007). The arrest of T cells in response to antigen does not result from the inactivation of the motile machinery, but is due to the generation of radial symmetry in the lamellar force generating apparatus such that the traction forces in the stable IS are balanced, resulting in no net motility and the generation of a very efficient system for gathering MHC–peptide complexes and moving them toward the center of the contact area (Kaizuka et al. 2007; Sims et al. 2007). While activated T cells form relatively stable IS over several hours, the naive T cells tend to form and break IS with a period of ~20 min (Sims et al. 2007). Protein kinase C-θ (PKC-θ) stimulated the symmetry breaking process and Wiskott Aldrich Syndrome Protein (WASp) stimulated symmetrization (Sims et al. 2007). The tendency of PKC-θ-deficient T cells to form hyperstable IS was confirmed by analysis of T cell motility *in vivo* (Sims et al. 2007). These experiments demonstrate that the pSMAC has a critical role in control of T cell motility. When the pSMAC is well formed, the cell is relatively immobile due to the balance of forces exerted inward by the integrins around the ring. When the symmetry of the pSMAC is disrupted, the cell will tend to migrate. Thus, a key function of the immunological synapse is to control T cell migration.

4.3 Importance of the pSMAC in Killing

Cytotoxic T lymphocytes (CTL) and natural killer cells are the classical effector cells for which an immunological synapse is proposed to be functionally mandated (Stinchcombe et al. 2001). This is based on the concept that the LFA-1–ICAM-1-

dependent annulus, first described in electron micrographs from NK cells (Schmidt et al. 1988) has been proposed to act as a gasket to contain cytotoxic agents in the synaptic cleft and prevent damage to bystander cells. Yet the evidence for an important role of the immunological synapse in killing has been largely circumstantial or even, in one study, against a critical role (Purbhoo et al. 2004). We have engaged in a systematic analysis of the importance of the pSMAC in CTL-mediated killing with Yuri Sykulev's laboratory. We have found that the CTL can form a pSMAC-like structure rapidly and without recognition of specific peptide–MHC recognition (Somersalo et al. 2004). Under these conditions, no granule polarization or release takes place. We posit that this tendency to form a pSMAC-like structure may increase the rate at which CTL can release granules when needed. The formation of this adhesion ring was dependent upon ICAM-1 density and was enhanced by NKG2d ligands (Markiewicz et al. 2005; Somersalo et al. 2004). In order to better understand the importance of the pSMAC, Sykulev's laboratory studied efficiency of CTL-mediated killing and IS formation with and without LFA-1–ICAM-1 interactions (Anikeeva et al. 2005). The CD2–CD58 adhesion system was utilized in both cellular targets and planar bilayers to provide alternative adhesion, and the density of ligands was sufficiently high to overcome defects in T cells sensitivity due to loss of LFA-1 engagement. Under these conditions, LFA-1–ICAM-1 interaction was essential for efficient killing as defined by the ability to kill efficiently with minimal granule release. Granule release was not altered by loss of LFA-1 engagement, but granules appeared to be poorly targeted toward the contact area, although they were polarized. This also demonstrated that other important adhesion systems do not substitute for LFA-1 in pSMAC formation. Recently, Beal et al. determined one mechanism for the well-described inefficiency of $CD4^+$ human CTL (Beal et al. 2008). When compared to efficient $CD8^+$ CTL the less efficient $CD4^+$ CTL tended to form incomplete pSMACs from which the granules appeared to leak out. Inspired by our earlier studies on the pSMAC destabilizing activity of PKC-θ, we used a specific PKC-θ inhibitor to stabilize the pSMACs of the $CD4^+$ CTLs, resulting in more efficient killing without altering granule release (Beal et al. 2008). This is the most direct evidence to date for an important role of the pSMAC in the efficiency of CTL-mediated killing. The dynamics of CTL *in vivo* are consistent with formation of stable immunological synapses prior to killing (Breart et al. 2008; Mempel et al. 2006). Histological analysis, particularly in the CNS, supports the generation of synapse-like structure by CTLs (Barcia et al. 2006; McGavern et al. 2002). *In vivo* analysis of immunological synapse in still in its infancy and it will be exciting to see if such mechanisms control efficiency of killing *in vivo* when such processes can be directly visualized.

5 Conclusions

Multiscale imaging from the "near-field" 100 nm range of TIRFM to the "far-field" 1 mm range of TPLSM have resulted in important discoveries about immune cell dynamics and interactions during immune responses. Biochemical and imaging studies demonstrate that T cells achieve single molecule sensitivity. The mechanisms by which T cell achieve this sensitivity are completely open to study. A major challenge is to study the molecular scale phenomena of T cell activation in cell–cell interfaces and *in vivo*. New breakthroughs in data collection and data processing are needed to take these next steps.

Acknowledgements I acknowledge Mark Philips for valuable discussions on issues of Ras activation in lymphocytes. I thank Rajat Varma, Santosh Vardhana, Julie Huang and Janelle Waite for contributions of ideas and observations that underpin some of the discussion and speculation in this discussion.

References

Alam SM, Travers PJ, Wung JL, Nasholds W, Redpath S, Jameson SC, Gascoigne NR (1996) T-cell-receptor affinity and thymocyte positive selection. Nature 381:616–620

Anikeeva N, Somersalo K, Sims TN, Thomas VK, Dustin ML, Sykulev Y (2005) Distinct role of lymphocyte function-associated antigen-1 in mediating effective cytolytic activity by cytotoxic T lymphocytes. Proc Natl Acad Sci USA 102:6437–6442

Anikeeva N, Lebedeva T, Clapp AR, Goldman ER, Dustin ML, Mattoussi H, Sykulev Y (2006) Quantum dot/peptide-MHC biosensors reveal strong CD8-dependent cooperation between self and viral antigens that augment the T cell response. Proc Natl Acad Sci USA 103:16846–16851

Axelrod D (1989) Total internal reflection fluorescence microscopy. Methods Cell Biol 30:245–270

Bachmann MF, McKall-Faienza K, Schmits R, Bouchard D, Beach J, Speiser DE, Mak TW, Ohashi PS (1997a) Distinct roles for LFA-1 and CD28 during activation of naive T cells: adhesion versus costimulation. Immunity 7:549–557

Bachmann MF, Oxenius A, Speiser DE, Mariathasan S, Hengartner H, Zinkernagel RM, Ohashi PS (1997b) Peptide-induced T cell receptor down-regulation on naive T cells predicts agonist/partial agonist properties and strictly correlates with T cell activation. Eur J Immunol 27:2195–2203

Bajenoff M, Egen JG, Koo LY, Laugier JP, Brau F, Glaichenhaus N, Germain RN (2006) Stromal cell networks regulate lymphocyte entry, migration, and territoriality in lymph nodes. Immunity 25:989–1001

Barcia C, Thomas CE, Curtin JF, King GD, Wawrowsky K, Candolfi M, Xiong WD, Liu C, Kroeger K, Boyer O, Kupiec-Weglinski J, Klatzmann D, Castro MG, Lowenstein PR (2006) In vivo mature immunological synapses forming SMACs mediate clearance of virally infected astrocytes from the brain. J Exp Med 203:2095–2107

Bayas MV, Kearney A, Avramovic A, van der Merwe PA, Leckband DE (2007) Impact of salt bridges on the equilibrium binding and adhesion of human CD2 and CD58. J Biol Chem 282:5589–5596

Beal AM, Anikeeva N, Varma R, Cameron TO, Norris PJ, Dustin ML, Sykulev Y (2008) Protein Kinase Cθ Regulates Stability of the Peripheral Adhesion Ring Junction and Contributes to the Sensitivity of Target Cell Lysis by CTL. J Immunol 181:4815–4824

Bell GI (1978) Models for the specific adhesion of cells to cells. Science 200:618–627

Breart B, Lemaitre F, Celli S, Bousso P (2008) Two-photon imaging of intratumoral CD8+ T cell cytotoxic activity during adoptive T cell therapy in mice. J Clin Invest 118:1390–1397

Bromley SK, Peterson DA, Gunn MD, Dustin ML (2000) Cutting edge: hierarchy of chemokine receptor and TCR signals regulating T cell migration and proliferation. J Immunol 165:15–19

Bromley SK, Iaboni A, Davis SJ, Whitty A, Green JM, Shaw AS, Weiss A, Dustin ML (2001) The immunological synapse and CD28-CD80 interactions. Nat Immunol 2:1159–1166

Castellino F, Huang AY, Altan-Bonnet G, Stoll S, Scheinecker C, Germain RN (2006) Chemokines enhance immunity by guiding naive CD8+ T cells to sites of CD4+ T cell-dendritic cell interaction. Nature 440:890–895

Celli S, Garcia Z, Bousso P (2005) CD4 T cells integrate signals delivered during successive DC encounters in vivo. J Exp Med 202:1271–1278

Cemerski S, Das J, Locasale J, Arnold P, Giurisato E, Markiewicz MA, Fremont D, Allen PM, Chakraborty AK, Shaw AS (2007) The stimulatory potency of T cell antigens is influenced by the formation of the immunological synapse. Immunity 26:345–355

Cemerski S, Das J, Giurisato E, Markiewicz MA, Allen PM, Chakraborty AK, Shaw AS (2008) The Balance between T Cell Receptor Signaling and Degradation at the Center of the Immunological Synapse Is Determined by Antigen Quality. Immunity 29(3):414–422

Chan PY, Lawrence MB, Dustin ML, Ferguson LM, Golan DE, Springer TA (1991) Influence of receptor lateral mobility on adhesion strengthening between membranes containing LFA-3 and CD2. J Cell Biol 115:245–255

Chen W, Evans EA, McEver RP, Zhu C (2008) Monitoring receptor-ligand interactions between surfaces by thermal fluctuations. Biophys J 94:694–701

Choudhuri K, Wiseman D, Brown MH, Gould K, van der Merwe PA (2005) T-cell receptor triggering is critically dependent on the dimensions of its peptide-MHC ligand. Nature 436:578–582

Daniels MA, Teixeiro E, Gill J, Hausmann B, Roubaty D, Holmberg K, Werlen G, Hollander GA, Gascoigne NR, Palmer E (2006) Thymic selection threshold defined by compartmentalization of Ras/MAPK signalling. Nature 444:724–729

Davis SJ, van der Merwe PA (1996) The structure and ligand interactions of CD2: implications for T-cell function. Immunol Today 17:177–187

De Magistris MT, Alexander J, Coggeshall M, Altman A, Gaeta FC, Grey HM, Sette A (1992) Antigen analog-major histocompatibility complexes act as antagonists of the T cell receptor. Cell 68:625–634

Delamarre L, Pack M, Chang H, Mellman I, Trombetta ES (2005) Differential lysosomal proteolysis in antigen-presenting cells determines antigen fate. Science 307:1630–1634

Demotz S, Grey HM, Sette A (1990) The minimal number of class II MHC-antigen complexes needed for T cell activation. Science 249:1028–1030

Dudziak D, Kamphorst AO, Heidkamp GF, Buchholz VR, Trumpfheller C, Yamazaki S, Cheong C, Liu K, Lee HW, Park CG, Steinman RM, Nussenzweig MC (2007) Differential antigen processing by dendritic cell subsets in vivo. Science 315:107–111

Dustin ML (1997) Adhesive bond dynamics in contacts between T lymphocytes and glass supported planar bilayers reconstituted with the immunoglobulin related adhesion molecule CD58. J Biol Chem 272:15782–15788

Dustin ML (2007) Cell adhesion molecules and actin cytoskeleton at immune synapses and kinapses. Curr Opin Cell Biol 19:529–533

Dustin ML (2008a) Hunter to Gatherer and Back: Immunological Synapses and Kinapses as Variations on the Theme of Amoeboid Locomotion. Annu Rev Cell Dev Biol 24:577–596

Dustin ML (2008b) T-cell activation through immunological synapses and kinapses. Immunol Rev 221:77–89

Dustin ML, Springer TA (1989) T-cell receptor cross-linking transiently stimulates adhesiveness through LFA-1. Nature 341:619–624

Dustin ML, Sanders ME, Shaw S, Springer TA (1987) Purified lymphocyte function-associated antigen 3 binds to CD2 and mediates T lymphocyte adhesion. J Exp Med 165:677–692

Dustin ML, Carpen O, Springer TA (1992) Regulation of locomotion and cell-cell contact area by the LFA-1 and ICAM-1 adhesion receptors. J Immunol 148:2654–2663

Dustin ML, Ferguson LM, Chan PY, Springer TA, Golan DE (1996) Visualization of CD2 interaction with LFA-3 and determination of the two-dimensional dissociation constant for adhesion receptors in a contact area. J Cell Biol 132:465–474

Dustin ML, Bromley SK, Kan Z, Peterson DA, Unanue ER (1997a) Antigen receptor engagement delivers a stop signal to migrating T lymphocytes. Proc Natl Acad Sci USA 94:3909–3913

Dustin ML, Golan DE, Zhu DM, Miller JM, Meier W, Davies EA, van der Merwe PA (1997b) Low affinity interaction of human or rat T cell adhesion molecule CD2 with its ligand aligns adhering membranes to achieve high physiological affinity. J Biol Chem 272:30889–30898

Dustin ML, Olszowy MW, Holdorf AD, Li J, Bromley S, Desai N, Widder P, Rosenberger F, van der Merwe PA, Allen PM, Shaw AS (1998) A novel adaptor protein orchestrates receptor patterning and cytoskeletal polarity in T-cell contacts. Cell 94:667–677

Dustin ML, Starr T, Coombs D, Majeau GR, Meier W, Hochman PS, Douglass A, Vale R, Goldstein B, Whitty A (2007) Quantification and modeling of tripartite CD2-, CD58FC chimera (alefacept)-, and CD16-mediated cell adhesion. J Biol Chem 282:34748–34757

Evavold BD, Sloan-Lancaster J, Hsu BL, Allen PM (1993) Separation of T helper 1 clone cytolysis from proliferation and lymphokine production using analog peptides. J Immunol 150:3131–3140

Fischer UB, Jacovetty EL, Medeiros RB, Goudy BD, Zell T, Swanson JB, Lorenz E, Shimizu Y, Miller MJ, Khoruts A, Ingulli E (2007) MHC class II deprivation impairs CD4 T cell motility and responsiveness to antigen-bearing dendritic cells in vivo. Proc Natl Acad Sci USA 104:7181–7186

Freiberg BA, Kupfer H, Maslanik W, Delli J, Kappler J, Zaller DM, Kupfer A (2002) Staging and resetting T cell activation in SMACs. Nat Immunol 3:911–917

Germain RN (1997) T-cell signaling: the importance of receptor clustering. Curr Biol 7:R640–R644

Giannone G, Dubin-Thaler BJ, Rossier O, Cai Y, Chaga O, Jiang G, Beaver W, Dobereiner HG, Freund Y, Borisy G, Sheetz MP (2007) Lamellipodial actin mechanically links Myosin activity with adhesion-site formation. Cell 128:561–575

Grakoui A, Bromley SK, Sumen C, Davis MM, Shaw AS, Allen PM, Dustin ML (1999) The immunological synapse: A molecular machine controlling T cell activation. Science 285:221–227

Gunzer M, Schafer A, Borgmann S, Grabbe S, Zanker KS, Brocker EB, Kampgen E, Friedl P (2000) Antigen presentation in extracellular matrix: interactions of T cells with dendritic cells are dynamic, short lived, and sequential. Immunity 13:323–332

Harding CV, Unanue ER (1990) Quantitation of antigen-presenting cell MHC class II/peptide complexes necessary for T-cell stimulation. Nature 346:574–576

Henrickson SE, Mempel TR, Mazo IB, Liu B, Artyomov MN, Zheng H, Peixoto A, Flynn MP, Senman B, Junt T, Wong HC, Chakraborty AK, von Andrian UH (2008) T cell sensing of antigen dose governs interactive behavior with dendritic cells and sets a threshold for T cell activation. Nat Immunol 9:282–291

Huang CT, Huso DL, Lu Z, Wang T, Zhou G, Kennedy EP, Drake CG, Morgan DJ, Sherman LA, Higgins AD, Pardoll DM, Adler AJ (2003) CD4+ T cells pass through an effector phase during the process of in vivo tolerance induction. J Immunol 170:3945–3953

Huang JH, Cardenas-Navia LI, Caldwell CC, Plumb TJ, Radu CG, Rocha PN, Wilder T, Bromberg JS, Cronstein BN, Sitkovsky M, Dewhirst MW, Dustin ML (2007) Requirements for T lymphocyte migration in explanted lymph nodes. J Immunol 178:7747–7755

Huse M, Lillemeier BF, Kuhns MS, Chen DS, Davis MM (2006) T cells use two directionally distinct pathways for cytokine secretion. Nat Immunol 7:247–255

Imai T, Hieshima K, Haskell C, Baba M, Nagira M, Nishimura M, Kakizaki M, Takagi S, Nomiyama H, Schall TJ, Yoshie O (1997) Identification and molecular characterization of fractalkine receptor CX3CR1, which mediates both leukocyte migration and adhesion. Cell 91:521–530

Irvine DJ, Purbhoo MA, Krogsgaard M, Davis MM (2002) Direct observation of ligand recognition by T cells. Nature 419:845–849

James JR, White SS, Clarke RW, Johansen AM, Dunne PD, Sleep DL, Fitzgerald WJ, Davis SJ, Klenerman D (2007) Single-molecule level analysis of the subunit composition of the T cell receptor on live T cells. Proc Natl Acad Sci USA 104:17662–17667

Kaizuka Y, Douglass AD, Varma R, Dustin ML, Vale RD (2007) Mechanisms for segregating T cell receptor and adhesion molecules during immunological synapse formation in Jurkat T cells. Proc Natl Acad Sci USA 104:20296–20301

Kaldjian EP, Gretz JE, Anderson AO, Shi Y, Shaw S (2001) Spatial and molecular organization of lymph node T cell cortex: a labyrinthine cavity bounded by an epithelium-like monolayer of

fibroblastic reticular cells anchored to basement membrane-like extracellular matrix. Int Immunol 13:1243–1253
Krogsgaard M, Li QJ, Sumen C, Huppa JB, Huse M, Davis MM (2005) Agonist/endogenous peptide-MHC heterodimers drive T cell activation and sensitivity. Nature 434:238–243
Krummel MF, Sjaastad MD, Wulfing C, Davis MM (2000) Differential clustering of CD4 and CD3zeta during T cell recognition. Science 289:1349–1352
Kuhn JR, Poenie M (2002) Dynamic polarization of the microtubule cytoskeleton during CTL-mediated killing. Immunity 16:111–121
Kupfer A, Mosmann TR, Kupfer H (1991) Polarized expression of cytokines in cell conjugates of helper T cells and splenic B cells. Proc Natl Acad Sci U S A 88:775–779
Lammermann T, Bader BL, Monkley SJ, Worbs T, Wedlich-Soldner R, Hirsch K, Keller M, Forster R, Critchley DR, Fassler R, Sixt M (2008) Rapid leukocyte migration by integrin-independent flowing and squeezing. Nature 453:51–55
Latek RR, Unanue ER (1999) Mechanisms and consequences of peptide selection by the I-Ak class II molecule. Immunol Rev 172:209–228
Lee SJ, Hori Y, Groves JT, Dustin ML, Chakraborty AK (2002) Correlation of a dynamic model for immunological synapse formation with effector functions: two pathways to synapse formation. Trends Immunol 23:492–499
Lillemeier BF, Pfeiffer JR, Surviladze Z, Wilson BS, Davis MM (2006) Plasma membrane-associated proteins are clustered into islands attached to the cytoskeleton. Proc Natl Acad Sci USA 103:18992–18997
Lindquist RL, Shakhar G, Dudziak D, Wardemann H, Eisenreich T, Dustin ML, Nussenzweig MC (2004) Visualizing dendritic cell networks in vivo. Nat Immunol 5:1243–1250
Lovitch SB, Pu Z, Unanue ER (2006) Amino-terminal flanking residues determine the conformation of a peptide-class II MHC complex. J Immunol 176:2958–2968
Markiewicz MA, Carayannopoulos LN, Naidenko OV, Matsui K, Burack WR, Wise EL, Fremont DH, Allen PM, Yokoyama WM, Colonna M, Shaw AS (2005) Costimulation through NKG2D Enhances Murine CD8+ CTL Function: Similarities and Differences between NKG2D and CD28 Costimulation. J Immunol 175:2825–2833
Mattheyses AL, Axelrod D (2006) Direct measurement of the evanescent field profile produced by objective-based total internal reflection fluorescence. J Biomed Opt 11:14006
McGavern DB, Christen U, Oldstone MB (2002) Molecular anatomy of antigen-specific CD8(+) T cell engagement and synapse formation in vivo. Nat Immunol 3:918–925
Mempel TR, Pittet MJ, Khazaie K, Weninger W, Weissleder R, von Boehmer H, von Andrian UH (2006) Regulatory T cells reversibly suppress cytotoxic T cell function independent of effector differentiation. Immunity 25:129–141
Miller MJ, Wei SH, Parker I, Cahalan MD (2002) Two-photon imaging of lymphocyte motility and antigen response in intact lymph node. Science 296:1869–1873
Miller MJ, Wei SH, Cahalan MD, Parker I (2003) Autonomous T cell trafficking examined in vivo with intravital two-photon microscopy. Proc Natl Acad Sci USA 100:2604–2609
Miller MJ, Hejazi AS, Wei SH, Cahalan MD, Parker I (2004) T cell repertoire scanning is promoted by dynamic dendritic cell behavior and random T cell motility in the lymph node. Proc Natl Acad Sci USA 101:998–1003
Molon B, Gri G, Bettella M, Gomez-Mouton C, Lanzavecchia A, Martinez AC, Manes S, Viola A (2005) T cell costimulation by chemokine receptors. Nat Immunol 6:465–471
Monks CR, Freiberg BA, Kupfer H, Sciaky N, Kupfer A (1998) Three-dimensional segregation of supramolecular activation clusters in T cells. Nature 395:82–86
Mor A, Campi G, Du G, Zheng Y, Foster DA, Dustin ML, Philips MR (2007) The lymphocyte function-associated antigen-1 receptor costimulates plasma membrane Ras via phospholipase D2. Nat Cell Biol 9:713–719
Negulescu PA, Krasieva TB, Khan A, Kerschbaum HH, Cahalan MD (1996) Polarity of T cell shape, motility, and sensitivity to antigen. Immunity 4:421–430
Norcross MA (1984) A synaptic basis for T-lymphocyte activation. Ann Immunol 135D:113–134

O'Rourke AM, Mescher MF (1992) Cytotoxic T-lymphocyte activation involves a cascade of signalling and adhesion events. Nature 138:253–255

O'Rourke AM, Rogers J, Mescher MF (1990) Activated CD8 binding to class I protein mediated by the T-cell receptor results in signalling. Nature 346:187–189

Oddos S, Dunsby C, Purbhoo MA, Chauveau A, Owen DM, Neil MA, Davis DM, French PM (2008) High-speed high-resolution imaging of intercellular immune synapses using optical tweezers. Biophys J 95(10):L66–L68

Okada T, Cyster JG (2007) CC chemokine receptor 7 contributes to Gi-dependent T cell motility in the lymph node. J Immunol 178:2973–2978

Paul WE, Seder RA (1994) Lymphocyte responses and cytokines. Cell 76:241–251

Peterson DA, DiPaolo RJ, Kanagawa O, Unanue ER (1999) Cutting edge: negative selection of immature thymocytes by a few peptide-MHC complexes: differential sensitivity of immature and mature T cells. J Immunol 162:3117–3120

Pu Z, Carrero JA, Unanue ER (2002) Distinct recognition by two subsets of T cells of an MHC class II-peptide complex. Proc Natl Acad Sci U S A 99:8844–8849

Pu Z, Lovitch SB, Bikoff EK, Unanue ER (2004) T cells distinguish MHC-peptide complexes formed in separate vesicles and edited by H2-DM. Immunity 20:467–476

Purbhoo MA, Irvine DJ, Huppa JB, Davis MM (2004) T cell killing does not require the formation of a stable mature immunological synapse. Nat Immunol 5:524–530

Qi S, Krogsgaard M, Davis MM, Chakraborty AK (2006) Molecular flexibility can influence the stimulatory ability of receptor-ligand interactions at cell-cell junctions. Proc Natl Acad Sci USA 103:4416–4421

Rabinowitz JD, Beeson C, Wulfing C, Tate K, Allen PM, Davis MM, McConnell HM (1996) Altered T cell receptor ligands trigger a subset of early T cell signals. Immunity 5:125–135

Redmond WL, Hernandez J, Sherman LA (2003) Deletion of naive CD8 T cells requires persistent antigen and is not programmed by an initial signal from the tolerogenic APC. J Immunol 171:6349–6354

Reich Z, Boniface JJ, Lyons DS, Borochov N, Wachtel EJ, Davis MM (1997) Ligand-specific oligomerization of T-cell receptor molecules. Nature 387:617–620

Robertson JM, Jensen PE, Evavold BD (2000) DO11.10 and OT-II T cells recognize a C-terminal ovalbumin 323-339 epitope. J Immunol 164:4706–4712

Roose JP, Mollenauer M, Ho M, Kurosaki T, Weiss A (2007) Unusual interplay of two types of Ras activators, RasGRP and SOS, establishes sensitive and robust Ras activation in lymphocytes. Mol Cell Biol 27:2732–2745

Santori FR, Arsov I, Vukmanovic S (2001) Modulation of CD8+ T cell response to antigen by the levels of self MHC class I. J Immunol 166:5416–5421

Schamel WW, Arechaga I, Risueno RM, van Santen HM, Cabezas P, Risco C, Valpuesta JM, Alarcon B (2005) Coexistence of multivalent and monovalent TCRs explains high sensitivity and wide range of response. J Exp Med 202:493–503

Schmidt RE, Caulfield JP, Michon J, Hein A, Kamada MM, MacDermott RP, Stevens RL, Ritz J (1988) T11/CD2 activation of cloned human natural killer cells results in increased conjugate formation and exocytosis of cytolytic granules. J Immunol 140:991–1002

Scholer A, Hugues S, Boissonnas A, Fetler L, Amigorena S (2008) Intercellular adhesion molecule-1-dependent stable interactions between T cells and dendritic cells determine CD8+ T cell memory. Immunity 28:258–270

Shaw AS, Dustin ML (1997) Making the T cell receptor go the distance: a topological view of T cell activation. Immunity 6:361–369

Sims TN, Soos TJ, Xenias HS, Dubin-Thaler B, Hofman JM, Waite JC, Cameron TO, Thomas VK, Varma R, Wiggins CH, Sheetz MP, Littman DR, Dustin ML (2007) Opposing Effects of PKCtheta and WASp on Symmetry Breaking and Relocation of the Immunological Synapse. Cell 129:773–785

Skokos D, Shakhar G, Varma R, Waite JC, Cameron TO, Lindquist RL, Schwickert T, Nussenzweig MC, Dustin ML (2007) Peptide-MHC potency governs dynamic interactions between T cells and dendritic cells in lymph nodes. Nat Immunol 8:835–844

Sloan-Lancaster J, Evavold BD, Allen PM (1993) Induction of T-cell anergy by altered T-cell-receptor ligand on live antigen-presenting cells. Nature 363:156–159
Sloan-Lancaster J, Shaw AS, Rothbard JB, Allen PM (1994) Partial T cell signaling: altered phospho-zeta and lack of zap70 recruitment in APL-induced T cell anergy. Cell 79:913–922
Somersalo K, Anikeeva N, Sims TN, Thomas VK, Strong RK, Spies T, Lebedeva T, Sykulev Y, Dustin ML (2004) Cytotoxic T lymphocytes form an antigen-independent ring junction. J Clin Invest 113:49–57
Springer TA (1990) Adhesion receptors of the immune system. Nature 346:425–434
Springer TA (1995) Traffic signals on endothelium for lymphocyte recirculation and leukocyte emigration. Annu Rev Physiol 57:827–872
Springer TA, Dustin ML, Kishimoto TK, Marlin SD (1987) The lymphocyte function-associated LFA-1, CD2, and LFA-3 molecules: cell adhesion receptors of the immune system. Annu Rev Immunol 5:223–252
Stefanova I, Dorfman JR, Germain RN (2002) Self-recognition promotes the foreign antigen sensitivity of naive T lymphocytes. Nature 420:429–434
Steinman RM, Hawiger D, Nussenzweig MC (2003) Tolerogenic dendritic cells. Annu Rev Immunol 21:685–711
Stinchcombe JC, Bossi G, Booth S, Griffiths GM (2001) The immunological synapse of CTL contains a secretory domain and membrane bridges. Immunity 15:751–761
Sykulev Y, Joo M, Vturina I, Tsomides TJ, Eisen HN (1996) Evidence that a single peptide-MHC complex on a target cell can elicit a cytolytic T cell response. Immunity 4:565–571
Trombetta ES, Mellman I (2005) Cell biology of antigen processing in vitro and in vivo. Annu Rev Immunol 23:975–1028
Tseng SY, Waite JC, Liu ML, Vardhana S, Dustin ML (2008) T cell-dendritic cell immunological synapses contain TCR dependent multiple CD28-CD80 clusters that recruit protein kinase C-θ. J Immunol 181(7):4852–4863
van der Merwe PA, McNamee PN, Davies EA, Barclay AN, Davis SJ (1995) Topology of the CD2-CD48 cell-adhesion molecule complex: implications for antigen recognition by T cells. Curr Biol 5:74–84
Varma R, Campi G, Yokosuka T, Saito T, Dustin ML (2006) T cell receptor-proximal signals are sustained in peripheral microclusters and terminated in the central supramolecular activation cluster. Immunity 25:117–127
Weiss A, Littman DR (1994) Signal transduction by lymphocyte antigen receptors. Cell 76:263–274
Wild MK, Cambiaggi A, Brown MH, Davies EA, Ohno H, Saito T, van der Merwe PA (1999) Dependence of T cell antigen recognition on the dimensions of an accessory receptor-ligand complex. J Exp Med 190:31–41
Woolf E, Grigorova I, Sagiv A, Grabovsky V, Feigelson SW, Shulman Z, Hartmann T, Sixt M, Cyster JG, Alon R (2007) Lymph node chemokines promote sustained T lymphocyte motility without triggering stable integrin adhesiveness in the absence of shear forces. Nat Immunol 8:1076–1085
Worbs T, Mempel TR, Bolter J, von Andrian UH, Forster R (2007) CCR7 ligands stimulate the intranodal motility of T lymphocytes in vivo. J Exp Med 204:489–495
Wu LC, Tuot DS, Lyons DS, Garcia KC, Davis MM (2002) Two-step binding mechanism for T-cell receptor recognition of peptide MHC. Nature 418:552–556
Wulfing C, Sumen C, Sjaastad MD, Wu LC, Dustin ML, Davis MM (2002) Costimulation and endogenous MHC ligands contribute to T cell recognition. Nat Immunol 3:42–47
Yachi PP, Lotz C, Ampudia J, Gascoigne NR (2007) T cell activation enhancement by endogenous pMHC acts for both weak and strong agonists but varies with differentiation state. J Exp Med 204:2747–2757
Zinkernagel RM, Doherty PC (1975) H-2 compatability requirement for T-cell-mediated lysis of target cells infected with lymphocytic choriomeningitis virus. Different cytotoxic T-cell specificities are associated with structures coded for in H-2K or H-2D. J Exp Med 141:1427–1436
Zipfel WR, Williams RM, Webb WW (2003) Nonlinear magic: multiphoton microscopy in the biosciences. Nat Biotechnol 21:1369–1377

T Cell Migration Dynamics Within Lymph Nodes During Steady State: An Overview of Extracellular and Intracellular Factors Influencing the Basal Intranodal T Cell Motility

Tim Worbs and Reinhold Förster

Contents

T. Worbs (✉) and R. Förster
Institute of Immunolgy, Hannover Medical School,
Carl-Neuberg-Strasse 1, 30625 Hannover, Germany
e-mail: worbs.tim@mh-hannover.de

D. McGavern and M. Dustin (eds.), *Visualizing Immunity,* Current Topics in Microbiology and Immunology 334,
DOI: 10.1007/978-3-540-93864-4_4,

Abstract Naive T lymphocytes continuously recirculate through secondary lymphoid organs such as lymph nodes until they are eventually activated by recognizing cognate peptide/MHC-complexes on the surface of antigen-presenting cells. The intranodal T cell migration behavior leading to these crucial – and potentially rare – encounters during the induction of an adaptive immune response could not be directly addressed until, in 2002, the use of two-photon microscopy also allowed the visualization of cellular dynamics deep within intact lymph nodes. Since then, numerous studies have confirmed that, by default, naive T cells are extremely motile, scanning the paracortical T cell zone for cognate antigen by means of an apparent random walk.

This review attempts to summarize the current knowledge of factors influencing the basal migration behavior of naive T lymphocytes within lymph nodes during steady state. Extracellular cues, such as the motility-promoting influence of CCR7 ligands and the role of integrins during interstitial migration, as well as intracellular signaling pathways involved in T cell motility, will be discussed. Particular emphasis is placed on structural features of the lymph node environment orchestrating T cell migration, namely the network of fibroblastic reticular cells serving as migration "highways." Finally, new approaches to simulate the cellular dynamics within lymph nodes in silico by means of mathematical modeling will be reviewed.

Abbreviations

3D	Three-dimensional
APC	Antigen-presenting cell
CTLA-4	Cytotoxic T lymphocyte antigen-4
DC	Dendritic cell
DOCK2	Dedicator of cytokinesis 2
FDC	Follicular dendritic cell
FRC	Fibroblastic reticular cell
GEF	Guanine nucleotide exchange factor
GFP	Green fluorescent protein
GPCR	G protein-coupled receptor
HEV	High endothelial venules
ICAM-1	Intercellular adhesion molecule-1
LFA-1	Lymphocyte function-associated antigen-1
LN	Lymph node
PI3K	Phosphoinositide 3-kinase
PIP3	Phosphatidylinositol-3,4,5-triphosphate
PTX	Pertussis toxin
S1P	Sphingosine-1-phosphate
SLO	Secondary lymphoid organ
TCR	T cell receptor
TCZ	T cell zone

TPM	Two-photon laser-scanning microscopy
VCAM-1	Vascular cell adhesion molecule-1
VLA-4	Very late activated antigen-4
WT	Wild type

1 Introduction

1.1 Cellular Dynamics Within Lymph Nodes During the Induction of an Adaptive Immune Response

The successful induction of a primary adaptive immune response requires naive T lymphocytes to recognize, by means of their specific T cell receptor (TCR), cognate antigenic peptides presented in the context of major histocompatibility complex (MHC) molecules on the surface of antigen-presenting cells (APC) such as dendritic cells (DC). This interaction leads to the activation and subsequent proliferation of antigen (Ag)-specific T lymphocytes and is thought to essentially determine the outcome of an ensuing adaptive immune response, depending on the immunogenic or tolerogenic context of Ag presentation (Banchereau et al. 2000).

During evolution, secondary lymphoid organs (SLO) such as lymph nodes (LN) have evolved to provide a highly organized functional environment suited for the efficient induction of adaptive immune responses. Strategically positioned at the interface between the blood and the lymphatic system, LN represent important meeting places, bringing together different migratory immune cell populations to allow for the intimate interaction and exchange of (antigenic) information between cells of the innate and the adaptive immune system (von Andrian and Mempel 2003). Whereas Ag-carrying DC are known to migrate into draining LN from peripheral tissues and organs via afferent lymph vessels, recirculating naive T lymphocytes enter LN from the blood circulation via specialized high endothelial venules (HEV). Both of these homing processes have been shown to crucially depend on signals mediated by the chemokine receptor CCR7 (Forster et al. 1999; Gunn et al. 1999). Migrating DC as well as naive T cells are known to express this chemokine receptor and have been described, after entry into LN, to predominantly localize to the paracortical T cell zone (TCZ) where CCR7 ligands are constitutively expressed (Forster et al. 1999; Luther et al. 2000).

Despite these common guidance cues, finding DC bearing the "right" antigenic peptides within the crowded cellular environment of a LN is probably like finding the "needle in a haystack" for incoming T lymphocytes. It has been estimated that the TCR repertoire contains at least 25 million different specificities (Arstila et al. 1999), with approximately only 1 in 10^5–10^6 T cells recognizing cognate Ag during the induction of a primary immune response (Blattman et al. 2002). The paracortex of a LN is therefore virtually packed with T cells of irrelevant TCR specificity competing with the few T lymphocytes specific for a given Ag presented by DC.

At the same time, only a few DC within the LN draining a site of infection will present the relevant antigenic peptides. Furthermore, in order to frequently patrol the various SLO of the body, recirculating naive T cells spend only a limited amount of time in any given LN (with an estimated average dwell time of 12–18 h; Cyster 2005).

Taking into account all these limitations, it becomes clear that strategies enhancing the probability of fast encounters between Ag-presenting DC and T cells of matching TCR specificity within LN are of utmost importance for the efficient induction of primary adaptive immune responses (Bajenoff et al. 2007). Several mechanisms seem to cooperate to achieve this goal. By exhibiting sweeping movements with their highly dynamic dendritic extensions, DC are able to efficiently cover large volumes of their surrounding, and naive T lymphocytes exhibit impressive velocities while scanning the DC repertoire within the paracortical TCZ by means of an apparent random walk migration (Miller et al. 2004a). Finally, the nonhematopoietic backbone of the LN paracortex, namely the network of fibroblastic reticular cells (FRC), has recently been shown to provide the structural matrix on which not only DC reside (Katakai et al. 2004b; Sixt et al. 2005) but also T lymphocytes move along during their journey in the LN (Bajenoff et al. 2006), thereby most probably further enhancing the likelihood of T cell–DC interactions.

1.2 Two-Photon Microscopy

For a long time, the structural organization of LN as well as the functional properties of DC–T cell interactions could only be studied by evaluating sections of fixed tissues. The resulting microscopic images provided high resolution but were inevitably static and therefore not well suited to address the inherently dynamic processes of immune cell migration and interaction. This situation changed dramatically with the application of two-photon imaging. Compared to conventional epifluorescence microscopy as well as single-photon confocal laser scanning microscopy, two-photon laser scanning microscopy (TPM) offers several unique advantages that allow for the imaging of cellular dynamics within the depth of nonmanipulated, intact lymphoid tissues (Cahalan et al. 2002; Wei et al. 2003; Germain et al. 2006; Cahalan and Parker 2008; Niesner et al. 2008): Most importantly, using ultra-short (femtosecond) pulsed infrared laser sources in TPM, fluorochromes are excited by the near-simultaneous absorption of two infrared photons. Because of its much longer wavelength, the infrared excitation light is scattered within tissues much less compared to visible light, resulting in dramatically increased penetration depths. As the two-photon excitation is inherently limited to the actual focal spot of the infrared laser beam, out-of-focus bleaching and photodamage are significantly reduced. Without the need for a pinhole to reject out-of-focus light as in conventional single-photon confocal microscopy, the light gathering efficiency and therefore the penetration depth is even further improved (Wei et al. 2003). Depending on the microscope and the experimental setup, penetration depths of up to 300 μm have been reported for the imaging of peripheral LN of mice (Huang et al. 2004), which is sufficient to also address the migration and interaction behavior of immune cells in the deeper paracortical TCZ.

1.3 Experimental Setups Used for TPM Imaging of Immune Cell Dynamics Within Lymph Nodes

Different experimental models have been described for studying the motility of murine T lymphocytes in their natural LN environment (Germain et al. 2006, Cahalan et al. 2003; Bousso and Robey 2004; Mempel et al. 2004b). In 2002, Miller et al. imaged adoptively transferred, ex vivo fluorescent-dye-labeled T and B lymphocytes within explanted LN bathed in medium perfused with 95% O_2 by TPM. Providing the first detailed quantitative analysis of the movement behavior of immune cells in a lymphoid tissue, they reported mean velocities of ~11 μm min^{-1} for T cells and ~6 μm min^{-1} for B cells (Miller et al. 2002). One year later, the same group developed an intravital TPM imaging approach to visualize the migration of T lymphocytes within exposed inguinal LN of anesthetized mice, finding a strikingly similar intranodal T cell motility *in vivo* (Miller et al. 2003). With their series of pioneering studies, Miller and coworkers established the general idea of the basal intranodal migration behavior of naive T and B lymphocytes in the absence of exogenous Ag or inflammatory stimuli to resemble a high motile apparently autonomous random walk (Miller et al. 2004a, 2002, 2003). Although analyzing different naive T cell populations [CD4$^+$, CD8$^+$ or total T cells, isolated either from wild type (wt) or different TCR-transgenic mice], all following TPM studies imaging LN during steady state have confirmed the random walk migration behavior of T lymphocytes with mean velocities of ~12 μm min^{-1} (range 9–15 μm min^{-1}) and peak velocities of >30 μm min^{-1} (Bajenoff et al. 2006, Bousso and Robey 2003; Miller et al. 2004; Mempel et al. 2004; Hugues et al. 2004; Shakhar et al. 2005; Zinselmeyer et al. 2005; Okada and Cyster 2007; Huang et al. 2007; Worbs et al. 2007; Hwang et al. 2007; Wei et al. 2007; Fischer et al. 2007; Matheu et al. 2007). Interestingly, a direct side-by-side comparison of the baseline motility of naive CD4$^+$ versus CD8$^+$ T lymphocytes within the very same LN sample has not so far been performed.

The two different LN imaging setups possess unique advantages and disadvantages which have been extensively discussed (Germain et al. 2005, 2006; Huang et al. 2004, Sumen et al. 2004; Halin et al. 2005a). Generally faster and easier to handle, LN explants provide a greater degree of accessibility: They allow the free choice of different imaging angles and therefore the collection of data sets throughout the whole tissue sample as well as the direct manipulation of the explanted LN during imaging. Furthermore, the explant preparation is usually more stable, lacking problems of possible motion artifacts resulting from respiratory or pulse movement of the anesthetized animal. On the other hand, only the intravital imaging of LN in situ ensures the analysis of cellular dynamics in a truly physiological environment. This includes crucial factors such as intact blood circulation, lymph flow, and innervation of the LN tissue, as well as physiological oxygenation. Finally, only in living animals, systemic processes such as LN homing as well as lymphocyte egress can be studied comprehensively.

Despite these obvious differences regarding LN status, intravital as well as LN explant imaging setups used in a large number of studies have yielded surprisingly similar results regarding various readouts of intranodal immune cell migration and interaction dynamics (Germain et al. 2006; Cahalan and Parker 2008). As a result, the LN explant preparation is now commonly accepted as a validated alternative technique to study cellular dynamics within LN, as long as certain critical environmental parameters apply. For example, intranodal T cell motility was indeed reported to be strongly compromised when the oxygen partial pressure or the perfusion rate within the imaging chamber dropped below certain threshold values (Huang et al. 2007).

Importantly, these two basic experimental setups for TPM imaging of LN have also been used with great success to intensively study the interaction dynamics of different T cell subsets and DC during priming and tolerance induction, as well as the migration and interaction behavior of B cells (reviewed in Cahalan and Parker 2008; Allen et al. 2007a; Celli et al. 2008; Bousso 2008; Germain et al. 2008).

2 Extracellular Factors Influencing the Intranodal Migration Behavior of T Lymphocytes

2.1 G Protein-Coupled Receptors are Crucially Required for T Cell Motility

Two important families of cell-surface receptors involved in immune cell migration, the large group of chemokines receptors and the sphingosine-1-phosphate (S1P) receptor family, both belong to the superfamily of G protein-coupled receptors (GPCR). Those members primarily implicated in the trafficking of naive T lymphocytes (CCR7, CXCR4 and $S1P_1$) are believed to predominantly signal via activating heterotrimeric G proteins containing pertussis toxin (PTX)-sensitive Gαi-subunits (Cyster 2005). Using different PTX treatment regimes, three groups have recently performed LN imaging studies to analyze the Gαi-dependency of the basal intranodal T cell migration behavior.

As a prolonged treatment with PTX completely abrogates the CCR7-dependent LN homing of naive T cells, Cyster et al. have developed a method of *in vitro* PTX-pulse loading, taking advantage of the fact that the full-scale PTX-effect occurs with a delay of at least 1 h. Cotransferring differentially labeled PTX pulse-loaded and saline-treated control T cells into untreated recipients, they evaluated the motility of these cells 2 h after transfer by TPM within explanted inguinal LN (Okada and Cyster 2007): The median velocity of PTX-treated T cells within the TCZ was reduced by approximately 50% compared to control T cells (~6 μm min^{-1} vs ~12 μm min^{-1}), whereas the mean motility coefficient, a parameter describing the propensity of a cell to move away from its initial starting position, was decreased by more than 90% (3.8 μm^{2} min^{-1} vs 44 μm^{2} min^{-1}). The more pronounced impairment of displacement resulted from the PTX-treated T cells additionally displaying a much sharper turning behavior (mean turning angle 72.4° vs 44.0°), i.e., a reduced directional persistence of movement (Okada and Cyster 2007). Interestingly, 8 h after adoptive transfer, the PTX-treated T cells were no longer distributed homogenously but localized preferentially to the peripheral TCZ, in line with the idea that (Gαi-dependent) chemokine-mediated positional cues contribute to the intranodal localization of lymphocytes (see Sect. 2.9), leading to the functional organization of SLO (Cyster 2005).

Using an experimental approach without internal control, Huang and coworkers adoptively transferred naive T cells 24 h before treating the recipient mice systemically with high doses of PTX (200 μg kg^{-1} body weight) to block Gαi-signaling. Naive T lymphocytes visualized by TPM within explanted LN of high dose PTX-treated animals displayed a more than 60% reduced median velocity, a much sharper turning behavior, and a significantly higher arrest rate compared to naive T cells migrating within explanted LN of untreated recipients (Huang et al. 2007).

Finally, Asperti-Boursin and colleagues made use of a conventional inverted epifluorescence microscope instead of TPM to image the cellular dynamics at depths of only 40–50 μm within explanted LN slices. Adding PTX-treated as well as control T cells directly to the surface of LN slices, the recruitment and proper positioning of T lymphocytes was observed to be severely impaired after continuous *in vitro* PTX treatment for 2 h (Asperti-Boursin et al. 2007), which is reminiscent of Okada and Cyster reporting the misplacement of PTX-pulse loaded T lymphocytes in explanted LN to increase over time. Therefore, adapting the PTX-pulse loading method, Asperti-Boursin et al. also found both the mean velocity as well as the mean motility coefficient of PTX-treated T cells within LN slices to be markedly reduced (Asperti-Boursin et al. 2007). Of note, however, is that all motility parameter values reported in this study are consistently lower than those resulting from the TPM observation of intranodal T cell migration in intravital imaging setups or within intact explanted LN (see Sect. 1.3), possibly owing to the special preparation method not completely preserving physiological conditions.

Whereas the PTX-treatment used in the three studies described above led to the blockade of all three known isoforms of the Gαi-subunit, Hwang et al. analyzed the intranodal motility of naive T cells deficient only for the isoform Gαi2 by intravital TPM. Displaying a significantly reduced cell polarity, the Gαi2-deficient T cells preferentially localized within outer parts of the TCZ, where they migrated with average velocities reduced by approximately 50% compared to wt control T cells (Hwang et al. 2007). This resulted in a severely lower displacement over time, whereas the directionality of Gαi2-deficient T cell migration was only slightly reduced. Interestingly, the overall motility reduction was less pronounced within deeper areas of the paracortical TCZ, where only a few Gαi2-deficient T cells could be imaged.

The limited residual, highly confined migration observed also after potentially complete PTX-blockade of Gαi function indicates the possibility that a minimal

Gαi-independent component of basal intranodal naive T lymphocyte motility exists. On the other hand, Okada and Cyster reported the occasional appearance of PTX-treated T cells showing high velocity migration with very significant displacement rates possibly reflecting the presence of a Gαi-independent migration mechanism being active in only a few cells (Okada and Cyster 2007).

In summary, however, these studies have clearly demonstrated an absolutely crucial requirement of Gαi-mediated signals for the basal intranodal motility of naive T lymphocytes, in particular implicating an important role for the Gαi2-subunit. Consequently, the question arises: which cell surface receptors upstream of Gαi are actually recognizing the extracellular motility cues necessary for T cells to maintain their migratory pace within the TCZ of LN?

2.2 *Chemokine Receptors: Chemotaxis, Chemokinesis, Haptokinesis*

2.2.1 Homeostatic Chemokines Orchestrate Lymphocyte Migration During Steady State

Named after their characteristic ability to induce chemotaxis, i.e., a directional migration in response to a concentration gradient, in a wide range of immune cells the family of chemokines (*chemo*tactic cyto*kines*) consists of more than 40 small, mostly secreted chemoattractant proteins (Cyster 2005). Whereas the first chemokines were discovered to mediate the recruitment of innate immune cells and effector T cells to sites of inflammation, it is now established that several chemokines act as central regulators of the homeostatic trafficking of immune cells, playing an important role in the development and functional organization of SLO (von Andrian and Mempel 2003). This group of homeostatic chemokines includes the CCR7 ligands CCL19 and CCL21, the CXCR5 ligand CXCL13, and the CXCR4 ligand CXCL12, and all these chemokines have been demonstrated to exert strong chemotactic effects on naive lymphocytes *in vitro* (Gunn et al. 1998a, b; Ngo et al. 1998; Bleul et al. 1996). Being expressed within different regions of resting LN (Cyster 2005), it initially seemed possible that these homeostatic chemokines might also function as chemotactical cues within the physiological LN environment *in vivo*, guiding naive lymphocytes in a mode of directional migration along potential intranodal chemokine gradients (see also Sect. 2.8). Indeed, once the development of TPM allowed the direct visualization of lymphocyte migration dynamics deep within intact LN, the "mode" of chemokine action during intranodal T cell movement quickly became a focus of research (von Andrian and Mempel 2003; Wei et al. 2003).

2.2.2 CCR7 Ligands Mediate T Cell Chemokinesis In Vitro

An alternative mode of action discussed for chemokines is the so-called chemokinesis, i.e., the nondirectional enhancement of baseline motility in the absence of a concentration gradient. Such an effect was observed *in vitro* for T cells during the analysis of T cell/DC-interactions in 3D collagen gel matrices once DC were present either in the gel matrix itself or in the adjacent fluid phase containing culture supernatant (Gunzer et al. 2000). The authors speculated that soluble factors produced by the DC might have enhanced T cell motility and suggested DC-derived migration-enhancing CC-chemokines as possible candidates. Several further *in vitro* studies have reported that the CCR7 ligands CCL19 and CCL21 are indeed able to elicit an increased motility of naive T lymphocytes in the absence of a concentration gradient, i.e., to induce chemokinesis. Okada and Cyster observed a chemokinetic effect of CCL21, but not of CXCL12, on the *in vitro* transwell migration of naive T cells. In this case, to test for chemokinesis instead of chemotaxis, the chemokines are added to the upper as well as to the lower chamber of the transwell chambers in equal concentrations (Okada and Cyster 2007). Kaiser et al. studied the polarization and motility of human naive T lymphocytes *in vitro* on a monolayer of mature DC by video-microscopy. Employing PTX-treatment as well as blocking antibodies against CCL19 and CCR7, they identified CCL19 produced by DC as an important factor eliciting random motility of naive T cells *in vitro*, thereby enhancing their scanning performance (Kaiser et al. 2005). Irvine and coworkers reported the chemokines CCL19 and CCL21 to induce long-lasting chemokinesis in the absence of concentration gradients in $CD4^+$ T cells migrating on several adhesion ligands such as ICAM-1, VCAM-1, and fibronectin *in vitro* (Stachowiak et al. 2006), whereas Woolf et al. also demonstrated sustained rapid random walk motility of naive T lymphocytes on immobilized CCL21 in the complete absence of integrin ligands (Woolf et al. 2007).

2.2.3 CCR7 Ligands Function as Haptokinetic Stimulators of the Basal Intranodal T Cell Motility In Vivo

Recently, four independent studies have directly addressed the influence of CCR7-mediated signals on the basal migration behavior of naive T cells within LN, reporting largely similar results (Okada and Cyster 2007; Huang et al. 2007; Worbs et al. 2007; Asperti-Boursin et al. 2007).

Imaging adoptively transferred $CD4^+$ T lymphocytes in the paracortical TCZ of popliteal LN by intravital TPM, we found a significant reduction of the median of average cell velocities by ~33% (9.7 $\mu m\ min^{-1}$ vs 14.6 $\mu m\ min^{-1}$) as well as of the mean motility coefficient by ~ 55% (27.7 $\mu m^2\ min^{-1}$ vs 61.0 $\mu m^2\ min^{-1}$) when comparing the migration behavior of CCR7-deficient and wt T cells, whereas parameters characterizing the directional persistence of movement, such as the turning angle distribution and the meandering index, were largely similar (Worbs et al. 2007). Importantly, an almost identical impairment of intranodal motility was observed for wt T cells migrating within LN of *plt/plt* (plt) mice, a naturally occurring mutant lacking the expression of CCL19 and CCL21 within lymphoid tissues (median of average cell velocities 9.4 $\mu m\ min^{-1}$, mean motility coefficient 30.7 $\mu m^2\ min^{-1}$). Systemic application of a large dose of recombinant CCL21 by intravenous injection during TPM imaging restored the velocity and motility coefficient of wt T cells migrating within the TCZ of popliteal LN of plt mice to wt levels within 2.5 h (Worbs et al. 2007). Furthermore, the overall motility of wt T lymphocytes within superficial areas of wt LN was much lower than in the deeper paracortical TCZ, which is reminiscent of earlier reports (Huang et al. 2004, 2007; Mempel et al. 2004a). Considering the promigratory effect of CCR7 ligands, their almost complete absence in superficial regions of the LN cortex (Luther et al. 2000; Worbs et al. 2007; Asperti-Boursin et al. 2007; Katakai et al. 2004a) might directly explain the observed correlation between T cell motility and distance to the LN surface. In fact, when artificially reverting the distribution of CCR7 ligands within the popliteal LN by subcutaneous injection of a large dose of recombinant CCL19 into the hind foot pad, wt T lymphocytes were seen to accumulate rapidly within the subcapsular sinus exhibiting very high velocities (Worbs et al. 2007).

Performing TPM imaging of explanted inguinal LN, Okada and Cyster found the median velocity of CCR7-deficient T cells in LN of wt recipients as well as of wt T cells in LN of plt recipients to be reduced by approximately 20% compared to the median velocity of wt T cell migrating in explanted LN of wt recipients. The mean motility coefficient, in both situations of impaired CCR7-signaling, was decreased by approximately 50% whereas the mean turning angle was again not significantly changed (Okada and Cyster 2007).

In contrast to the aforementioned two studies, Huang and coworkers, also imaging explanted inguinal LN by TPM, reported comparatively smaller differences between CCR7-deficient and wt naive T cells regarding most of the motility parameters analyzed, namely median velocity, arrest coefficient, and confinement index. Interestingly, they instead found a significant difference in the median turning angle, with CCR7-deficient T cells expressing sharper turns, i.e., exhibiting a lower directional persistence (Huang et al. 2007).

Finally, Asperti-Boursin and colleagues found cellular velocities as well as mean motility coefficients of naive T lymphocytes migrating within plt LN slices to be significantly reduced compared to the motility observed in wt LN slices. The same held true for the direct comparison of the motility parameters of differentially labeled $CCR7^{-/-}$ and wt T cells either overlaid onto a wt LN slice or transferred into wt recipients before explantation and imaging (Asperti-Boursin et al. 2007).

Together, these data strongly indicate that the presence of CCR7 ligands within LN is required to maintain the highly motile random walk of naive T lymphocytes observed within the paracortical TCZ during steady state. As the motility-stimulating chemokines are most probably not present within LN in solution, but immobilized on FRC of the stromal cell network (see Sect. 2.9), it might be more adequate to refer to this effect as haptokinesis, i.e., the nondirectional stimulation of motility due to physical contact with a ligand-covered substrate (Mempel et al. 2006). Importantly, all these studies found the motility reduction in the absence of CCR7-mediated signals to be significantly lower than the more severe impairment caused by the PTX-induced complete blocking of the Gαi-signaling pathway (Okada and Cyster 2007; Huang et al. 2007; Asperti-Boursin et al. 2007), strongly suggesting that further Gαi-dependent receptors contribute to the basal intranodal motility of naive T lymphocytes.

2.2.4 The Chemokine Receptor CXCR4 and Its Ligand CXCL12

In the search for further motility-promoting GPCR besides CCR7, CXCR4 was immediately considered a candidate as this homeostatic chemokine receptor is also expressed on naive T lymphocytes (albeit at relatively low levels) and its ligand CXCL12 (SDF-1) is constitutively present in LN, although most prominently in the medullary and subcapsular regions and not in the paracortical TCZ (Asperti-Boursin et al. 2007; Hargreaves et al. 2001). However, using specific CXCR4-antagonists, two independent studies were unable to demonstrate a significant influence of a block of CXCR4-mediated signals on the movement behavior of naive T cells, either alone or in combination with deficiency for CCR7-signaling (Okada and Cyster 2007; Asperti-Boursin et al. 2007). Therefore, although it remains possible that CXCR4-signaling might influence the migration of T cells within the area of the medullary cords prior to egress from LN, an important contribution to the basal motility within the TCZ seems rather unlikely.

2.3 The Sphingosine-1-Phosphate Receptor $S1P_1$ Regulates T Cell Egress

$S1P_1$, a member of the sphingosine-1-phosphate receptor family of Gαi-dependent GPCR, is highly expressed on naive T cells and has been shown to be crucially required for the egress of T lymphocytes from LN, spleen, and thymus during steady state (Mandala et al. 2002; Matloubian et al. 2004; Allende et al. 2004; Halin et al. 2005; Pappu et al. 2007). Furthermore, upregulation of CD69 on recently activated lymphocytes has been correlated with a simultaneous downregulation of $S1P_1$ surface levels leading to the efficient "lymph node trapping" of antigen-specific lymphocytes within inflamed LN (Shiow et al. 2006). As S1P was described to elicit T cell chemotaxis *in vitro* and $S1P_1$ surface levels seem to gradually increase with LN dwell time of T lymphocytes (Matloubian et al. 2004, Graeler and Goetzl 2002; Lo et al. 2005; Schwab et al. 2005), it was hypothesized that the mechanism of LN egress might in fact represent a $S1P_1$-mediated chemotactic response of T lymphocytes within medullary regions directed towards high levels of the ligand S1P present in the efferent lymph fluid.

Recognizing the tremendous importance of the S1P–$S1P_1$-axis in the orchestration of lymphocyte trafficking, several TPM imaging studies have been performed to investigate if $S1P_1$-mediated signals also influence the interstitial migration of naive T lymphocytes within the paracortex of LN, using either $S1P_1$-deficient lymphocytes or pharmacologically interfering with $S1P_1$-signaling (Huang et al. 2007; Halin et al. 2005b; Wei et al. 2005). However, none of these experiments revealed a significant contribution of $S1P_1$-mediated signals to the basal intranodal random walk motility of T lymphocytes within the TCZ.

As the medullary cords had long been regarded to represent the main LN exit compartment for lymphocytes, two TPM studies have tried to visualize the lymphocyte migration behavior within medullary areas of explanted LN, reporting an overall reduced baseline T cell motility compared to that observed in the paracortical TCZ, as well as examples of "log-jammed" T cells accumulating along the borders of medullary sinuses in the presence of $S1P_1$-agonists (Wei et al. 2005; Sanna et al. 2006). Consequently, these studies also support an alternative mechanistic model for a role of S1P in egress regulation, indicating the presence of potential "egress portals" in the medullary lymphatic endothelium that can be closed due to agonistic engagement of $S1P_1$ on endothelial cells, thereby blocking the transendothelial migration of lymphocytes into the medullary sinuses and the efferent lymphatics.

In the light of these findings, it seems possible that the molecular cues governing lymphocyte motility might indeed be significantly different between different sub-compartments of LN. In such a scenario, the high abundance of CCR7 ligands within paracortical T cell areas would help to maintain the observed high intranodal random walk motility of T lymphocytes, whereas their relative lack in the medullary region would result in a much lower baseline T cell motility. According to the first model, S1P-gradients might then take over in the direct vicinity of efferent lymph vessels, possibly leading to a chemotactic T cell response promoting egress from LN within the medullary cords (Nombela-Arrieta et al. 2007). In this context, it is interesting to note that S1P was found to either enhance or inhibit chemokine-induced chemotaxis of naive but not activated T lymphocytes *in vitro* in a concentration-dependent manner (Graeler et al. 2002), suggesting an intimate cross-talk between these two important receptor families regulating lymphocyte migration in a Gαi-dependent manner.

Finally, in a recent study, Pham et al. provide strong evidence that the presence of high levels of CCL21 on FRC within the paracortical TCZ does not only provide haptokinetic promigratory stimuli but additionally actively inhibits T cell egress, effectively acting as a LN retention signal that competes with $S1P_1$-mediated cues (Pham et al. 2008). First, the authors identify the so-called cortical sinusoids, lined by LYVE-1^+ lymphatic endothelium and connected to the efferent lymphatics, as an additional site of lymphocyte egress from LN, which is actually much closer to the TCZ compared to the medullary sinuses classically regarded to be the main LN exit portal. Second, they show that CCR7-deficient T lymphocytes egress faster from LN than wt T cells, whereas T cells overexpressing CCR7 are retained longer within LN. Importantly, CCR7 deficiency also led to a partial recovery of T cell egress in mice treated with the functional $S1P_1$-antagonist FTY720, and PTX-treatment (abrogating CCR7- as well as $S1P_1$-mediated signals) was found to enhance the egress of already $S1P_1$-deficient T cells (Pham et al. 2008). Although the latter finding might suggest LN exit (potentially via cortical sinusoids), and not random walk migration on the paracortical FRC network, to be the "default" behavior of T lymphocytes in the absence of this competitive signaling (Dustin and Chakraborty 2008), it is somewhat difficult to envision T cell egress to take place at all under these conditions, given the extremely low residual T cell motility observed after complete PTX-induced Gαi-blockade (Okada and Cyster 2007).

In this scenario, while migrating within the paracortical TCZ constantly receiving haptokinetic CCL21 stimuli, a T cell moving into the vicinity of a cortical sinusoid during random walk would additionally experience the influence of $S1P_1$-mediated signals (high local S1P concentration or a – probably rather short-ranged – S1P-gradient).

The T cell would then have to make a decision to either continue with its random walk migration driven by the promigratory CCR7-signal, thereby staying on the CCL21 expressing FRC "highways," or to react to the $S1P_1$-mediated egress signal, thereby localizing within the exit compartment of the cortical sinusoid, subsequently leaving the LN. In the absence of CCR7-signals, T cells consequently seem to have a higher chance of egress, whereas blocking of both pathways by PTX-treatment leads to a reduction of motility as well as egress compared to the wt situation, possibly favoring a more rapid LN egress due to T cells quickly encountering cortical sinusoids after entering via neighboring HEV.

In the future, it will be highly interesting to see how this competitive interplay of migration-determining GPCR is actually regulated: (1) by features of the LN architecture, most importantly the stromal cell networks (Bajenoff et al. 2006), (2) by potentially variable expression levels of CCR7- as well as $S1P_1$-ligands within different compartments of either resting or inflamed LN (Pappu et al. 2007; Mueller et al. 2007), (3) by the T cell activation state, as T cells that have divided several times after Ag-recognition seem to simultaneously upregulate $S1P_1$ and downregulate CCR7 (Pham et al. 2008), and, particularly, (4) at the level of intracellular signaling, as both receptors share major signaling pathways including Gαi-recruitment (Dustin and Chakraborty 2008). However, in order to directly address by TPM the relative contribution of cortical sinusoids versus medullary cords to T lymphocyte egress, further tissue markers are needed as "landmarks" to precisely delineate the exact position of cells within LN subcompartments and relative to structural features such as the FRC network and lymphatic sinuses during imaging.

2.4 Analysis of the Role of Further GPCR in Intranodal T Cell Motility

Based on diverse findings, several further candidate GPCR as well as their ligands have recently been tested for their impact on the intranodal motility of T lymphocytes using TPM imaging of explanted LN (Huang et al. 2007): Analyzing T cells deficient for the thromboxane prostanoid receptor (TP) as well as generally interfering with the arachidonic acid metabolism by systemic treatment of the recipient mice with the cyclooxygenase/lipoxygenase-inhibitor ETYA for 18 h before LN explantation revealed no evidence for a contribution of arachidonic acid metabolites to the basal intranodal motility of T lymphocytes. Furthermore, T cells deficient for the adenosine receptor $A_{2A}R$ or the lysophosphatidylcholine receptor G2A also showed an intranodal migration behavior comparable to wt T lymphocytes (Huang et al. 2007).

In summary, the quest to identify all extracellular cues contributing to the high motility phenotype of naive T lymphocytes within LN is far from being solved. First, the "CCR7-independent part" of Gαi-dependent T cell motility has not yet been elucidated, and second, an explanation for the (very limited) Gαi-independent residual motility – GPCR coupling via non Gαi-subunits, motility-stimulating non-GPCR or a T cell intrinsic motility program – is also still missing.

2.5 The Role of Integrins in Steady State T Cell Migration

Adhesion molecules such as intercellular adhesion molecule-1 (ICAM-1) and vascular cell adhesion molecule-1 (VCAM-1) together with their corresponding integrin receptors LFA-1 (lymphocyte function-associated antigen-1) and VLA-4 (very late activated antigen-4) have been shown to be of crucial importance for the homing of leukocytes into lymphoid as well as nonlymphoid compartments during steady state and inflammation (von Andrian and Mackay 2000). Following activation by chemokine signals, integrins mediate the firm arrest of circulating leukocytes in the presence of strong shear forces to allow for the subsequent extravasation from the vascular lumen (Shamri et al. 2005).

Over recent years, several *in vitro* studies have additionally implicated integrins, and in particular LFA-1, in the migration of T cells (Smith et al. 2007). However, despite the expression of ICAM-1 and VCAM-1 by FRC as well as LN-resident DC (Katakai et al. 2004b; Wilson et al. 2003), it might turn out that the role of these important adhesion molecules in supporting basal T cell motility within the physiological 3D environment of the paracortical TCZ is in fact rather limited (Cahalan and Parker 2008).

Woolf and coworkers have recently reported T cells to exhibit a highly motile and sustained random walk on immobilized CCL21 *in vitro*, irrespective of the presence or absence of coimmobilized ICAM-1 and/or VCAM-1. They further demonstrate that, under shear-free conditions comparable to the LN parenchyma, LFA-1 and VLA-4 are actually kept in a nonadhesive state, effectively being "silenced" in the presence of this motility-promoting homeostatic chemokine (Woolf et al. 2007). In line with such a "mechanoregulation" of integrin responsiveness, the additional application of shear stress led to the chemokine-mediated activation of LFA-1 *in vitro* and consequently resulted in firm T cell arrest on surfaces coated with ICAM-1 and CCL21. Importantly, the basal intranodal motility of CD18-deficient T cells, lacking functional LFA-1 (CD11a/CD18), was only slightly reduced compared to wt control T cells as observed by TPM imaging of explanted inguinal LN (Woolf et al. 2007) as well as intravital microscopy of popliteal LN (Worbs et al., unpublished observation). Furthermore, blockade of VLA-4 resulted in a similar motility reduction of wt T cells migrating within the paracortical TCZ and did not further reduce the motility of CD18-deficient T cells (Woolf et al. 2007).

A recent study by Lämmermann and coworkers, analyzing the migration of DC rather than T cells, might provide further insight into the function of integrins during rapid interstitial migration of immune cells (Lammermann et al. 2008): Generating DC deficient for all 24 known integrin heterodimers expressed on the surface of leukocytes, the authors confirmed that integrins are indeed necessary for the *in vitro* migration on integrin ligand-coated 2D surfaces. Strikingly, however, integrin-deficient DC exhibited a completely normal migration behavior in all 3D environments tested, including 3D collagen gel matrices and skin explants. Importantly, the intranodal migration of integrin-deficient DC from the subcapsular sinus towards the paracortical TCZ was also comparable to wt DC as observed by intravital TPM imaging of popliteal LN (Lammermann et al. 2008).

Despite potential differences in the composition of the molecular machinery utilized by T lymphocytes and DC during interstitial migration, these findings suggest that, in the shear-free environment of the LN parenchyma, the integrin receptors of naive T cells migrating on a FRC scaffold highly decorated with motility-stimulating CCR7 ligands might indeed largely remain in an inactive, nonadhesive state in order to allow for the rapid random walk migration of T lymphocytes observed in the absence of cognate Ag. On the other hand, the small reduction in velocity as well as directionality observed for T lymphocytes migrating in the absence of either LFA-1 or VLA-4 function could be indicative for an alternative scenario: Low-level integrin-mediated adhesion forces might help T cells to "stay on track," i.e., to remain in direct contact with the surface of FRC to constantly receive haptokinetic motility cues as well as to retain the migration directionality imprinted by the FRC scaffold.

2.6 The Inhibitory Coreceptor CTLA-4 – the Reverse Stop Signal Model

It has been observed in a number of systems that a T cell recognizing cognate Ag by its TCR receives a so-called "stop signal," leading to intracellular Ca^{2+} signaling, reversal of T cell polarity, and an increase in T cell adhesiveness. It seems clear that this stop signal allows for the establishment of an intimate contact to the Ag-presenting APC by high-affinity LFA-1-ICAM-1 interactions and subsequent formation of the so-called immunological synapse (IS), although the details of these highly dynamic interactions leading to T cell activation remain to be further elucidated (Dustin 2008). In this context, cytotoxic T lymphocyte antigen-4 (CTLA-4)-expression on T lymphocytes might act as a special motility-enhancing factor, effectively overriding the TCR-mediated stop signal (Schneider et al. 2006). Performing TPM imaging, the authors describe preactivated antigen-specific T cells expressing the inhibitory coreceptor CTLA-4 to continue their basal random migration within LN also during Ag challenge, whereas those negative for CTLA-4 displayed a reduced motility and higher arrest rates under these conditions.

2.7 Endogenous MHC Class II Molecules Sustain $CD4^+$ T Cell Motility

In a recent study, Fischer and colleagues provide evidence that, also in the absence of cognate antigen, TCR-mediated signals might actually influence basal $CD4^+$ T cell motility (Fischer et al. 2007). Transferring TCR-transgenic OT-II T cells into MHC II-deficient recipients, the authors found a progressive impairment of the *in vivo* proliferative capacity as well as of the basal motility of $CD4^+$ T cells over time. TPM imaging of explanted LN revealed a normal intranodal T cell motility at day

2 after transfer into MHCII-deficient recipients, whereas after 7 days the transferred OT-II T cells were almost completely immotile. Further results indicated a reduced activity of the small GTPases Rac and Rap1, but not Ras, to be the underlying mechanism of the observed motility defect. Importantly, the reduced motility of CD4$^+$ T cells lacking self-peptide/MHCII-contacts also seemed to severely impair their capability to scan the repertoire of LN-resident DC, thereby contributing to the observed diminished *in vivo* proliferative response of OT II T cells (Fischer et al. 2007). Therefore, constant tonic TCR signaling due to nonspecific recognition of endogenous self-peptide/MHC class II complexes might not only promote the survival and proliferative capacity of peripheral CD4$^+$ T cells (Delon et al. 2002), but also preserve their basal motility status during migration within SLO.

2.8 Examples of Directional Lymphocyte Migration Within Lymphoid Organs: The Role of Chemotactic Guidance Cues

As described, numerous studies have confirmed that the default migration behavior of naive T lymphocytes within resting LN in the absence of cognate antigen represents an apparently autonomous random walk (see Sect. 1.3). According to this model, naive T cells would scan the DC repertoire present within the paracortical TCZ in a series of nondirectional "lunges," making the encounter of a T cell with a DC presenting cognate antigen effectively a stochastic event (Wei et al. 2003).

However, evidence for highly directional lymphocyte movement, possibly representing actual chemotaxis in response to chemokine gradients, comes from TPM studies imaging nonnaive or developing cell populations as well as situations of Ag-specific priming:

In 2005, Witt et al. reported a rapid directional movement of positively selected thymocytes towards the medulla of intact explanted thymi, speculating that CCR7-signaling could be chemotactically regulating this intrathymic migration (Witt et al. 2005b). By TPM imaging of intact explanted LN, a CCR7-dependent chemotactic migration of Ag-activated B cells was shown to occur along a CCL21 gradient directed towards the B/T border (Okada et al. 2005). Focusing on T cell priming, Castellino and colleagues published a study demonstrating the recruitment of naive CD8$^+$ T lymphocytes to sites of previous antigen-specific CD4$^+$ T cell–DC interactions. This recruitment was indeed found to be independent of antigen recognition by the naive CD8$^+$ T cells, but seemed to involve directional migration towards the CCR5 ligands CCL3 (MIP-1α) and CCL4 (MIP-1β) (Castellino et al. 2006). Statically evaluating DC–T cell-clusters during priming, Beuneu and coworkers report similar findings that existing Ag-specific interaction of CD4$^+$ T cells and DC led to a preferential recruitment of (in this case Ag-specific) CD8$^+$ T cells (Beuneu et al. 2006). Finally, Hugues et al. recently demonstrated by TPM analysis of explanted LN that naive CD8$^+$ T cells, independent of Ag specificity, "preferentially" interacted with DC already presenting cognate Ag to Ag-specific CD8$^+$ T cells (Hugues et al. 2007). With an average duration of around 5 min, these "preferential" interactions were much shorter than actual Ag-dependent T cell arrest, but significantly longer than observed in the absence of an ongoing Ag-specific CD8$^+$ T cell response. Although it remained unclear if these "preferential" interactions actually resulted from a directional chemotactic recruitment towards the DC or instead reflected prolonged contact times, the effect obviously required CCR5 expression by the naive CD8$^+$ T lymphocytes (Hugues et al. 2007).

Therefore, under conditions of priming, the directional chemotactic recruitment of T cells might indeed represent an important strategy to enhance the likelihood of productive encounters with rare Ag-presenting DC. Taking into account the role of the FRC network as the primary physical substrate of intranodal T cell movement (see Sect. 2.9), this "fine-tuning" of the basal migration behavior might involve chemokines immobilized on the surface of FRC increasing the probability for T cells to take certain routes on these stromal cell "highways," preferentially leading towards, e.g., mature DC presenting cognate Ag (Bajenoff et al. 2007; Mempel et al. 2006). In contrast, it remains unlikely that (at least long-range) chemokine gradients orchestrate the intranodal T cell migration behavior during steady state, as this should result in a highly synchronized and persistent directional movement of all lymphocytes present in one LN region. However, it has been hypothesized that even the random walk migration of lymphocytes observed within resting LN in the absence of cognate Ag might theoretically still result from chemokine gradients as long as the chemotactic effect of directional guidance was sufficiently short-ranged (Beauchemin et al. 2007).

2.9 *The Fibroblastic Reticular Cell Network Serves as a Scaffold for Intranodal T Cell Migration*

Long before the advent of TPM imaging, histological analysis of LN sections indicated that lymphocytes migrating within LN parenchyma would experience a densely packed as well as strategically compartmentalized microenvironment (von Andrian and Mempel 2003). Sophisticated immunofluorescence studies have consequently revealed the presence of intricate cellular networks of nonhematopoietic stroma cells forming the structural backbone of LN (Katakai et al. 2004a, b; Gretz et al. 1996; Kaldjian et al. 2001).

Within B cell follicles, follicular dendritic cells (FDC) represent the most important population of stromal cells of mesenchymal origin (Bajenoff et al. 2006). Being the main producers of the CXCR5–ligand CXCL13, FDC have been described to play a crucial role in Ag presentation to B cells and to support B cell homing and survival (Cyster et al. 2000).

In contrast, the stroma cell network within paracortical T cell areas was found to consist almost exclusively of FRC (Bajenoff et al. 2006; Katakai et al. 2004a), which produce and ensheath the extracellular matrix proteins of the basement membrane of the so-called "conduit system" (Sixt et al. 2005; Gretz et al. 2000). Based on the 3D reticular network of collagen fibers supposed to confer essential mechanical stability to the LN, the conduit system connects subcapsular and paracortical sinuses with the perivascular spaces of HEV. By this means, lymph-borne molecules below a certain molecular-weight threshold are channeled through the paracortex towards the HEV without percolating through the interstitial space of the LN parenchyma (Gretz et al. 2000). Furthermore, resident DC were found to tightly associate with FRC sampling the fluid content of the conduits for lymph-borne Ag (Sixt et al. 2005). T cells were shown to migrate *in vitro* on fibronectin-coated surfaces as well as within 3D collagen gel matrices (Gunzer et al. 2000; Stachowiak et al. 2006); however, these molecules are most probably "invisible" to migrating T lymphocytes within the paracortical TCZ as they are located within the lumen of the FRC-ensheathed conduit system. On the other hand, the FRC of the paracortical TCZ have been described to produce and present the CCR7 ligands CCL19 and CCL21 on their surface (Luther et al. 2000; Link et al. 2007), thereby potentially forming a chemokine-rich migration substrate for the apparent random walk of T lymphocytes within the paracortiocal TCZ of LN *in vivo*. Finally, FRC have also been reported to express the integrin ligands ICAM-1 and VCAM-1 (Katakai et al. 2004b); however, the influence of these adhesion molecules during rapid interstitial migration of T lymphocytes within the LN parenchyma may actually be rather limited (see Sect. 2.5).

In a groundbreaking study, Bajenoff and colleagues verified this hypothesis by intravital TPM imaging of the popliteal LN of special chimeric mice (Bajenoff et al. 2006): They generated bone-marrow chimeras by transferring wt bone marrow cells into irradiated ubiquitin promotor-GFP transgenic recipient animals in which the fluorescent protein is expressed by all nucleated cells (Schaefer et al. 2001). Eight weeks after reconstitution, hematopoietic cells were almost completely derived from the (nonfluorescent) wt bone marrow graft, whereas the nonhematopoietic cells constituting the LN stromal cell networks remained of GFP-transgenic host origin. Imaging the intranodal migration behavior of adoptively transferred fluorescently labeled T lymphocytes within the outer paracortex, they found a close correlation between the turning behavior of T cells and the presence of T cell-associated GFP^+ branches of the FRC network running at the corresponding angle (Bajenoff et al. 2006). Demonstrating that T cells almost exclusively move on the pathways provided by the FRC network, the authors could convincingly show that the 3D structure of the FRC network itself, with its characteristic dimensions and frequent crossing points (average distance between two intersections 17 ± 7 μm, range 5–37 μm), creates the observed random walk migration of T lymphocytes within the boundaries of the paracortical TCZ (Bajenoff et al. 2006).

Consequently, the constitutive presentation of the CCR7 ligands on the surface of FRC seems to provide important haptokinetic cues to T cells migrating on the FRC network, stimulating the basal intranodal motility of naive T lymphocytes (Okada and Cyster 2007; Worbs et al. 2007). At the same time, the preferential association of LN-resident DC with the FRC network (Sixt et al. 2005; Lindquist et al. 2004) should increase the chance of Ag-specific T cell activation by physically bringing together (largely sessile) DC and (highly motile) T cells on a shared structural matrix within LN (Lammermann and Sixt 2008).

Furthermore, the FRC network, extending towards and surrounding the perivascular space of HEV, was found to restrict the access of immigrating T lymphocytes into the paracortical zone to discrete HEV regions aptly termed "exit ramps" (Bajenoff et al. 2006). On the other hand, newly immigrated dermal DC bearing tissue-derived Ag have been reported to localize near to HEV within the area of the cortical ridge thereby again potentially enhancing the likelihood for direct encounters with newly immigrated T cells entering via HEV (Katakai et al. 2004a; Bajenoff et al. 2003).

The strands of the FDC stromal cell network, although possessing a much higher inherent motility than the FRC of the paracortical T cell, seem to play a very similar role for the steady state migration behavior of B cells within B cell follicles. Bajenoff and colleagues visualized B lymphocytes actively following the directions of the more irregular, moving FDC branches while migrating through B cell follicles by means of an apparent random walk (Bajenoff et al. 2006). Additionally, recent studies provided evidence that indeed CXCL13 presented on these FDC might be an important haptokinetic promigratory signal for B cells, showing that the motility of germinal center (GC) B cells is partially reduced within CXCL13-deficient animals (Allen et al. 2007b), and B cells deficient for the Gαi2-subunit exhibit a reduced motility within LN follicles (Han et al. 2005).

Finally, the mutually exclusive presence of the two different stromal cell networks within the two primary subcompartments of LN provides an elegant explanation for the strictly confined migration behavior observed for B and T lymphocytes (Bajenoff et al. 2006): Rather than following the chemotactic influence of soluble chemokine gradients, B and T lymphocytes seem to be confined to their respective compartments by a strong preference to remain in contact with either FDC or FRC during random walk migration, receiving haptokinetic motility signals by different surface-bound chemokines in the process (Cahalan and Parker 2008). In line with this hypothesis, T cells moving towards the T–B boundary and into the area of the follicular edge – the zone where B cells and helper T cells interact after antigen priming – are observed to turn sharply away from the B cell follicle without slowing down, most probably because they are still migrating on FRC which extend into the follicular edge region, potentially overlapping with the FDC network in this region (Bajenoff et al. 2006; Wei et al. 2003).

3 Intracellular Factors Contributing to Intranodal T Cell Motility

3.1 Cell Polarity and Directional Migration: Rho GTPases and the Actomyosin Cytoskeleton

The establishment and maintenance of a polarized cell shape is generally assumed to be an absolute prerequisite for directional migration, involving the differential distribution of several molecular systems between the front (leading edge) and the back (uropod) of a moving cell. In T cells, this includes the actomyosin cytoskeleton, the microtubule system, several classes of surface receptors such as integrins, and the vesicular transport as well as a set of evolutionary highly conserved "polarity proteins" (Krummel and Macara 2006). In particular, the continuous rapid remodeling of the actomyosin cytoskeleton might be absolutely crucial for the successful polarization and translocation of T lymphocytes, as they move much faster than most nonhematopoietic cells, lacking the formation of stress fibers or focal adhesions while exhibiting a characteristic "hand mirror" morphology during their amoeboid migration (Jacobelli et al. 2004; Friedman et al. 2005; Burkhardt et al. 2008).

For a large range of cells, including T lymphocytes, the intracellular signaling controlling actin dynamics during migration has been shown to rely on a set of interlinked feedback loops involving several members of the Rho family of small GTPases such as Rac, Cdc42, and Rho (Krummel and Macara 2006; Ward 2006). Whereas the formation of pseudopodia protrusions at the leading edge of migrating T cells by actin polymerization is largely controlled by Rac, Rho is thought to actively prevent a counterproductive formation of protusions at the uropod, and Cdc42 seems to control the overall cell polarity as a kind of directional "compass" (Burkhardt et al. 2008). Furthermore, MyH9, the heavy chain of the nonmuscle myosin IIA, was found to be enriched at the uropod, in particular the head–tail junction, of crawling T cells, critically contributing to uropod formation and motility of T lymphocytes *in vitro* (Jacobelli et al. 2004). The fast amoeboid migration of naive T cells, reaching mean velocities of ~12 μm min^{-1} and peak velocities of >30 μm min^{-1} *in vivo* (see Sect. 1.3), might therefore result from two independent remodeling processes of the actomyosin cytoskeleton acting in concert. Rac-controlled actin polymerization at the leading edge promotes the formation of protusions, and in parallel, myosin II-mediated contraction at the uropod might generate extrusive forces that effectively propel the bulk of the cytoplasm forward into the leading edge (Jacobelli et al. 2004). Finally, the authors speculate that inactivation of the myosin II machinery at the uropod might be a key step during depolarization and synapse formation after a T cell has received a TCR-mediated "stop signal" due to recognition of cognate Ag (Jacobelli et al. 2004).

On the other hand, a recent study by Morin et al. suggests a different function of myosin II during T cell migration: Within T lymphocytes crawling on integrin ligand-coated surfaces *in vitro*, binding of LFA-1 to ICAM-1 led to the recruitment of MyH9 to the uropod, where it might actually generate the critical force required for the successful cellular de-adhesion by rapid rupture of LFA-1-ICAM-1 bonds (Morin et al. 2008). Interfering with MyH9-function in the presence of immobilized ICAM-1 consequently resulted in a dramatically aberrant elongation of the uropod because of the defective tail detachment and retraction, whereas the motility of the leading edge, driven by actin-polymerization, was found to be largely unaffected (Morin et al. 2008).

3.2 Class IA, but not IB, Isoforms of Phosphoinositide 3-Kinase Contribute to the Basal Motility Level of T Lymphocytes Within LN

The family of phosphoinositide 3-kinases (PI3K) consists of four classes of enzymes (IA, IB, II, and III). Class IA and IB PI3K are heterodimers composed of different regulatory and catalytic subunits and have been described to mediate the intracellular production of the important second messenger phosphatidylinositol-3,4,5-triphosphate (PIP3) in response to different extracellular signals (Ward 2006). Class IA PI3K, containing one of the catalytic subunits p110α, p110β, or p110δ, are thought to act primarily downstream of immune cell receptors such as the TCR that are phosphorylated by specific tyrosine kinases upon cognate stimulus (Deane and Fruman 2004). On the other hand, class IB PI3K containing the catalytic subunit p110γ are regarded to be predominantly involved in chemotactic responses, as they are recruited to the inner surface of the plasma membrane upon chemokine receptor activation coupling through βγ-subunits of heterotrimeric G proteins (Ward 2006). However, whereas chemotaxis of neutrophils, monocytes and macrophages in response to inflammatory chemokines indeed crucially requires class IB PI3K, chemokine receptor signaling in lymphocytes seems to be much more complex (Oak et al. 2007). The *in vitro* migration towards the chemokines CCL19, CCL21, and CXCL12 was found to be only partially reduced in T cells deficient for p110γ, therefore lacking class IB PI3K (Nombela-Arrieta et al. 2004; Reif et al. 2004), and chemotactic T cell responses were completely unaffected by a lack of the p110δ catalytic isoform (i.e., the class IA isoenzyme PI3Kδ). In contrast, *in vitro* chemotaxis of B cells towards the CXCR5 ligand CXCL13 seems to be mainly PI3Kδ-dependent, but in turn largely independent of 110γ (Nombela-Arrieta et al. 2004; Reif et al. 2004). A similar differential dependency on class IA and IB PI3K was reported for lymphocyte homing to SLO, with p110γ deficiency completely abrogating the residual LN homing of DOCK2-deficient ($DOCK2^{-/-}$) T cells, whereas lack of p110δ, but not p110γ, abrogated B cell homing to Peyer's patches and the splenic white pulp (Nombela-Arrieta et al. 2004; Reif et al. 2004).

In order to clarify the actual contribution of different PI3K isoenzymes to the random walk motility of naive T lymphocytes within the LN environment, three groups have recently performed imaging studies using animals deficient for certain catalytic or regulatory subunits of class IA or IB PI3K as well as the pan-PI3K-inhibitor Wortmannin (Matheu et al. 2007; Asperti-Boursin et al. 2007; Nombela-Arrieta et al. 2007). Performing intravital TPM of popliteal LN, Nombela-Arrieta et al. analyzed the relative contribution of DOCK2, as well as of the class IB catalytic subunit of PI3K, p110γ, to basal intranodal T cell motility, finding that DOCK2 plays a dominant role (see Sect. 3.3). Imaging lymphocytes within explanted LN by TPM, Matheu and coworkers report the intranodal velocity of B and T cells treated with PI3K-specific concentrations of Wortmannin to be reduced by ~20–25% compared to untreated control cells (Matheu et al. 2007). Interestingly, the loss of different class IA regulatory subunits due to genetic deletion resulted in graded effects on the basal intranodal motility levels. Whereas T cells single-deficient for p85α or p85β exhibited a velocity reduction of 12 and 26%, respectively, T lymphocytes deficient for four out of five known class IA regulatory subunits (p85α, p85β, p50α, p55α) moved ~37% slower than control cells within the same LN, and the motility coefficient of these cells, as a measure for efficient cellular displacement, was even more strongly decreased in a similarly cumulative fashion (Matheu et al. 2007). In contrast, p85α proved to be the dominant isoform responsible for basal motility in B cells, with a lack of p85α additionally resulting in a strikingly altered B cell morphology featuring dendritic-like extensions. Taken together, although clearly implicating class IA PI3K in the maintenance of intranodal lymphocyte motility, these experiments do not distinguish if the impaired migration behavior is actually a direct consequence of an abrogated enzymatic activity of class IA PI3K and therefore PI3K-signaling in T and B cells or due to the loss of an adaptor functionality of class IA regulatory subunits independent of their usual interaction with class IA catalytic subunits (Matheu et al. 2007).

A third group utilized the LN slices model described earlier to image the migration behavior of differentially labeled Wortmannin-treated and untreated control T cells within areas corresponding to the paracortical TCZ by epifluorescence microscopy (Asperti-Boursin et al. 2007). In contrast to the aforementioned study, Asperti-Boursin et al. did not find any evidence for an influence of PI3K-inhibition by Wortmannin on the motility of naive T lymphocytes in the LN slice model. They therefore speculate that the actual relevance of PI3K-dependent intracellular signaling events during intranodal migration might be determined by the set of Rac guanine exchange factors (GEF) expressed within a specific cell type (Asperti-Boursin et al. 2007). On the other hand, again the particular preparation method of LN slices might result in a lymphocyte migration behavior differing from the one observed in a more physiological setting of intact LN imaging by TPM.

3.3 DOCK2 is Crucially Required for T Cell Motility During Steady State

It had been shown that Rac activity within chemokine-stimulated lymphocytes is predominantly controlled by DOCK2 (Fukui et al. 2001; Sanui et al. 2003), a hematopoietic cell-specific member of the CDM family of Rac GEF proteins (Reif and Cyster 2002). Indeed, Nombela-Arrieta and coworkers had previously demonstrated DOCK2 to play a crucial role during chemokine-induced F-actin polymerization, directed *in vitro* migration to chemokine gradients, and short-term homing to LN of T (as well as B) lymphocytes (Nombela-Arrieta et al. 2004; Fukui et al. 2001). Using intravital TPM to directly visualize the migration of adoptively transferred T lymphocytes within the popliteal LN of living anesthetized recipient mice, they recently analyzed the contribution of DOCK2, as well as of the class IB catalytic subunit of PI3K, p110γ, to basal intranodal motility during steady state conditions (Nombela-Arrieta et al. 2007). DOCK2$^{-/-}$ T cells displayed a massively impaired intranodal motility, characteristically only being able to oscillate around their initial tracking positions. The median instantaneous 3D velocity of T cells was almost 2-fold reduced whereas the mean motility coefficient was more than 15-fold decreased compared to wt control cells, and T cells exhibited a significantly broader distribution of turning angles (Nombela-Arrieta et al. 2007). On the other hand, T lymphocytes deficient for both DOCK2 as well as p110γ (DOCK2$^{-/-}$p110$\gamma^{-/-}$) did not display any further motility reduction compared to DOCK2$^{-/-}$ T cells, and, similar to the report of Asperti-Boursin et al. (Asperti-Boursin et al. 2007), p110γ-single deficient T lymphocytes (p110$\gamma^{-/-}$) were observed to display an intranodal migration behavior largely comparable to wt control cells except for a slightly broader distribution of turning angles and a slightly reduced mean motility coefficient (Nombela-Arrieta et al. 2007). Acting downstream of Gαi-dependent chemokine receptors such as CCR7, DOCK2 therefore seems to play a dominant nonredundant role in the intracellular signal transduction of environmental promigratory signals promoting an efficient intranodal T cell scanning behavior, whereas p110γ might partially contribute to directional persistence during interstitial T cell migration (see Sects. 2.1 and 2.2.3).

Furthermore, Nombela-Arrieta and coworkers found the egress of DOCK2$^{-/-}$ and DOCK2$^{-/-}$p110$\gamma^{-/-}$, but not p110$\gamma^{-/-}$, T lymphocytes from LN to be delayed and demonstrated by TPM imaging of medullary regions of explanted LN that DOCK2-mediated signals are substantially contributing to the motility of T lymphocytes within medullary cords (Nombela-Arrieta et al. 2007). Taking into account that the *in vitro* transwell migration of T (as well as B) cells towards S1P is almost completely abrogated in the absence of functional DOCK2 (Nombela-Arrieta et al. 2007), these findings suggest that DOCK2 might also be required for the intracellular signaling downstream of potential S1P$_1$-mediated egress signals within the LN medulla (also refer to Sect. 2.3).

3.4 *Velocity Fluctuations During Basal Intranodal T Cell Migration: The Role of Intracellular Ca^{2+} Signaling*

During their first TPM studies, Miller and coworkers observed naive T lymphocytes migrating within LN to exhibit large velocity fluctuations, effectively traveling in a series of "lunges." These were found to be separated by pauses during which the T cell was hypothesized to reorientate its cellular motility machinery of the actin cytoskeleton in order to be able to move on in a new direction (Wei et al. 2003; Miller et al. 2002). The apparent periodicity of these velocity fluctuations (of ~ 2 min) led to the speculation that naive T lymphocytes might actually follow an "intrinsic motility program," autonomously governing their intranodal migration behavior (Miller et al. 2003).

However, recent studies have made it clear that Gαi-dependent external promigratory signals are absolutely necessary for T lymphocytes to maintain their dynamic migratory behavior (Sect. 2.1), and that CCR7 ligands are important haptokinetic stimulators of the intranodal motility of naive T cells (Sect. 2.2.3). Furthermore, results from a computational approach to simultaneously model the intranodal movement behavior of naive T cells and DC in silico (Beltman et al. 2007b) suggest that the large and seemingly "rhythmic" velocity fluctuations described by Miller et al. are in fact random, most probably resulting from encounters between migrating T cells and physical "obstacles" within the LN TCZ such as the FRC network, DC, or other lymphocytes (see also Sect. 2.9).

Importantly, the repeated pausing of T lymphocytes during intranodal migration has been assumed to correlate with transient intracellular Ca^{2+} elevations

(Cahalan and Parker 2008), potentially elicited by very short-lived, Ag-independent encounters with DC (Miller et al. 2004a, b; Mempel et al. 2004a), inducing low-level TCR-signals in naive T cells that might also contribute to T cell survival (Revy et al. 2001) or even long-term maintenance of intranodal motility (see Sect. 2.7). Indeed, Asperti-Boursin et al. reported naive T cells loaded with the Ca^{2+} indicator fura-2 to occasionally exhibit transient elevations of intracellular Ca^{2+} levels in the absence of exogenous Ag while migrating within LN slices, and these Ca^{2+} elevations seemed to occur more frequently when the cells were arrested (Asperti-Boursin et al. 2007), consistent with the idea that an increase in the intracellular Ca^{2+} induces cell immobilization (Donnadieu et al. 1994; Bhakta et al. 2005). Wei and colleagues simultaneously monitored cytosolic Ca^{2+} levels as well as the motility of T lymphocytes loaded with the Ca^{2+} indicator indo-1 within intact explanted lymph nodes by TPM (Wei et al. 2007). They found infrequent, relatively small Ca^{2+} spikes, lasting less than 2 min and sharply rising to peak levels of about 200 mM, to be associated with the onset of a temporary slowing of T cells that migrated under steady state conditions in the absence of inflammatory or antigenic stimuli. However, under these experimental conditions, the averaged T cell velocity remained decreased for a few minutes even after intracellular Ca^{2+} levels had already returned to baseline (Wei et al. 2007). Therefore, the observed fluctuations of intracellular Ca^{2+} levels do not directly correlate with the ongoing stop-and-go behavior of naive T cells, and further TPM imaging studies will be necessary to elucidate the origin of the "fine-structure" of the velocity fluctuations during basal T lymphocyte migration within LN.

4 Computational Simulation of T Cell Migration and Interaction Dynamics: The In Silico Lymph Node Model

During the last 6 years, TPM imaging has been used extensively for the analysis of T cell migration and interaction processes within intact lymph nodes both ex vivo and *in vivo* (see Sect. 1.3). The large data sets generated in the process usually contain the 3D coordinates of individual T cells as a function of time as well as information on the duration of cellular contacts of T lymphocytes and, e.g., DC or FRC. However, while TPM is well suited for the direct visualization of relatively low numbers of labeled cells within lymph nodes, several (technical) limitations of present TPM studies have become apparent: The tissue volume being imaged is relatively small compared to the biological compartments of interest, the maximum duration of a single imaging session is normally restricted to 30–60 min, and, importantly, the density of cellular events that can be resolved and tracked efficiently during subsequent image analysis is clearly limited, prohibiting the simultaneous visualization of more than a few cellular players within the lymph node. The opportunity to circumvent at least some of these limitations by complementing TPM imaging with computational simulations was recognized early (Chakraborty et al. 2003; Witt et al. 2005a), and, consequently, several groups have developed mathematical models approximating the cellular dynamics within the paracortical TCZ of LN based on existing TPM imaging data (Beauchemin et al. 2007; Beltman et al. 2007a, b; Meyer-Hermann and Maini 2005; Preston et al. 2006; Riggs et al. 2008). While employing completely different algorithms, all computational models were designed to most accurately match the migration and interaction behavior of real and in silico T lymphocytes.

Utilizing the first available set of T cell motility data (Miller et al. 2002), Meyer-Herrman and Maini started from the principal idea of intranodal T cell migration as a random walk with persistence of orientation of individual T cells in the range of 2 min. They programmed a modified 2D cellular automaton-type Potts model (Graner and Glazier 1992; Glazier and Graner 1993), in which the in silico T cell is represented by a collection of subunits on a lattice. The migration of a T cell is then simulated as the independent movement of its cell membrane subunits which have to maintain coherence as well as a constant cell volume during the movement step (Meyer-Hermann and Maini 2005). Therefore, movement and shape changes (deformation) of the migrating in silico T cell are directly correlated. After each period of movement within this model, the simulated cell undergoes a phase of reshaping to regain spherical appearance before continuing the movement into a new direction. Importantly, this model yields a distribution of in silico T lymphocyte velocities that fits well with the variation observed *in vivo* by TPM imaging. As a further result, the authors confirmed the assumption of a motility-induced cell elongation during lymphocyte migration and explicitly reinforced the general model of the intranodal migration of lymphocytes being best described as a random walk with persistence of orientation (Meyer-Hermann and Maini 2005).

In a very different approach, Preston et al. modeled the encounters between T cells and DC within the LN quantitatively as a transport-limited chemical reaction with the dynamics being dependent on physical contact between randomly moving reactants, i.e., cellular players (Preston et al. 2006). Using asymptotic methods and Monte Carlo simulations, they derived a time distribution characterizing the delay before individual T cells encounter antigen-bearing DC. Testing their model, they found that the density of DC presenting a specific target antigen has to exceed 35 DC mm^{-3} for a T cell to have a more than 50% chance of an antigen-specific encounter to occur within the first 24 h (Preston et al. 2006). This value is actually much larger than previous estimations based on calculations not taking into account the transport process dynamics had predicted and nicely illustrates the potential of computational simulations to generate new hypothesis regarding the requirements and limitations of the cellular dynamics during the induction of an adaptive immune response within LN.

Beauchemin and colleagues have applied a very simple mathematical model, describing T cell motion as an unhindered 3D random walk within a 2D projection using only three independent variables (Beauchemin et al. 2007): After a period t_{free} of movement in a straight line with a fixed speed v_{free}, each in silico T cell remains immotile for a pause time t_{pause} (in which it is assumed to reorganize its actin cytoskeleton, allowing it to turn). After that, a new movement direction is assigned to each cell at random and the next round of straight movement is calculated, effectively generating trajectories representing a 2D random walk (Beauchemin et al. 2007). The parameter triplet fitting best to the pooled experimental data sets of four publications (Mempel et al. 2004b; Miller et al. 2002, 2003, 2004b) predicted a pause time of 0.5 min, a period of free movement of 2 min and an effective mean 3D velocity of 11.8 μm min^{-1} (Beauchemin et al. 2007). Interestingly, the resulting mean free path length d_{free} of the best fitting triplet is 38 μm, which is approximately twice the distance being reported as the average distance between two intersections of the FRC network (~17 ± 7 μm, range 5–37 μm; Bajenoff et al. 2006). Beauchmin et al. speculate that the simulation results might be indicative for migrating T cells arriving at an intersection of the FRC network to turn onto a new FRC fiber in 50% of the cases, a prediction which should be relatively easy to test by TPM imaging.

Beltman and coworkers have simulated the intranodal movement behavior of T cells and their interaction with DC in the absence of antigen applying a 3D cellular Potts model (Graner and Glazier 1992; Glazier and Graner 1993), additionally integrating (static) FRC as simple structural barrier elements (Beltman et al. 2007b). Similar to the model of Meyer-Herrman and Maini (see above), the in silico cells are composed of independently motile subunits occupying space within a 3D lattice. The principal assumption of this computational approach is that a migrating T cell has to polarize, forming a leading edge that determines the approximate direction of its movement (Friedman et al. 2005). Each in silico T cell is therefore assigned a preferred direction of movement which is repeatedly updated after certain time intervals according to its recent displacement. Realistic motility parameters (motility coefficients of 50–100 μm^2 min^{-1} and persistence times of ~2 min; Miller et al. 2003, 2004b; Mempel et al. 2004a) could be achieved by updating the "target direction" of each simulated T cell every 15–20 s. Importantly, the self-adjusting random walk migration behavior of the in silico T cells, closely resembling actual experimental data (Miller et al. 2004b), resulted exclusively from the interplay of the basic motility principle of directional persistence with the simulated, densely packed LN environment. Consequently, Beltman and colleagues also postulate that the intranodal T cell migration *in vivo*, being characterized by persistent motion on a short timescale but random walk migration on a long timescale as well as large velocity fluctuations with (apparently) periodic pausing, might be rather environmentally determined than being the result of an intrinsic T cell "motility program" (Beltman et al. 2007b).

Although the "directional propensity" parameter allowed the adjustment of in silico T cell velocities to realistic values, the authors report in a second study (Beltman et al. 2007a) that it also results in a directional persistence of the simulated T cells that is significantly higher than observed *in vivo* [data from Mempel et al. (2004a) evaluated in Preston et al. (2006)], leading to smaller in silico turning angles while the motility coefficients actually seem similar. Beltman et al. suggest that these differences might be explained by a strong preference of *in vivo* T lymphocytes to remain in contact with cells of the FRC network, representing the structural matrix of T lymphocyte migration within the paracortical TCZ (Bajenoff et al. 2006). The shape of the FRC network might therefore force T cells to exhibit larger turns (resulting in a lower directional persistence observed *in vivo*) while at the same time ensuring large displacements leading to high motility coefficients. However, in the first study, Beltman et al. also report the characteristics of in silico T cell migration to be highly similar in simulations with or without preferred adherence to the simulated FRC elements (Beltman et al. 2007b). Taken together, these discrepancies definitely call for refined computational models employing simulated FRC networks with realistic topology, dimensions, and adhesion properties.

An interesting prediction of the simulations of Beltman et al. was that migrating T cells would self-organize into highly dynamic streams (Beltman et al. 2007b). Although earlier TPM studies had not revealed any evidence for a globally organized large-scale bulk movement of T cells within the paracortical TCZ (Miller et al. 2003), Beltman and coworkers speculated that small or especially dynamic T cell streams could have escaped detection because of the small proportion of (labeled) T cells actually being imaged. Indeed, when adoptively transferring much larger numbers of labeled T lymphocytes for TPM imaging studies, Miller and colleagues could also detect some indications for the presence of short-lived and highly dynamic T cell streams in the TCZ of real LN (Beltman et al. 2007b). Furthermore, the simulation results suggested that these small streams would convert into a global bulk movement of the in silico T cells once the elements of the reticular network were omitted from the simulation routine indicating a further possible function of the FRC network as a "flow breaker," preventing undesirable large-scale bulk movements of T cells which could interfere with an efficient scanning of resident DC.

The calculated DC–T cell contact rates are largely comparable to previous estimations based on experimental results (Miller et al. 2004a; Bousso and Robey 2003). At any given time point, a simulated DC was found to be in contact with ~75 T cells, and most of these contacts were very short with an average duration of only 1.3 min (compared to approximately 3 min as described in previous TPM studies; Miller et al. 2004a, b; Mempel et al. 2004a). The in silico DC scanning rate equaled 3,000 T cells per hour of which 2,000 contacts were classified as unique (first contact of the same T cell–DC pair). In silico T cells, on the other hand, were estimated to scan 150 DC per hour, of which 100 contacts were considered unique. The computational model also confirmed the hypothesis that the scanning rates of both T cells and DC critically depend on the degree of T cell motility (Miller et al. 2004a; Bousso and Robey 2003), yielding extremely low scanning rates in silico when the simulated T cells were largely immotile (Beltman et al. 2007b).

In a second study, Beltman et al. investigated the interaction dynamics of T cells and DC in the presence of cognate Ag using the 3D cellular Potts model in combination with a modified, more elaborate algorithm to describe the dendrite dynamics of DC (Beltman et al. 2007a). They found that the initial decrease in T cell motility observed in the presence of activated Ag-loaded DC ["phase 1" behavior as reported in Mempel et al. (2004a)] could not be achieved by simulating an increased adhesion between Ag-specific T cells and Ag-bearing DC but required a dedicated "stop signal" delivered to the in silico T cell during direct contact with the DC (leading to the immediate loss of the directional propensity in the Ag-specific T cell). In order to correctly model long-lasting T cell–DC contacts observed at later time points *in vivo* ("phase 2" interactions), it was necessary to implement such a "stop signal"-mechanism as well as an increase in the adhesion strength between Ag-specific T cells and DC presenting cognate Ag (Beltman et al. 2007a). Furthermore, the simulation results suggested that DC might promote long-lasting contacts mechanistically by retracting those dendrites more slowly that are engaged with Ag-specific T cells, thereby preventing an accidental premature loss of contact, a hypothesis that again could be easily tested by TPM imaging. Finally, performing very long simulation runs of up to 2.5 days (after fitting the parameters to short timescale experimental data), Beltman and colleagues determined the average duration of Ag-specific contacts in silico to be approximately 1.5 h. This is much shorter than initially predicted, possibly because the short time window of TPM imaging (typically only 30–60 min due to technical limitations) leads to contact duration distributions biased against intermediate duration times (Beltman et al. 2007a). However, one should keep in mind that in the computational simulation of T cell–DC interaction dynamics all losses of contact were strictly stochastically, not taking into account any possible "termination signal" that might be required *in vivo* for the T cell to actively end an established contact (e.g., once a certain threshold of intracellular signaling is reached during activation).

Riggs and coworkers recently developed a new agent-based computational model of cellular dynamics within LN to compare random walk motility and chemotaxis-driven contacts during the scanning of DC by T cells in silico (Riggs et al. 2008). Their simulations closely mirror the *in vivo* situation as observed by TPM regarding T cell speed, short-term directional persistence of motion, and cell motility. Additionally, data regarding the density of T cells and DC within LN as well as values characterizing the LN topology (distance between HEV and efferent lymphatics) were integrated into the mathematical model. Strikingly, chemotaxis-driven repertoire scanning increases the total number of T cell–DC-contacts while at the same time decreasing unique contacts (first-time contacts between a given T cell–DC pair), effectively leading to the generation of fewer activated T lymphocytes (Riggs et al. 2008). As a result, Riggs et al. suggest that random walk motility might actually be the optimal search strategy for rare Ag-specific T cells to find the few DC presenting the matching cognate Ag within the LN TCZ, thereby maximizing the production of activated T cells within a given time frame.

5 Conclusion

Within the 6 years since the technique of TPM was introduced into the field of immunology, our view of how immune cells move, meet, and communicate during an immune response has changed dramatically. Going beyond the static analysis of literally "frozen" pictures of LN cryosections, TPM has revealed the interior of these SLO to be a densely packed place of truly bustling activity, invoking the conception of LN as dynamic "interaction enhancement machines" for immune cells: Although seemingly chaotic, the intranodal random walk migration of T lymphocytes represents a highly effective search strategy. It allows naive T cells to quickly scan large volumes of the LN interior for cognate Ag in an unbiased fashion and without hampering each other's pace, as might happen in a scenario of strictly guided directional mass movement. Mechanistically, the apparent random walk actually seems to result from a series of stochastic "route choices" as T cells repeatedly encounter bifurcation on the intranodal FRC "highways." Constantly receiving external motility-stimulating signals, the T cells are even more efficiently distributed within the paracortical TCZ while moving along the strands of the FRC network, at the same time inevitably encountering Ag-presenting DC residing on the same structural backbone. The induction phase of an immune response is often a close race between the pathogen and the immune system, and it has become clear that the extremely high intranodal motility of T lymphocytes is essential to increase the probability of early productive encounters between rare T cells specific for relevant pathogen-derived peptides and potentially equally rare APC presenting these peptides in LN draining an infection site.

Whereas several factors influencing the basal T cell motility within LN have been characterized during recent years, many open questions remain, including the identity of the – possibly redundant – promigratory Gαi-dependent signals besides the demonstrated haptokinetic effect of CCR7 ligands. Furthermore, it will be highly interesting to see if and how adhesion molecules, in particular integrins and integrin ligands, really contribute to the interstitial migration behavior of lymphocytes in the truly physiological 3D environment of intact LN. Similar to elucidating the precise role of intracellular Ca^{2+} signals during intranodal T cell migration, this task will most probably require TPM providing a higher temporal and spatial resolution in order to resolve the "fine-structure" of the cellular shape, the migratory path, and the interaction history of individual T lymphocytes. At the same time, it will be of utmost importance to visualize different cellular constituents of the LN interior simultaneously by cell-type-specific expression of fluorescent proteins or labeling with specific antibodies. Often mistakenly perceived to represent void space, the characteristic "black background" of current TPM imaging studies might finally be illuminated, revealing cellular and structural "landmarks." These features in turn will facilitate the intranodal "navigation" within different LN regions during imaging, which is critical, e.g., for the further analysis of lymphocyte migration events during LN egress. Importantly, TPM studies imaging nonnaive T lymphocyte populations (such as regulatory T cells or cytotoxic T lymphocytes) as well as compartments other than LN (e.g., tumors) provide new insights into the mechanism regulating T cell motility (Germain et al. 2008; Pittet and Mempel 2008; Ng et al. 2008). Finally, computational simulations of the cellular dynamics within LN might prove to be valuable tools supplementing TPM studies in the future, as many technical limitations encountered during the in situ imaging of real lymphocytes within intact LN do not apply in silico (Worbs et al. 2008). In the end, the growing knowledge about how T lymphocytes migrate and interact within LN might hopefully lead to the development of therapeutical strategies selectively targeting the intranodal motility of certain T cell subsets to either enhance or dampen a specific immune response.

References

Allen CD, Okada T, Cyster JG (2007a) Germinal-center organization and cellular dynamics. Immunity 27:190–202

Allen CD, Okada T, Tang HL et al (2007b) Imaging of germinal center selection events during affinity maturation. Science 315:528–531

Allende ML, Dreier JL, Mandala S et al (2004) Expression of the sphingosine 1-phosphate receptor, S1P1, on T-cells controls thymic emigration. J Biol Chem 279:15396–15401

Arstila TP, Casrouge A, Baron V et al (1999) A direct estimate of the human alphabeta T cell receptor diversity. Science 286:958–961

Asperti-Boursin F, Real E, Bismuth G et al (2007) CCR7 ligands control basal T cell motility within lymph node slices in a phosphoinositide 3-kinase-independent manner. J Exp Med 204:1167–1179

Bajenoff M, Granjeaud S, Guerder S (2003) The strategy of T cell antigen-presenting cell encounter in antigen-draining lymph nodes revealed by imaging of initial T cell activation. J Exp Med 198:715–724

Bajenoff M, Egen JG, Koo LY et al (2006) Stromal cell networks regulate lymphocyte entry, migration, and territoriality in lymph nodes. Immunity 25:989–1001

Bajenoff M, Egen JG, Qi H et al (2007) Highways, byways and breadcrumbs: directing lymphocyte traffic in the lymph node. Trends Immunol 28:346–352

Banchereau J, Briere F, Caux C et al (2000) Immunobiology of dendritic cells. Annu Rev Immunol 18:767–811

Beauchemin C, Dixit NM, Perelson AS (2007) Characterizing T cell movement within lymph nodes in the absence of antigen. J Immunol 178:5505–5512

Beltman JB, Maree AF, de Boer RJ (2007a) Spatial modelling of brief and long interactions between T cells and dendritic cells. Immunol Cell Biol 85:306–314

Beltman JB, Maree AF, Lynch JN et al (2007b) Lymph node topology dictates T cell migration behavior. J Exp Med 204:771–780

Beuneu H, Garcia Z, Bousso P (2006) Cutting edge: cognate CD4 help promotes recruitment of antigen-specific CD8 T cells around dendritic cells. J Immunol 177:1406–1410

Bhakta NR, Oh DY, Lewis RS (2005) Calcium oscillations regulate thymocyte motility during positive selection in the three-dimensional thymic environment. Nat Immunol 6:143–151

Blattman JN, Antia R, Sourdive DJ et al (2002) Estimating the precursor frequency of naive antigen-specific CD8 T cells. J Exp Med 195:657–664

Bleul CC, Fuhlbrigge RC, Casasnovas JM et al (1996) A highly efficacious lymphocyte chemoattractant, stromal cell-derived factor 1 (SDF-1). J Exp Med 184:1101–1109

Bousso P (2008) T-cell activation by dendritic cells in the lymph node: lessons from the movies. Nat Rev Immunol 8:675

Bousso P, Robey E (2003) Dynamics of CD8+ T cell priming by dendritic cells in intact lymph nodes. Nat Immunol 4:579–585

Bousso P, Robey EA (2004) Dynamic behavior of T cells and thymocytes in lymphoid organs as revealed by two-photon microscopy. Immunity 21:349–355

Burkhardt JK, Carrizosa E, Shaffer MH (2008) The actin cytoskeleton in T cell activation. Annu Rev Immunol 26:233–259

Cahalan MD, Parker I (2008) Choreography of cell motility and interaction dynamics imaged by two-photon microscopy in lymphoid organs. Annu Rev Immunol 26:585–626

Cahalan MD, Parker I, Wei SH et al (2002) Two-photon tissue imaging: seeing the immune system in a fresh light. Nat Rev Immunol 2:872–880

Cahalan MD, Parker I, Wei SH et al (2003) Real-time imaging of lymphocytes in vivo. Curr Opin Immunol 15:372–377

Castellino F, Huang AY, Altan-Bonnet G et al (2006) Chemokines enhance immunity by guiding naive CD8+ T cells to sites of CD4+ T cell-dendritic cell interaction. Nature 440:890–895

Celli S, Garcia Z, Beuneu H et al (2008) Decoding the dynamics of T cell-dendritic cell interactions in vivo. Immunol Rev 221:182–187

Chakraborty AK, Dustin ML, Shaw AS (2003) In silico models for cellular and molecular immunology: successes, promises and challenges. Nat Immunol 4:933–936
Cyster JG (2005) Chemokines, sphingosine-1-phosphate, and cell migration in secondary lymphoid organs. Annu Rev Immunol 23:127–159
Cyster JG, Ansel KM, Reif K et al (2000) Follicular stromal cells and lymphocyte homing to follicles. Immunol Rev 176:181–193
Deane JA, Fruman DA (2004) Phosphoinositide 3-kinase: diverse roles in immune cell activation. Annu Rev Immunol 22:563–598
Delon J, Stoll S, Germain RN (2002) Imaging of T-cell interactions with antigen presenting cells in culture and in intact lymphoid tissue. Immunol Rev 189:51–63
Donnadieu E, Bismuth G, Trautmann A (1994) Antigen recognition by helper T cells elicits a sequence of distinct changes of their shape and intracellular calcium. Curr Biol 4:584–595
Dustin ML (2008) T-cell activation through immunological synapses and kinapses. Immunol Rev 221:77–89
Dustin ML, Chakraborty AK (2008) Tug of war at the exit door. Immunity 28:15–17
Fischer UB, Jacovetty EL, Medeiros RB et al (2007) MHC class II deprivation impairs CD4 T cell motility and responsiveness to antigen-bearing dendritic cells in vivo. Proc Natl Acad Sci USA 104:7181–7186
Forster R, Schubel A, Breitfeld D et al (1999) CCR7 coordinates the primary immune response by establishing functional microenvironments in secondary lymphoid organs. Cell 99:23–33
Friedman RS, Jacobelli J, Krummel MF (2005) Mechanisms of T cell motility and arrest: deciphering the relationship between intra- and extracellular determinants. Semin Immunol 17:387–399
Fukui Y, Hashimoto O, Sanui T et al (2001) Haematopoietic cell-specific CDM family protein DOCK2 is essential for lymphocyte migration. Nature 412:826–831
Germain RN, Castellino F, Chieppa M et al (2005) An extended vision for dynamic high-resolution intravital immune imaging. Semin Immunol 17:431–441
Germain RN, Miller MJ, Dustin ML et al (2006) Dynamic imaging of the immune system: progress, pitfalls and promise. Nat Rev Immunol 6:497–507
Germain RN, Bajenoff M, Castellino F et al (2008) Making friends in out-of-the-way places: how cells of the immune system get together and how they conduct their business as revealed by intravital imaging. Immunol Rev 221:163–181
Glazier JA, Graner F (1993) Simulation of the differential adhesion driven rearrangement of biological cells. Phys Rev E Stat Phys Plasmas Fluids Relat Interdiscip Top 47:2128–2154
Graeler M, Goetzl EJ (2002) Activation-regulated expression and chemotactic function of sphingosine 1-phosphate receptors in mouse splenic T cells. FASEB J 16:1874–1878
Graeler M, Shankar G, Goetzl EJ (2002) Cutting edge: suppression of T cell chemotaxis by sphingosine 1-phosphate. J Immunol 169:4084–4087
Graner F, Glazier JA (1992) Simulation of biological cell sorting using a two-dimensional extended Potts model. Phys Rev Lett 69:2013–2016
Gretz JE, Kaldjian EP, Anderson AO et al (1996) Sophisticated strategies for information encounter in the lymph node: the reticular network as a conduit of soluble information and a highway for cell traffic. J Immunol 157:495–499
Gretz JE, Norbury CC, Anderson AO et al (2000) Lymph-borne chemokines and other low molecular weight molecules reach high endothelial venules via specialized conduits while a functional barrier limits access to the lymphocyte microenvironments in lymph node cortex. J Exp Med 192:1425–1440
Gunn MD, Ngo VN, Ansel KM et al (1998a) A B-cell-homing chemokine made in lymphoid follicles activates Burkitt's lymphoma receptor-1. Nature 391:799–803
Gunn MD, Tangemann K, Tam C et al (1998b) A chemokine expressed in lymphoid high endothelial venules promotes the adhesion and chemotaxis of naive T lymphocytes. Proc Natl Acad Sci USA 95:258–263
Gunn MD, Kyuwa S, Tam C et al (1999) Mice lacking expression of secondary lymphoid organ chemokine have defects in lymphocyte homing and dendritic cell localization. J Exp Med 189:451–460

Gunzer M, Schafer A, Borgmann S et al (2000) Antigen presentation in extracellular matrix: interactions of T cells with dendritic cells are dynamic, short lived, and sequential. Immunity 13:323–332

Halin C, Rodrigo Mora J, Sumen C et al (2005a) In vivo imaging of lymphocyte trafficking. Annu Rev Cell Dev Biol 21:581–603

Halin C, Scimone ML, Bonasio R et al (2005b) The S1P-analog FTY720 differentially modulates T-cell homing via HEV: T-cell-expressed S1P1 amplifies integrin activation in peripheral lymph nodes but not in Peyer patches. Blood 106:1314–1322

Han SB, Moratz C, Huang NN et al (2005) Rgs1 and Gnai2 regulate the entrance of B lymphocytes into lymph nodes and B cell motility within lymph node follicles. Immunity 22:343–354

Hargreaves DC, Hyman PL, Lu TT et al (2001) A coordinated change in chemokine responsiveness guides plasma cell movements. J Exp Med 194:45–56

Huang AY, Qi H, Germain RN (2004) Illuminating the landscape of in vivo immunity: insights from dynamic in situ imaging of secondary lymphoid tissues. Immunity 21:331–339

Huang JH, Cardenas-Navia LI, Caldwell CC et al (2007) Requirements for T lymphocyte migration in explanted lymph nodes. J Immunol 178:7747–7755

Hugues S, Fetler L, Bonifaz L et al (2004) Distinct T cell dynamics in lymph nodes during the induction of tolerance and immunity. Nat Immunol 5:1235–1242

Hugues S, Scholer A, Boissonnas A et al (2007) Dynamic imaging of chemokine-dependent CD8+ T cell help for CD8+ T cell responses. Nat Immunol 8:921–930

Hwang IY, Park C, Kehrl JH (2007) Impaired trafficking of Gnai2+/- and Gnai2-/- T lymphocytes: implications for T cell movement within lymph nodes. J Immunol 179:439–448

Jacobelli J, Chmura SA, Buxton DB et al (2004) A single class II myosin modulates T cell motility and stopping, but not synapse formation. Nat Immunol 5:531–538

Kaiser A, Donnadieu E, Abastado JP et al (2005) CC chemokine ligand 19 secreted by mature dendritic cells increases naive T cell scanning behavior and their response to rare cognate antigen. J Immunol 175:2349–2356

Kaldjian EP, Gretz JE, Anderson AO et al (2001) Spatial and molecular organization of lymph node T cell cortex: a labyrinthine cavity bounded by an epithelium-like monolayer of fibroblastic reticular cells anchored to basement membrane-like extracellular matrix. Int Immunol 13:1243–1253

Katakai T, Hara T, Lee JH et al (2004a) A novel reticular stromal structure in lymph node cortex: an immuno-platform for interactions among dendritic cells, T cells and B cells. Int Immunol 16:1133–1142

Katakai T, Hara T, Sugai M et al (2004b) Lymph node fibroblastic reticular cells construct the stromal reticulum via contact with lymphocytes. J Exp Med 200:783–795

Krummel MF, Macara I (2006) Maintenance and modulation of T cell polarity. Nat Immunol 7:1143–1149

Lammermann T, Sixt M (2008) The microanatomy of T-cell responses. Immunol Rev 221:26–43

Lammermann T, Bader BL, Monkley SJ et al (2008) Rapid leukocyte migration by integrin-independent flowing and squeezing. Nature 453:51–55

Lindquist RL, Shakhar G, Dudziak D et al (2004) Visualizing dendritic cell networks in vivo. Nat Immunol 5:1243–1250

Link A, Vogt TK, Favre S et al (2007) Fibroblastic reticular cells in lymph nodes regulate the homeostasis of naive T cells. Nat Immunol 8:1255–1265

Lo CG, Xu Y, Proia RL et al (2005) Cyclical modulation of sphingosine-1-phosphate receptor 1 surface expression during lymphocyte recirculation and relationship to lymphoid organ transit. J Exp Med 201:291–301

Luther SA, Tang HL, Hyman PL et al (2000) Coexpression of the chemokines ELC and SLC by T zone stromal cells and deletion of the ELC gene in the plt/plt mouse. Proc Natl Acad Sci USA 97:12694–12699

Mandala S, Hajdu R, Bergstrom J et al (2002) Alteration of lymphocyte trafficking by sphingosine-1-phosphate receptor agonists. Science 296:346–349

Matheu MP, Deane JA, Parker I et al (2007) Class IA phosphoinositide 3-kinase modulates basal lymphocyte motility in the lymph node. J Immunol 179:2261–2269
Matloubian M, Lo CG, Cinamon G et al (2004) Lymphocyte egress from thymus and peripheral lymphoid organs is dependent on S1P receptor 1. Nature 427:355–360
Mempel TR, Henrickson SE, Von Andrian UH (2004a) T-cell priming by dendritic cells in lymph nodes occurs in three distinct phases. Nature 427:154–159
Mempel TR, Scimone ML, Mora JR et al (2004b) In vivo imaging of leukocyte trafficking in blood vessels and tissues. Curr Opin Immunol 16:406–417
Mempel TR, Junt T, von Andrian UH (2006) Rulers over randomness: stroma cells guide lymphocyte migration in lymph nodes. Immunity 25:867–869
Meyer-Hermann ME, Maini PK (2005) Interpreting two-photon imaging data of lymphocyte motility. Phys Rev E Stat Nonlin Soft Matter Phys 71:061912
Miller MJ, Wei SH, Parker I et al (2002) Two-photon imaging of lymphocyte motility and antigen response in intact lymph node. Science 296:1869–1873
Miller MJ, Wei SH, Cahalan MD et al (2003) Autonomous T cell trafficking examined in vivo with intravital two-photon microscopy. Proc Natl Acad Sci U S A 100:2604–2609
Miller MJ, Hejazi AS, Wei SH et al (2004a) T cell repertoire scanning is promoted by dynamic dendritic cell behavior and random T cell motility in the lymph node. Proc Natl Acad Sci USA 101:998–1003
Miller MJ, Safrina O, Parker I et al (2004b) Imaging the single cell dynamics of CD4+ T cell activation by dendritic cells in lymph nodes. J Exp Med 200:847–856
Morin NA, Oakes PW, Hyun YM et al (2008) Nonmuscle myosin heavy chain IIA mediates integrin LFA-1 de-adhesion during T lymphocyte migration. J Exp Med 205:195–205
Mueller SN, Hosiawa-Meagher KA, Konieczny BT et al (2007) Regulation of homeostatic chemokine expression and cell trafficking during immune responses. Science 317:670–674
Ng LG, Mrass P, Kinjyo I et al (2008) Two-photon imaging of effector T-cell behavior: lessons from a tumor model. Immunol Rev 221:147–162
Ngo VN, Tang HL, Cyster JG (1998) Epstein-Barr virus-induced molecule 1 ligand chemokine is expressed by dendritic cells in lymphoid tissues and strongly attracts naive T cells and activated B cells. J Exp Med 188:181–191
Niesner RA, Andresen V, Gunzer M (2008) Intravital two-photon microscopy: focus on speed and time resolved imaging modalities. Immunol Rev 221:7–25
Nombela-Arrieta C, Lacalle RA, Montoya MC et al (2004) Differential requirements for DOCK2 and phosphoinositide-3-kinase gamma during T and B lymphocyte homing. Immunity 21:429–441
Nombela-Arrieta C, Mempel TR, Soriano SF et al (2007) A central role for DOCK2 during interstitial lymphocyte motility and sphingosine-1-phosphate-mediated egress. J Exp Med 204:497–510
Oak JS, Matheu MP, Parker I et al (2007) Lymphocyte cell motility: the twisting, turning tale of phosphoinositide 3-kinase. Biochem Soc Trans 35:1109–1113
Okada T, Cyster JG (2007) CC chemokine receptor 7 contributes to Gi-dependent T cell motility in the lymph node. J Immunol 178:2973–2978
Okada T, Miller MJ, Parker I et al (2005) Antigen-engaged B cells undergo chemotaxis toward the T zone and form motile conjugates with helper T cells. PLoS Biol 3:e150
Pappu R, Schwab SR, Cornelissen I et al (2007) Promotion of lymphocyte egress into blood and lymph by distinct sources of sphingosine-1-phosphate. Science 316:295–298
Pham TH, Okada T, Matloubian M et al (2008) S1P1 receptor signaling overrides retention mediated by G alpha i-coupled receptors to promote T cell egress. Immunity 28:122–133
Pittet MJ, Mempel TR (2008) Regulation of T-cell migration and effector functions: insights from in vivo imaging studies. Immunol Rev 221:107–129
Preston SP, Waters SL, Jensen OE et al (2006) T-cell motility in the early stages of the immune response modeled as a random walk amongst targets. Phys Rev E Stat Nonlin Soft Matter Phys 74:011910

Reif K, Cyster J (2002) The CDM protein DOCK2 in lymphocyte migration. Trends Cell Biol 12:368–373

Reif K, Okkenhaug K, Sasaki T et al (2004) Cutting edge: differential roles for phosphoinositide 3-kinases, p110gamma and p110delta, in lymphocyte chemotaxis and homing. J Immunol 173:2236–2240

Revy P, Sospedra M, Barbour B et al (2001) Functional antigen-independent synapses formed between T cells and dendritic cells. Nat Immunol 2:925–931

Riggs T, Walts A, Perry N et al (2008) A comparison of random vs. chemotaxis-driven contacts of T cells with dendritic cells during repertoire scanning. J Theor Biol 250:732–751

Sanna MG, Wang SK, Gonzalez-Cabrera PJ et al (2006) Enhancement of capillary leakage and restoration of lymphocyte egress by a chiral S1P1 antagonist in vivo. Nat Chem Biol 2:434–441

Sanui T, Inayoshi A, Noda M et al (2003) DOCK2 regulates Rac activation and cytoskeletal reorganization through interaction with ELMO1. Blood 102:2948–2950

Schaefer BC, Schaefer ML, Kappler JW et al (2001) Observation of antigen-dependent CD8+ T-cell/ dendritic cell interactions in vivo. Cell Immunol 214:110–122

Schneider H, Downey J, Smith A et al (2006) Reversal of the TCR stop signal by CTLA-4. Science 313:1972–1975

Schwab SR, Pereira JP, Matloubian M et al (2005) Lymphocyte sequestration through S1P lyase inhibition and disruption of S1P gradients. Science 309:1735–1739

Shakhar G, Lindquist RL, Skokos D et al (2005) Stable T cell-dendritic cell interactions precede the development of both tolerance and immunity in vivo. Nat Immunol 6:707–714

Shamri R, Grabovsky V, Gauguet JM et al (2005) Lymphocyte arrest requires instantaneous induction of an extended LFA-1 conformation mediated by endothelium-bound chemokines. Nat Immunol 6:497–506

Shiow LR, Rosen DB, Brdickova N et al (2006) CD69 acts downstream of interferon-alpha/beta to inhibit S1P1 and lymphocyte egress from lymphoid organs. Nature 440:540–544

Sixt M, Kanazawa N, Selg M et al (2005) The conduit system transports soluble antigens from the afferent lymph to resident dendritic cells in the T cell area of the lymph node. Immunity 22:19–29

Smith A, Stanley P, Jones K et al (2007) The role of the integrin LFA-1 in T-lymphocyte migration. Immunol Rev 218:135–146

Stachowiak AN, Wang Y, Huang YC et al (2006) Homeostatic lymphoid chemokines synergize with adhesion ligands to trigger T and B lymphocyte chemokinesis. J Immunol 177:2340–2348

Sumen C, Mempel TR, Mazo IB et al (2004) Intravital microscopy: visualizing immunity in context. Immunity 21:315–329

von Andrian UH, Mackay CR (2000) T-cell function and migration. Two sides of the same coin. N Engl J Med 343:1020–1034

von Andrian UH, Mempel TR (2003) Homing and cellular traffic in lymph nodes. Nat Rev Immunol 3:867–878

Ward SG (2006) T lymphocytes on the move: chemokines, PI 3-kinase and beyond. Trends Immunol 27:80–87

Wei SH, Parker I, Miller MJ et al (2003) A stochastic view of lymphocyte motility and trafficking within the lymph node. Immunol Rev 195:136–159

Wei SH, Rosen H, Matheu MP et al (2005) Sphingosine 1-phosphate type 1 receptor agonism inhibits transendothelial migration of medullary T cells to lymphatic sinuses. Nat Immunol 6:1228–1235

Wei SH, Safrina O, Yu Y et al (2007) Ca2+ signals in CD4+ T cells during early contacts with antigen-bearing dendritic cells in lymph node. J Immunol 179:1586–1594

Wilson NS, El-Sukkari D, Belz GT et al (2003) Most lymphoid organ dendritic cell types are phenotypically and functionally immature. Blood 102:2187–2194

Witt C, Raychaudhuri S, Chakraborty AK (2005a) Movies, measurement, and modeling: the three Ms of mechanistic immunology. J Exp Med 201:501–504

Witt CM, Raychaudhuri S, Schaefer B et al (2005b) Directed migration of positively selected thymocytes visualized in real time. PLoS Biol 3:e160

Woolf E, Grigorova I, Sagiv A et al (2007) Lymph node chemokines promote sustained T lymphocyte motility without triggering stable integrin adhesiveness in the absence of shear forces. Nat Immunol 8:1076–1085

Worbs T, Bernhardt G, Forster R (2008) Factors governing the intranodal migration behavior of T lymphocytes. Immunol Rev 221:44–63

Worbs T, Mempel TR, Bolter J et al (2007) CCR7 ligands stimulate the intranodal motility of T lymphocytes in vivo. J Exp Med 204:489–495

Zinselmeyer BH, Dempster J, Gurney AM et al (2005) In situ characterization of CD4+ T cell behavior in mucosal and systemic lymphoid tissues during the induction of oral priming and tolerance. J Exp Med 201:1815–1823

Chemoattractant Receptor Signaling and Its Role in Lymphocyte Motility and Trafficking

John H. Kehrl, Il-Young Hwang, and Chung Park

Contents

Abstract Intravital microscopy has provided extraordinary glimpses of lymphocytes crossing high endothelial venules, detailed the movements and interactions of lymphocytes within lymph organs, and recorded lymphocytes crossing the lymphatic endothelium into the efferent lymph. Helping to orchestrate these movements are signals generated by the engagement of chemoattractants with

---

J.H. Kehrl (✉), Il-Y. Hwang, and C. Park
B-cell Molecular Immunology Section, Laboratory of Immunoregulation, NIAID, NIH, 10 Center Dr. Bethesda, MD20892-1876, USA
e-mail: jkehrl@niaid.nih.gov

D. McGavern and M. Dustin (eds.), *Visualizing Immunity,* Current Topics in Microbiology and Immunology 334,
DOI: 10.1007/978-3-540-93864-4_5,

their cognate receptors. Chemokines present on high endothelial venules and within lymph organs, and the high levels of sphingosine 1-phosphate in the lymph provide signposts to help guide lymphocytes and provide intracellular signals that affect lymphocyte polarity and motility. Within lymph nodes, T and B lymphocytes migrate along networks of fibroblastic reticular cells and follicular dendritic, respectively, which provide an adhesive platform and solid phased chemokines. Illustrating the importance of chemoattractant receptors in these processes, lymphocytes that lack CXCR4, CXCR5, CCR7, or S1PR1, or which lack crucial signaling molecules activated by these receptors, exhibit defects in lymph node entrance, positioning, polarity, motility, and/or lymph node egress. This review will focus on the contributions of *in vivo* imaging of lymphocytes from various mouse mutants to our understanding of the roles chemoattractants play in lymphocyte entrance into and exit from lymph nodes, and in coordinating and facilitating the movements of lymphocytes within lymph nodes.

1 Introduction

The roles of chemokines and chemokine receptor signaling have been subject to numerous reviews as has lymphocyte trafficking (Bromley et al. 2008; Cyster 2005; Sumen et al. 2004). However, to provide a perspective for the discussion of the imaging studies that have broadened our understanding of chemokines and other chemoattractants in lymphocyte trafficking and motility, this review includes a overview of the major chemoattractant receptors important for naive T and B cells lymph node homing, interstitial motility, and lymph node egress. In addition, a summary is provided of the major signaling pathways that emanate from these receptors known to be relevant to lymphocytes. This is followed by a discussion of intravital microscopy of lymph nodes and lymphocyte motility. Finally, a more detailed analysis of the imaging studies examining the consequences of disrupting receptor expression or the downstream signaling pathways on the behavior of lymphocytes in lymph nodes will be presented.

2 Major Chemoattractant Receptors that Function in Lymph Node Homing, Motility, and Egress

2.1 CCR7 is a Critical Chemokine Receptor for Naive T cells

CCR7 is highly expressed on naive T cells and important for their normal trafficking (Forster et al. 1999). One of its ligands, CCL21, is expressed on the luminal side of HEVs where it promotes the firm arrest of recirculating naive T lymphocytes and facilitates their entrance into the lymph node parenchyma (Gretz et al. 2000). CCR7 ligands are present on the surface of the fibroreticular cells, which predominant in T cells zone, but not in the B cell follicle. Such a distribution provides a contact-dependent enhancement of cellular motility helping to maintain basal intranodal T cell motility and provides the positioning cues that restrict T cells to the T cell zones (Bajenoff et al. 2006; Forster et al. 1999). B lymphocytes, which express CCR7 at significantly lower levels than do T cells, can raise their CCR7 expression following engagement of the B cell antigen receptor. This facilitates T–B cell interactions by providing chemoattractant signal for B cells within a follicle to relocate to the border between the B and T cell zones in lymph nodes (Okada et al. 2005).

2.2 CXCR5 is a Critical Chemokine Receptor for Follicular B Lymphocytes

The chemokine receptor CXCR5 is of primary importance for B cells although T cells destined for the lymph node follicle express CXCR5 to do so (Ansel et al. 2000; Forster et al. 1996; Gunn et al. 1998). CXCL13, the sole CXCR5 ligand, is also expressed on the luminal side of HEVs, where it serves as an arrest chemokine for B cells particularly in Peyer's patches (Kanemitsu et al. 2005). In contrast, CCL19 and CXCL12 are the most important as arrest chemokines for B cell entrance into lymph nodes. The distribution of CXCL13 within lymph nodes is a mirror image of the distribution of CCR7 ligands, being restricted to lymph node follicles and undetectable in the T cell zones. CXCL13 is likely associated with follicular dendritic cells (FDCs), which are the major stromal cell in the lymph node follicle. CXCL13 distributed on the surface of FDCs delivers the signals that fuel B cell motility, provides the positioning cues that restricts B cells to follicles, and facilitates the entrance of CXCR5 positive T cells from the T cell zone (Bajenoff et al. 2006).

2.3 S1PR1 Functions in Lymphocyte Egress from Primary and Secondary Immune Organs

Sphingosine 1-Phosphate (S1P) is a lysophospholipid that is a potent regulator of a wide variety of cellular responses. Erythrocytes are a central source of the micromolar levels of S1P found in the blood while an unknown source accounts for the S1P present in lymph (Pappu et al. 2007). The localized concentration of S1P is regulated by cell surface lipid phosphate phosphatases and by S1P lyases. Five G-protein coupled receptors termed S1PR1–5 have been identified as targets of S1P mediating the cellular responses that have been attributed to it (Cyster 2005; Rosen et al. 2008). Lymphocytes prominently express S1PR1 and S1PR4 and to a lesser extent S1PR3 and S1PR5. S1PR4 signals by activating heterotrimeric G-protein Gi. Implicating S1P receptor signaling in the regulation of lymphocyte trafficking were studies with the pro-drug FTY720, which is converted to an active moiety by phosphorylation *in vivo*. FTY720-P exhibits agonistic activity at all of the S1P receptors with the exception of S1PR2. Treatment of mice with FTY720 led to the retention of lymphocytes in lymphoid organs (Rosen et al. 2008). Further studies specifically examining the role of S1PR1 signaling in lymphocyte trafficking noted that the absence of S1PR4 expression causes a thymocyte egress defect and the retention of lymphocytes in lymph organs (Matloubian et al. 2004). The authors concluded that the S1PR1 acts as a sensor for S1P to regulate the egress of lymphocyte from primary and secondary lymphatic organs. However, other studies supported a major role for S1PR1 signaling at the level of lymphatic endothelium (Sanna et al. 2006; Wei et al. 2005). Multiphoton imaging has provided much of the data supporting this alternative view (see below).

3 Major Mediators of Chemoattractant Receptor Signaling

3.1 Heterotrimeric G-Proteins Transduce Chemoattractant Receptor Signals

Chemokine and chemoattractant receptors use heterotrimeric G-proteins as signal transducers (Cyster 2005, Kehrl 2004). Heterotrimeric G-proteins are composed of α, β, and γ subunits (Neer 1995; Offermanns and Simon 1996). While G_α subunits can exist independently or in a trimeric complex, G_β and G_γ subunits exist as either a free heterodimer or as part of the trimeric complex, but not independently. The G_γ subunits undergo a lipid modification that targets $G_{\beta\gamma}$ subunits to the plasma membrane. The different G_α subunits have been used to grouped heterotrimeric G-proteins into subfamilies Gi, Gq, Gs, and G12. Members of Gi and perhaps Gq and G12 subfamilies mediate chemokine receptor signaling. In lymphocytes, one member of the Gi subfamily $G_{i\alpha2}$ plays a critical role (see below). Upon ligand binding, a GPCR functions as a guanine nucleotide exchange factor for G_α subunits resulting in the release of GDP and the binding of GTP. The binding of GTP to G_α allows G_α-GTP and $G_{\beta\gamma}$ to independently activate downstream effectors. Lymphocyte chemotaxis is triggered largely by the release of $G_{\beta\gamma}$ subunits from Gi proteins (Arai et al. 1997; Neptune et al. 1999). Also important are mechanisms that specifically regulate heterotrimeric G-proteins. G_α subunits have an intrinsic GTPase activity that hydrolyzes GTP returning the G_α subunit to a GDP-bound state and triggering its reassembly with $G_{\beta\gamma}$ to form the inactive heterotrimer. However, the intrinsic GTPase activity of G_α subunits cannot account for the rapid modulation of their GTP and GDP bound status. The discovery of regulators of G protein signaling (RGS) proteins, which accelerate the G_α GTPase reaction, resolved this issue (Druey et al. 1996; Koelle and Horvitz 1996). In addition, through a mechanism not yet fully understood, RGS proteins can accelerate the activation kinetics of the signal transduction pathway (Doupnik et al. 1997). Thus, RGS proteins may help G-protein linked signaling pathways decipher rapidly changing agonist profiles, thereby preserving temporal information. Since lymphocytes express a broad array of RGS proteins, they are likely to have significant roles in regulating chemoattractant receptor signaling (Kehrl 1998, 2004).

3.2 Small GTPases are Critical Downstream Molecules in the Chemokine Receptor Signaling Pathway

Small GTPases have all emerged as important intermediates in lymphocyte chemotaxis (see review in Thelen and Stein 2008). Unfortunately, how heterotrimeric G-protein signaling activates small GTPase remains poorly understood. This is in part because of the large number of guanine nucleotide exchange factors (GEFs) capable of activating these molecules. An early consequence of chemokine receptor signaling is the acquisition of a polarized shape with a leading edge, where chemokine receptors cluster, and a trailing edge or uropod, where adhesion molecules such as CD44, ICAM-1, and ICAM-3 accumulate (Krummel and Macara 2006; Manes et al. 2005). Helping to control some of these morphological changes are the small GTPases Rac, Cdc42, RhoA, and Rap1. The generation of actin filled lamellipodial protrusions depends upon Rac activation. Of the three Rac isoforms, Rac2 is exclusively expressed in myeloid and lymphoid cells and likely most important for lymphocyte chemotaxis (Croker et al. 2002). The mechanism by which Gi activation leads to Rac2 activation is still poorly understood. One important mediator is DOCK2. The lack of DOCK2 (SH3–SH2 adaptor protein dreadlocks), which functions in conjunction with ELMO1 (hematopoietic cell specific CDM family protein) and/or ELMO2 as an exchange factor for Rac, resulted in a substantial defect in B and T lymphocyte migration (Fukui et al. 2001). Cdc42 functions to regulate the extension of filopodia and its activation may help cells maintain a persistent leading edge in a migrating cell perhaps by limiting actin polymerization along the lateral edge of the pseudopodia. Tec kinases may function to activate Cdc42 in chemokine-induced lymphocyte migration (Takesono et al. 2004). RhoA functions in cell retraction and release of the uropod from substrates. Its activation in lymphocytes can occur via the chemoattractant receptor for lysophosphatic acid and those S1P receptors that couple to $G_{12/13}$, which activates RhoA through the exchange factor Lsc (Lee et al. 2004). Activated RhoA via its interaction with downstream effectors helps controls the high affinity state of LFA-1 and its lateral mobility, thereby regulating lymphocyte arrest on HEVs. The RhoA effector mDia is induced during T cell activation and regulates actin polymerization (Vicente-Manzanares et al. 2003). Furthermore, the RhoA effector p160 Rho-coiled coil kinase (ROCK) triggers myosin light chain phosphorylation and actomyosin contraction, which is required for lymphocyte polarization (Bardi et al. 2003). Since RhoA has been localized both at the leading edge and in the uropod of migrating cells, RhoA may be controlled by several distinct signaling pathways. Rap1 is implicated in integrin activation, lymphocyte polarization, and lymphocyte motility (Gerard et al. 2007; Kinashi and Katagiri 2005; Shimonaka et al. 2003).

A crucial downstream effector of Rap1 is termed regulator of adhesion and cell polarization enriched in lymphoid tissues (RAPL or Rassf5). Both B and T lymphocytes from these mice exhibited poor adhesion and homed poorly to lymphoid tissues, and *Rassf5* deficient thymocytes exhibited defective emigration from the thymus (Katagiri et al. 2003, 2004). A mediator of RASSF5 is the protein kinase Mst1, which functions in the induction of lymphocyte polarization and LFA-1 clustering (Katagiri et al. 2006). Furthermore, a leukocyte adhesion syndrome has been linked to impaired expression of a Rap1 activator calDAG-GEFI in lymphocytes, neutrophils, and platelets (Pasvolsky et al. 2007).

3.3 The Phosphoinositide 3-Kinase (PI3K) Family Members Function in Lymphocyte Trafficking and Motility

The PI3K lipid kinases regulate cell motility in a variety of cell types and have been recently implicated in lymphocyte motility (Matheu et al. 2007; Oak et al. 2007). Class IA and class IB PI3Ks trigger changes in intracellular phosphatidylinositol-3,4,5-triphosphate (PIP3) levels, which has been implicated as a fundamental mediator in chemoattractant gradient sensing, facilitating the recruitment of PH-domain containing proteins. Class IA PI3K exists as a heterodimer composed of one of three catalytic forms, p110α, p110β, or p110δ and one of five regulatory forms p85α (two alternative splice products p55α and p50α), p85β, and p55γ. Class IB PI3K has a p101/p87 regulatory subunit and p110γ catalytic subunit and is known as a downstream effector of $G_{\beta\gamma}$ signaling. As such it has been thought to be the major isoform activated by chemokine receptor signaling (Li et al. 2000). However, while p100γ has a prominent role in T cell responses to chemokine, it is much less important for B cells, and class IA isoforms are more important (Hirsch et al. 2000; Nombela-Arrieta et al. 2004). Although DOCK2 has a PIP3 binding site, PI3K activity is apparently not required for DOCK2-mediated lymphocyte migration. Recently, several studies using PI3K inhibitors and PI3K isoform-deficient mice have examined the behavior of lymphocytes in lymph nodes by TP-LSM or intravital imaging (Asperti-Boursin et al. 2007; Matheu et al. 2007; Ortolano et al. 2006).

3.4 Actin Dynamics and Myosin IIA

Chemokine receptor signaling ultimately leads to remodeling of the lymphocyte actin cytoskeleton, which accounts for the morphologic changes observed following chemokine exposure and the induction of cell motility. Actin nucleation is catalyzed by various assembly factors, several of which have been identified as important in lymphocytes (Burkhardt et al. 2008; Goley and Welch 2006). The activity of the assembly factor ARP2/3 is triggered by WASP and N-WASP, which are in turn activated by Cdc42. While lymphocytes deficient in WASP, N-WASP, and the WASP interacting protein WIP show defects in their chemotactic responses to specific chemokines, no imaging studies examining their intranodal motility have been reported (Gallego et al. 2006; Haddad et al. 2001; Snapper et al. 2005). Lymphocytes also express WAVE2, which acts downstream of Rac to activate ARP2/3; however, mice lacking WAVE2 die in utero impeding assessment of its role in lymphocyte chemotaxis and motility. Formins are actin assembly factors that act downstream of RhoA, which regulates the balance between filamentous and monomeric actin. Mouse T lymphocytes lacking the formin isoform mDia1 have impaired chemotaxis and reduced F-actin formation, while B lymphocytes are less affected (Sakata et al. 2007). Again, no imaging studies have been reported. Finally, the nonmuscle myosin heavy chain IIA, MyH9, the only class II myosin expressed in T cells, is associated with the uropod during crawling. Using D10.G4-IL2 CD4 T cell, MyH9 function was shown to be required for uropod maintenance and for the motility of the cells on a plate (Jacobelli et al. 2004). Since disruption of the MyH9 in mice results in early embryonic lethality, the analysis of MyH9 deficient lymphocytes will require a conditional knockout.

4 Imaging Lymphocyte Trafficking and Motility

4.1 Imaging T and B Lymphocytes in Lymph Nodes

The interactions of adoptively transferred lymphocytes with microvessels in lymph nodes were first observed by wide field fluorescence microscopy using a microsurgical preparation of the murine inguinal lymph node (von Andrian 1996). To observe the hemodynamics within HEVs and to capture lymphocyte–HEV endothelium interactions fluorescent images must be rapidly acquired. In general, the acquistion rate of laser scanning confocal microscopy limits its suitability for these types of studies. Two-photon laser scanning microscopy (TP-LSM) of a similar lymph node preparation does allow an examination of lymphocytes entering into the lymph node parenchyma. Within 10 min of their adoptive transfer, B and T lymphocytes can be observed entering into the lymph node cortical ridge region, and within 2 h, the cells have begun to move away from the cortical ridge region, B cells entering the lymph node follicle and T cells accessing the deeper cortical regions of the lymph node. Figure 1 shows several intravital TP-LSM images from the cortical ridge region of the inguinal lymph node of adoptively transferred fluorescently labeled lymphocytes entering through HEVs. The B and T lymphocytes entered with similar kinetics although they exhibited some preference for the particular HEV through which they entered. This suggests a nonuniform distribution of the arrest chemokines among the individual HEVs.

Two-photon microscopy imaging of explanted lymph nodes in medium perfused with oxygen provided the first detailed analysis of the movement of fluorescently labeled adoptively transferred lymphocytes within the lymph node (Miller et al. 2003). A year later, this methodology was adapted to imaging lymph nodes intravitally in

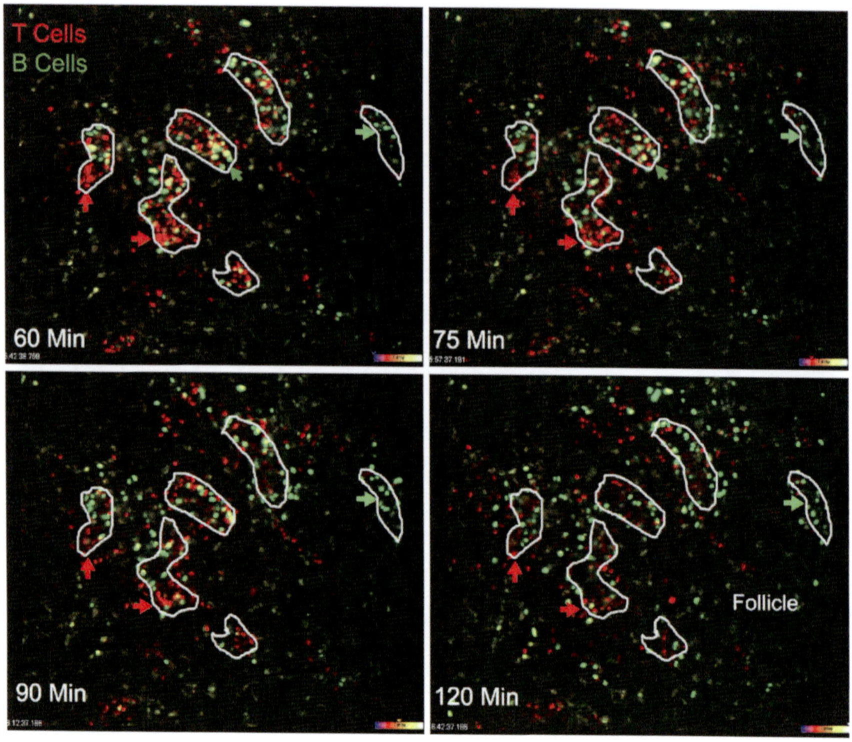

Fig. 1 Entrance of B and T cells into the cortical ridge region of the the inguinal lymph node detected by intravital microscopy. Fluroscently labeled B (*green*) and T (*red*) lymphocytes were adoptively transfered by intravenous injection into the tail vein of a recepient mouse. The entrance of the labeled cells was monitored by TP-LSM of an intravital preparation of the inguinal lymph node of the anesthetized recepient mouse. Single images of multiple images were acquired between 1 h and 2 h after transfer. The location of the high endothelial venules (*outlined in white*) was verified by intravenous injection of fluorescently labeled dextran, which is confined to the blood vessels, following the imaging (not shown). Show of the high endothelial venules showed apparent preference for T cells versus B cells

anesthetized mice (Miller et al. 2002). The results achieved by examining explanted lymph nodes versus those obtained by intravital microscopy have yielded remarkably consistent results. Several critical features emerged as important factors for the observation of lymphocyte motility including the level of oxygenation provided to explanted lymph nodes and the ambient temperature, which is critical for both ex vivo and *in vivo* imaging (Huang et al. 2007; Sumen et al. 2004). Within the T cell zone of the inguinal lymph node T lymphocytes are polarized moving at an average velocity of 10–12 μm min^{-1} in an apparent random walk pattern. The cortical T cells exhibited significant velocity fluctuations, often pausing followed by rapid cell movement described as lunges with peak velocities of 25 μm min^{-1}. Following each pause, there is a high probability that the T cell will reorient itself and move off in a different direction (Cahalan and Parker 2008; Wei et al. 2003). In contrast, T cells located more superficially in the lymph node in the region of the cortical ridge were noted to be much less motile. B lymphocytes exhibit a similarly polarized shape to that of T cells; however, they move more uniformly without the significant pausing and lunges of cortical T cells and at a slower average velocity (6–7 μm min^{-1}). The mechanisms underlying the different interstitial velocities and motility patterns of B and T lymphocytes within the lymph node are not known. While the maximum migration velocity of cells has been proposed to depend upon their levels of non-muscle myosin-II expression and the deformability of the cell's nucleus (Friedl and Weigelin 2008), follicular B cells have levels of *myh9* expression that exceed that of CD4 and CD8 T cells (Kehrl, unpublished observation).

Although the initial studies of the movement of B and T lymphocytes within the lymph node suggested that cells moved randomly, recent imaging studies have shown that fibroreticular cells in the T cell zone and follicular dendritic cells in the lymph node follicle help direct their movements (Bajenoff et al. 2006). Fibroreticular cells are fibroblast-like cells that provide a structural scaffold, which is decorated with cell surface ICAM-1 and chemokines. T cells follow the supporting fibers of the fibroreticular cells, rarely leaving them, as they migrate within the T cell zone. The fibers appear to dictate the directional movement of the cells. B cells are similarly associated with follicular dendritic processes. *In vitro*, it has been shown that chemokines can synergize with adhesion ligands to trigger T and B lymphocyte chemokinesis (Stachowiak et al. 2006). However, surprisingly, it seems that integrins are not required for lymphocyte and dendritic cell interstitial motility (Lammermann et al. 2008; Woolf et al. 2007) Superimposed upon the random movements of lymphocytes along the fibroreticular network are likely signals that generate nonrandom directed cell migration. For example, after antigen capture B cells located with the B cell follicle upregulate their expression of CCR7 and migrate along a chemotatic gradient towards the CCR7 ligand-rich T cell zone where they encounter activated CD4 T cells (Okada et al. 2005).

4.2 Imaging *Ccr7*$^{-/-}$ Lymphocytes

Lymph nodes from *Ccr7* deficient mice are devoid of naive T cells, but the T cell populations in the blood, the red pulp of the spleen, and in the bone marrow are greatly expanded (Forster et al. 1999). Secondary lymph organs exhibit morphological abnormalities. Several detailed analyses of adoptively transferred *Ccr7* deficient naive T cells have been performed by intravital TP-LSM, which have been recently reviewed (Worbs et al. 2008). One study showed that the deficient T -cells exhibit a very low motility in the superficial lymph node cortex, which could be rescued by the subcutaneous transfer of a CCR7 ligand. Imaging deeper into the T cell cortex, the velocity of the *Ccr7*$^{-/-}$ T cells was reduced by approximately 33% when compared to wild type cells, and the mean motility coefficient decreased by 55%. Other motility parameters such as turning angle distribution were not significantly altered suggesting that the CCR7 signaling contributed to basal motility of the cells and was not providing signals for directed migration (Worbs et al. 2007). Another group using an explanted lymph node approach also documented a reduction in interstitial velocity of *Ccr7*$^{-/-}$ T cells although the reduction was slightly smaller than by intravital imaging, 20 versus 33% (Okada and Cyster 2007). In contrast, another study also using explanted lymph nodes failed to observe a significant reduction in velocity, but did note increased median turning angles by the *Ccr7*$^{-/-}$ T cells as compared to wild type cells (Huang et al. 2007). Finally, a fourth group used lymph node slices from wild type mice and transferred fluorescently labeled wild type and *Ccr7*$^{-/-}$ T cells into the slice in order to compare their behavior. The authors found that the mean velocity of the *Ccr7* deficient T cells was reduced approximately 25% compared to that of wild type cells (Asperti-Boursin et al. 2007). *Ccr7*$^{-/-}$ B cells have been imaged in the context of an antigen driven B cell response. In contrast to wild type B cells from hen egg lysozyme transgenic mice, *Ccr7*-deficient hen egg lysozyme transgenic B cells failed to directionally migrate toward the lymph node follicle-T cell border following antigen challenge (Okada et al. 2005).

4.3 Imaging *Cxcr5*$^{-/-}$ Lymphocytes

Animals that are deficient for *Cxcr5* are devoid of inguinal lymph nodes and most Peyer's patches (Forster et al. 1996). The preserved PP exhibit an abnormal architecture. When *Cxcr5*-deficient B cells are adoptively transferred into wild type recipients they migrate into the T cell area, but fail to enter the B cell follicle. In contrast to the abundant imaging studies examining *Ccr7*$^{-/-}$ T cells, *Cxcr5*-deficient B cells have received less attention. Intravital and whole mount microscopy was used to examine B cell adhesion to high endothelial venules (HEVs). In Peyer's patches, B cell entry was found to be dependent upon CXCR5 although CCR7 and CXCR4 also contributed (Okada et al. 2002). In contrast, CXCR5 seemed to have little role in lymph node entry. No multiphoton studies of interstitial motility of *Cxcr5*$^{-/-}$ B cells have been reported.

4.4 S1PR1 Receptors and Lymph Node Egress

Genetic deletion studies examining the physiological implications associated with S1PR1 receptor signaling revealed that *S1pr1* null mice died in utero between E12.5 and E14.5. However, fetal liver transfers allowed the reconstitution of the hematopoietic system of lethally irradiated mice with *S1pr1* null lymphocytes. These mice lacked peripheral T cells because of impaired exit of T cells from the thymus, and have reduced numbers of B cells in the blood and the lymph. Adoptive cell transfer experiments established an intrinsic requirement for S1PR1 receptor expression by T and B cells for lymphoid organ egress (Matloubian et al. 2004).

This phenocopied the results observed by treating mice with the drug FTY720 and suggested that the drug behaved as a physiologic antagonist by hyperstimulating the S1PR1 triggering internalization and termination of S1P-generated promigratory signals (Cyster 2005). However, enigmatically, S1PR1 specific agonists are very poor chemoattractants and, in contrast to what would have been predicted the administration of *in vivo* of S1PR1 antagonists, do not induce lymphopenia (Rosen et al. 2008). This raises the possibility that the lymphocyte S1PR1 function in lymph organ egress by a mechanism distinct from the generally held hypothesis that they act as chemoattractant receptors. Unfortunately, no published studies have reported the imaging of *S1pr1* null lymphocytes crossing or attempting to cross the lymphatic endothelium of lymph nodes. Such studies are possible using either explanted lymph nodes or by intravital imaging B and T lymphocytes entering the cortical sinusoids, which provide a site for B and T cells egress within the lymph node cortex. Figure 2 shows an intravital image acquired in the region of the cortical sinusoid, which reveals the presence of adoptively transferred normal B cells within the sinusoid (imaging done by Park).

Besides lymphocytes, the lymphatic endothelium also expresses S1PR1 (Rosen et al. 2008). Evidence linking S1PR1 signaling to the regulation of barrier function of lymphatic endothelium initially arose from studies that associated signaling of this receptor with embryonic vascular maturation. S1P is known to induce the reorganization of the endothelial cytoskeleton and to stabilize adherens junctions and focal adhesion complexes at the cell membrane (McVerry and Garcia 2005). This offered an alternative hypothesis to account for the potent effect of FTY720 on lymphocyte trafficking. Two imaging studies have directly examined the impact of modulating S1PR signaling on lymphocytes in the medullary region of the lymph node. The first study used explanted lymph nodes and ex vivo application of wheat germ agglutinin. In fixed tissues, wheat germ agglutinin stained a stromal component subjacent to a known marker of the lymphatic endothelium (Wei et al. 2005). The second study addressed the criticism that the explanted lymph node preparations disrupted the normal flow of lymph by performing intravital microscopy. However, in the second study, the lymphatic endothelium was not directly visualized and only the behavior of the T cells in the medullary region of the lymph node was analyzed (Sanna et al. 2006). These studies showed that S1PR1 antagonists did not alter the interstitial movements of T and B cells in the lymph node cortex and follicle, respectively, nor did they alter the egress of lymphocytes into the lymph in the medullary region. However, S1PR1 agonists such as SEW2871 induced lymphopenia, slowed T cells in the medullary region, and inhibited T lymphocyte medullary egress. The reduction in T cell motility and egress inhibition could be rapidly reversed by the administration of a S1PR1 competitive antagonist. The reduction in T cell motility by the S1PR1 agonists was attributed to "log jamming" due to block of lymphocyte egress at the level of the endothelium, However, a direct suppression of lymphocyte motility by S1PR1 agonism remains a possibility. This possibility is supported by a study that showed S1PR1 agonism inhibited the entry of tissue T cells into afferent lymphatics, in part by triggering integrin activation (Ledgerwood et al. 2008).

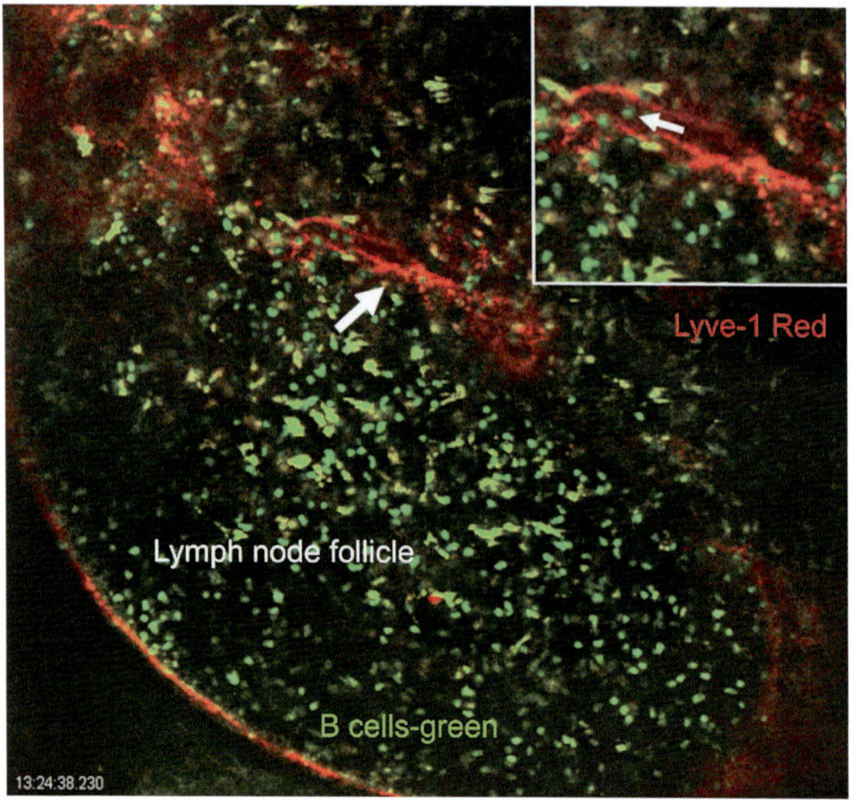

Fig. 2 Adoptively transferred B cells in the cortical sinusoid of the inguinal lymph node. Fluorescently labeled B lymphocytes (*green*) were adoptively transferred to a recipient mouse previously injected with a fluorescent LYVE-1 antibody to outline the cortical sinusoid. TP-LSM of an intravital preparation of the inguinal lymph node of the anesthestized recepient mouse 24 h after transfer. B lymphocytes are found in the lymph node follicle and several cells can be seen in the the cortical sinusoid as outlined by the LYVE-1 antibody staining. *Inset* 2× zoom of the cortical sinusoid region. *Breaks* in the LYVE-1 positive staining may present sites of entrance into the coritcal sinusoid

4.5 Imaging Mouse Lymphocytes Defective in Chemokine Receptor Signaling

An alternative approach to using lymphocytes that lack specfic chemoattractant receptors is to use lymphocytes from mice that lack an essential component in the chemoattractant signaling pathway. In addition, pharmacologic inhibitors of the chemoattractant-induced signaling pathway can be employed. However, since chemokine receptor signaling is required for lymph node entrance, if chemokine receptor-mediated integrin activation is completely blocked, lymphocytes cannot cross HEVs and no cells will be available to be imaged within the lymph node. For example, if 2 h prior to adoptive transfer lymphocytes are exposed to pertussis toxin, which adenosine diphosphate ribosylates $G_{\alpha i}$ proteins and prevents GPCRs-triggered nucleotide exchange, essentially none of the transferred lymphocytes enter into lymph nodes. Therefore, for intravital imaging, the genetically modified or pharmacologically treated lymphocytes must retain some capability to cross HEVs or the signaling pathway must be inactivated after the cells have entered the lymph node. Imaging lymphocytes in lymph node slices can avoid the requirement to cross HEVs since B and T lymphocytes added to a lymph node slice in culture will enter into their appropriate compartment. This should be a powerful techniques for examing the signaling pathways important for lymphocyte motility (see below).

4.5.1 Imaging *Gnai2*$^{-/-}$ Lymphocytes and Pertussis Toxin-Treated Lymphocytes

Gnai2$^{-/-}$ mice often lack multiple peripheral lymph nodes (Han et al. 2005) and likely have a defect in thymic egress (Rudolph et al. 1995). Splenic T and B lymphocytes isolated from these mice respond poorly to chemoattractants and home inefficiently to lymph nodes when adoptively transferred to wild type mice. This is despite their enhanced expression of another Gi protein, $G_{i\alpha 3}$ (Han et al. 2005). Imaging of wild type and *Gnia2*$^{-/-}$ B lymphocytes in HEVs revealed that, while they both rolled on the HEV endothelium, the *Gnai2*-deficient B cells arrested at a frequency 25% of that of wild type cells (Kehrl, unpublished data). Despite the reduction in homing, the transfer of larger numbers of *Gnai2*$^{-/-}$ cells does allow sufficient cells to enter the inguinal lymph node for intravital microscopy. Once the *Gnai2*$^{-/-}$ B cells crossed HEVs they entered lymph node follicles poorly and moved 25% slower than did wild type cells. They often oscillated rather than employing the typical ameboid movements observed with wild type cells (Han et al. 2005). Adoptively transferred *Gnai2*$^{-/-}$ T cells also exhibited defects in positioning as they failed to enter the lymph node cortex in usual numbers, and they often failed to adopt the normal polarized morphology typical of T cells residing in the lymph node (Hwang et al. 2007). Many of the cells remained at the lymph node cortical ridge region, close to HEVs through which they had entered the lymph node. Superficially, in lymph nodes, the *Gnai2*$^{-/-}$ T cells showed a 50% reduction in their mean velocity compared to wild type cells. The few cells observed deeper in the lymph node cortex had a 25% reduction in their mean velocity. Those *Gnai2*$^{-/-}$ T cells that did enter the cortex may be the cells that have best compensated for the loss of *Gnia2* (Hwang et al. 2007). Figure 3 illustrates the tracking of wild type and *Gnai2*-deficient T cells superficially in the lymph node and shows an image acquired during intravital microscopy of wild type T cells and *Gnai2*$^{-/-}$ T cells in the lymph node cortex. Surprisingly, despite the markedly impaired migration of *Gnia2*$^{-/-}$ T and B cells to S1P (Kehrl, unpublished observation), *Gnai2*-deficient T and B cells did not show a significant defect in lymph node egress. In contrast to the *Gnai2*-deficient T and B lymphocytes, *Gnai3*-deficient lymphocytes showed no chemotactic defect (Hwang, unpublished observation). Imaging studies to address their *in vivo* motility are in progress.

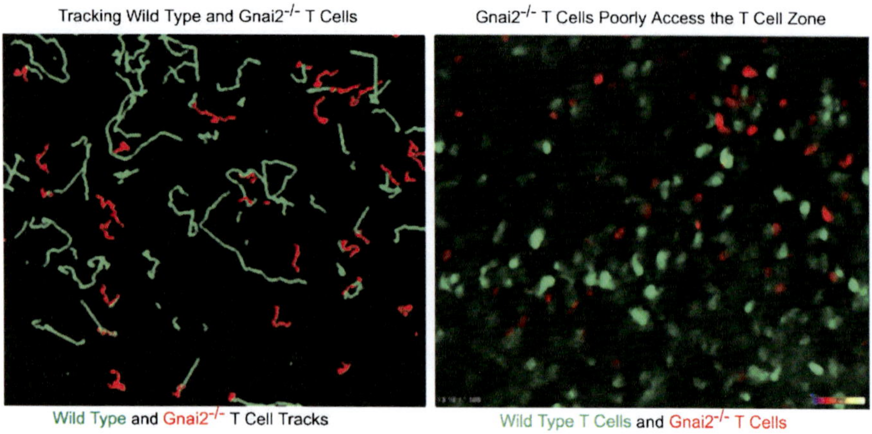

Fig. 3 Intravital microscopy of adoptively transferred wild type and *Gnai2*$^{-/-}$ lymphocytes in the inguinal lymph of a recepient mouse. Fluorescently labeled wild type T lymphocytes (*green*) and and *Gnai2*$^{-/-}$ T lymphocytes (*red*) were adoptively transferred to a recipient mouse. TP-LSM of an intravital preparation of the inguinal lymph node of the anesthestized recepient mouse was performed 24 h after transfer. The *left panel* shows the tracking of wild type T lymphocytes (*green*) and and *Gnai2*$^{-/-}$ T lymphocytes (*red*) approximately 100 μm beneath the surface of the inguinal lymph node. The shorter tracks of the *Gnai2*$^{-/-}$ T lymphocytes is indicative of their reduced interstitial velocity compared to the wild type cells. Individual cells were tracked from the acquired images using Imaris. The *right panel* shows an image acquired deeper from the surface of the lymph node at a depth of approximately 300 μm in a region within the lymph node cortex. Relatively few *Gnai2*$^{-/-}$ T lymphocytes (*red*) can be observed

Another approach to analyze the role of Gi proteins in lymphocyte motility has been to use pertussis toxin, which prevents Gi activation. Treatment of mice with high doses of pertussis toxin following lymphocyte transfer allowed the cells to cross HEVs prior to Gi inactivation. This reduced the mean velocity of B cells within the lymph node follicle by 25% as assessed by intravital TP-LSM (Han et al. 2005). Another approach was used to examine the effect of pertussis toxin treatment on T cell motility. In this case, T cells were treated with pertussis toxin and then immediately transferred. This allowed some of the cells to enter the lymph node before the drug deactivated the $G_{\alpha i}$ subunits. Subsequent imaging revealed a 50% decrease in interstitial velocity, a reduction of 90% in the motility coefficient, and sharper turning angles (Okada and Cyster 2007). In addition, similar to the *Gnai2*$^{-/-}$ T cells, the pertussis toxin-treated cells were mostly located superficially in the lymph node. Pertussis toxin was also used to treat mice prior to explanting the lymph nodes for imaging T cell motility, which again showed a significant decrease in T cell velocity, sharper turning angles, and in addition a higher arrest rate compared to cells from untreated mice (Huang et al. 2007). Finally, pertussis toxin-treated lymphocytes in lymph node sections also showed reduced motility as compared to control cells (Asperti-Boursin et al. 2007).

Together, these studies of *Gnai2*-deficient lymphocytes and those employing pertussis toxin demonstrated a marked impairment in the basal motility of T and

B lymphocytes in lymph nodes, which argues that chemokine receptors such as CCR7 and CXCR5 as well as perhaps other GPCRs provide the signals necessary to maintain lymphocyte migration through lymph node organs. However, since pertussis toxin completely blocks lymphocyte chemotaxis and lymph node homing, but only partially inhibits lymphocyte motility, other signals may also have roles in regulating lymphocyte motility. Arguing that may be the case, naive CD4 T-cells deprived of MHC class II molecules demonstrate a progressive and profound defect in motility (Fischer et al. 2007). It will be of interest to determine whether these cells retain normal responsiveness to chemokines.

4.5.2 Imaging *Rgs1*$^{-/-}$ B Lymphocytes

Murine B cell express *Rgs1*, which is significantly upregulated following activation. Follicular B cells from *Rgs1*$^{-/-}$ mice respond excessively and desensitize improperly to the chemokines. Many of the B cell follicles in the spleens of *Rgs1*$^{-/-}$ mice have germinal centers, and immunization of these mice leads to exaggerated germinal center formation, partial disruption of the normal lymph organ architecture, and abnormal lymphocyte trafficking (Moratz et al. 2004). Intravital microscopy of adoptively transferred *Rgs1*$^{-/-}$ B cells in the HEVs of normal mice revealed an increase in the percentage of B cells adhering to the endothelium (Han et al. 2005). Intravital TP-LSM imaging of adoptively transferred wild type and *Rgs1*$^{-/-}$ B cells in the inguinal lymph node showed that the *Rgs1*$^{-/-}$ B cells more rapidly entered lymph node follicles and that in the follicle they had an enhanced basal motility compared to wild type B cells (appproximately 25% increase). Thus, the lack of an RGS protein gave the opposite phenotype of the loss of *Gnai2* again arguing that the output from chemoattractant receptor signaling directly impact the intranodal motility of lymphocytes. *Rgs1* deficiency did not obviously impact B cell transit times through the inguinal lymph node.

4.5.3 Imaging *Dock2*$^{-/-}$ Lymphocytes

Dock2-deficient mice have T lymphocytopenia, atrophic lymphoid follicles, and a reduction in marginal zone B cells (Fukui et al. 2001). *Dock2*$^{-/-}$ lymphocytes have impaired chemotaxis and chemokine-induced Rac activation. Intravital TP-LSM showed a reduction in the intranodal motility of *Dock2*$^{-/-}$ T and B cells. Most *Dock2*$^{-/-}$ T cells oscillated, unable to initiate movement in any direction despite continuous shape changes and attempts at forming a leading edge. Mean velocity of control T cells was 12.6 μm min^{-1} while *Dock2*$^{-/-}$ T cells moved at 7.0 μm min^{-1}. Frame to frame variations in the centroid position of the *Dock2*$^{-/-}$ lymphocytes likely augmented their mean instantaneous velocities leading to a less dramatic decrease than appeared from the 3D tracking of the cells (Nombela-Arrieta et al. 2007). A similar situation occurred with the analysis of the tracking data from the

Gnai2-deficient B and T lymphocytes. Also similar to the *Gnai2*$^{-/-}$ lymphocytes, S1P-induced cell migration was significantly reduced in T and B cells lacking *Dock2*. However, in contrast to the *Gnai2*$^{-/-}$ T cells, the *Dock2*$^{-/-}$ T cells failed to exit lymph nodes as well as did wild type cells (Nombela-Arrieta et al. 2007). Imaging lymph node explants using wheat germ agglutinin to outline the lymphatic endothelium showed the *Dock2*$^{-/-}$ T cells in close proximity to efferent lymphatic vessels had markedly impaired motility. Whether this accounts for their decreased egress remains to be established. The reduction in the lymph node egress of *Dock2*$^{-/-}$ T cells is consistent with a study that examined T cell lymph node egress using the pertussis toxin treatment regimen described above (Pham et al. 2008). Such treatment modestly reduced the egress of normal T cells, but curiously augmented the egress of T cells in FTY720-treated mice. These data, along with the noted decrease in lymph node egress of either adoptively transferred *S1pr1*$^{-/-}$ or *Ccr7*$^{-/-}$ T,cells, was used to argue that a balance between S1PR1 and chemokine receptor signaling controls T lymphocyte egress.

4.5.4 Imaging *Pik3r1*$^{-/-}$, *Pik3r2*$^{-/-}$, and *Pik3cg*$^{-/-}$ Lymphocytes

While PI3Ks have emerged as important mediators of chemoattractant signaling particularly in studies of the model organism *Dictyostelium discoideum*, their precise roles in lymphocyte chemotaxis and motility have been more difficult to delineate (Nombela-Arrieta et al. 2007). Surprisingly, the consequences of disrupting various PI3K isoforms revealed some cell type specific effects as T and B lymphocytes had different phenotypes. Disruption of the class IB PI3K *Pik3cg* results in T cells that are hyporesponsive to chemokines and which have a lymph node homing defect (Nombela-Arrieta et al. 2004). In contrast, *Pik3cg*$^{-/-}$ B cells have not been shown to have any defects in their responsiveness to chemokines (Nombela-Arrieta et al. 2004). However, several studies do implicate PI3Ks in B cell responses to chemokines. B cells deficient in the Class IA PI3K *Pik3cd* showed diminished chemotactic responses, especially to CXCL13 (Reif et al. 2004). Adoptive transfer experiments of B cells treated with wortmannin, an inhibitor of PI3K activity, revealed diminished homing to Peyer's patches and splenic white pulp cords. Similar results were found with *Pik3cd*-deficient B cells. Another study using a slightly different protocol to treat the B cells with wortmannin showed that the treatment reduced mouse B cell chemotaxis and impaired lymph node homing (Ortolano et al. 2006). Intravital imaging of adoptively transferred B cells in high endothelial venules showed that wortmannin treatment did not impair endothelial rolling, but it did reduce the number of firmly adherent cells. In addition, imaging a wortmannin-treated B cell line *in vitro* showed a reduction in chemokine-induced motility (Ortolano et al. 2006). Several TP-LSM studies have provided additional data on the role of PI3Ks in T and B cell motility. A recent review has summarized the results of these studies in detail so only a brief summary is included here (Oak et al. 2007). Wortmannin treatment was shown to reduce the velocity of T and B cells in explanted lymph nodes by 21 and 26%, respectively while another study using lymph node slices noted that wortmannin treatment did not affect the velocity of T cells in the lymph node slice, although it did note a minor alteration in the positioning of the cells (Asperti-Boursin et al. 2007; Matheu et al. 2007). Intravital TP-LSM imaging of adoptively transferred *Pik3cg*$^{-/-}$ T cells revealed normal motility with the exception of slightly broader turning angles. The lack of *Pik3cg* in the setting of *Dock2* deficiency while further reducing T cells chemotactic responses to CCR7 ligands had little effect on the already significantly impaired motility of the *Dock2*$^{-/-}$ T cells (Nombela-Arrieta et al. 2004). Analysis of *Pik3r1*$^{-/-}$ lymphocytes in explanted lymph nodes found a 24 and 12% reduction in the velocity of B and T cells, respectively, while a similar analysis of *Pik3r2*$^{-/-}$ lymphocytes showed a 5 and 26% reduction in the B and T cells, respectively (Matheu et al. 2007). Further studies are needed to verify that the reduction in motility noted with wortmannin treatment is specifically related to its inhibition of PI3K isoforms and not any of the nonPI3K targets of the drug. Also, it remains unclear whether the reduction in motility of the *Pik3r1*$^{-/-}$ and *Pik3r2*$^{-/-}$ lymphocytes arises from a loss of enzymatic activity or from a loss of adaptor function of PIK3R1 or PIK3R2. Besides potential roles in basal motility and homing to immune organs, it will be of considerable interest to determine whether the PI3Ks function in lymphocyte gradient sensing *in vivo*.

5 Conclusions

Chemokine receptor signaling is required for lymphocytes to cross HEVs in order to enter lymphoid organs and for the proper positioning of lymphocytes within those organs. TP-LSM of explanted lymph nodes, thymus explants, and intravital preparations has revealed a new role for chemokine receptor signaling: that is to help maintain lymphocyte interstitial motility within lymphoid organs. This conclusion is largely based on the imaging of genetically or pharmacologically manipulated lymphocytes. Chemokine maintained movement of T and B lymphocytes along the fibroreticular network within lymphoid organs is required for the appropriate compartmentalization of lymph nodes, likely enhances the interactions between lymphocytes and antigen presenting cells bearing cognate antigen, and may eventually facilitate lymphocyte egress into efferent lymphatics. Signaling by the S1PR1 helps lymphocytes cross the lymphatic endothelium to enter the efferent lymph. S1PR1 signaling on both the endothelial cells and on lymphocytes likely contributes to the control of lymphocyte egress from lymph nodes. Direct imaging of lymphocyte egress is now possible and is contributing to our understanding of this process. While TP-LSM has allowed the visualization of cell dynamics and motility within lymph nodes providing a wealth of new information, many questions remain. For example, the mechanisms by which lymphocytes are organized in complex immunological reactions remain largely ill defined. While the recent application of TP-LSM to studying germinal center reactions has providing some new insights it has raised many new ones (Allen et al. 2007, Hauser et al. 2007, Schwickert et al. 2007). The answers to those questions will in part depend upon the development of improved imaging technology and the application of an array of sophisticated fluorescent sensors that can monitor the status of signaling pathways.

Acknowledgements The authors are supported by the intramural research program of the National Institute of Allergy and Infectious Diseases.

References

Allen CD, Okada T, Tang HL et al (2007) Imaging of germinal center selection events during affinity maturation. Science 315:528–531

Ansel KM, Ngo VN, Hyman PL et al (2000) A chemokine-driven positive feedback loop organizes lymphoid follicles. Nature 406:309–314

Arai H, Tsou CL, Charo IF (1997) Chemotaxis in a lymphocyte cell line transfected with C-C chemokine receptor 2B: evidence that directed migration is mediated by betagamma dimers released by activation of Galphai-coupled receptors. Proc Natl Acad Sci USA 94:14495–14499

Asperti-Boursin F, Real E, Bismuth G et al (2007) CCR7 ligands control basal T-cell motility within lymph node slices in a phosphoinositide 3-kinase-independent manner. J Exp Med 204:1167–1179

Bajenoff M, Egen JG, Koo LY, et al (2006) Stromal cell networks regulate lymphocyte entry, migration, and territoriality in lymph nodes. Immunity 25:989–1001

Bardi G, Niggli V, Loetscher P (2003) Rho kinase is required for CCR7-mediated polarization and chemotaxis of T lymphocytes. FEBS Lett 542:79–83

Bromley SK, Mempel TR, Luster AD (2008) Orchestrating the orchestrators: chemokines in control of T-cell traffic. Nat Immunol 9:970–980

Burkhardt JK, Carrizosa E, Shaffer MH (2008) The actin cytoskeleton in T-cell activation. Annu Review Immunol 26:233–259

Cahalan MD, Parker I (2008) Choreography of cell motility and interaction dynamics imaged by two-photon microscopy in lymphoid organs. Annu Rev Immunol 26:585–626

Croker BA, Tarlinton DM, Cluse LA et al (2002) The Rac2 guanosine triphosphatase regulates B lymphocyte antigen receptor responses and chemotaxis and is required for establishment of B-1a and marginal zone B lymphocytes. J Immunol 168:3376–3386

Cyster JG (2005) Chemokines, sphingosine-1-phosphate, and cell migration in secondary lymphoid organs. Annu Rev Immunol 23:127–159

Doupnik CA, Davidson N, Lester HA, Kofuji P (1997) RGS proteins reconstitute the rapid gating kinetics of gbetagamma-activated inwardly rectifying K+ channels. Proc Natl Acad Sci USA 94:10461–10466

Druey KM, Blumer KJ, Kang VH, Kehrl JH (1996) Inhibition of G-protein-mediated MAP kinase activation by a new mammalian gene family. Nature 379:742–746

Fischer UB, Jacovetty EL, Medeiros RB et al (2007) MHC class II deprivation impairs CD4 T-cell motility and responsiveness to antigen-bearing dendritic cells in vivo. Proc Natl Acad Sci USA 104:7181–7186

Forster R, Mattis AE, Kremmer E et al (1996) A putative chemokine receptor, BLR1, directs B-cell migration to defined lymphoid organs and specific anatomic compartments of the spleen. Cell 871037–1047

Forster R, Schubel A, Breitfeld D et al (1999) CCR7 coordinates the primary immune response by establishing functional microenvironments in secondary lymphoid organs. Cell 99:23–33

Friedl P, Weigelin B (2008) Interstitial leukocyte migration and immune function. Nat Immunol 9:960–969

Fukui Y, Hashimoto O, Sanui T et al (2001) Haematopoietic cell-specific CDM family protein DOCK2 is essential for lymphocyte migration. Nature 412:826–831

Gallego MD, de la Fuente MA, Anton IM et al (2006) WIP and WASP play complementary roles in T-cell homing and chemotaxis to SDF-1alpha. Int Immunol 18:221–232
Gerard A, Mertens AE, van der Kammen RA, Collard JG (2007) The Par polarity complex regulates Rap1- and chemokine-induced T-cell polarization. J Cell Biol 176(6):863–875
Goley ED, Welch MD (2006) The ARP2/3 complex: an actin nucleator comes of age. Nat Rev Mol Cell Biol 7:713–726
Gretz JE, Norbury CC, Anderson AO et al (2000) Lymph-borne chemokines and other low molecular weight molecules reach high endothelial venules via specialized conduits while a functional barrier limits access to the lymphocyte microenvironments in lymph node cortex. J Exp Med 192:1425–1440
Gunn MD, Ngo VN, Ansel KM et al (1998) A B-cell-homing chemokine made in lymphoid follicles activates Burkitt's lymphoma receptor-1. Nature 391:799–803
Haddad E, Zugaza JL, Louache F et al (2001) The interaction between Cdc42 and WASP is required for SDF-1-induced T-lymphocyte chemotaxis. Blood 97:33–38
Han SB, Moratz C, Huang NN et al (2005) Rgs1 and Gnai2 regulate the entrance of B lymphocytes into lymph nodes and B-cell motility within lymph node follicles. Immunity 22:343–354
Hauser AE, Junt T, Mempel TR et al (2007) Definition of germinal-center B-cell migration in vivo reveals predominant intrazonal circulation patterns. Immunity 26:655–667
Hirsch E, Katanaev VL, Garlanda C et al (2000) Central role for G protein-coupled phosphoinositide 3-kinase gamma in inflammation. Science 287:1049–1053
Huang JH, Cardenas-Navia LI et al (2007) Requirements for T lymphocyte migration in explanted lymph nodes. J Immunol 178:7747–7755
Hwang IY, Park C, Kehrl JH (2007) Impaired trafficking of Gnai2$^{+/-}$ and Gnai2$^{-/-}$ T lymphocytes: implications for T-cell movement within lymph nodes. J Immunol 179:439–448
Jacobelli J, Chmura SA, Buxton DB et al (2004) A single class II myosin modulates T-cell motility and stopping, but not synapse formation. Nat Immunol 5:531–538
Kanemitsu N, Ebisuno Y, Tanaka T et al (2005) CXCL13 is an arrest chemokine for B-cells in high endothelial venules. Blood 106:2613–2618
Katagiri K, Imamura M, Kinashi T (2006) Spatiotemporal regulation of the kinase Mst1 by binding protein RAPL is critical for lymphocyte polarity and adhesion. Nat Immunol 7:919–928
Katagiri K, Maeda A, Shimonaka M, Kinashi T (2003) RAPL, a Rap1-binding molecule that mediates Rap1-induced adhesion through spatial regulation of LFA-1. Nat Immunol 4:741–748
Katagiri K, Ohnishi N, Kabashima K et al (2004) Crucial functions of the Rap1 effector molecule RAPL in lymphocyte and dendritic cell trafficking. Nat Immunol 5:1045–1051
Kehrl JH (1998) Heterotrimeric G protein signaling: roles in immune function and fine-tuning by RGS proteins. Immunity 8:1–10
Kehrl JH (2004) G-protein-coupled receptor signaling, RGS proteins, and lymphocyte function. Crit Rev Immunol 24:409–423
Kinashi T, Katagiri K (2005) Regulation of immune cell adhesion and migration by regulator of adhesion and cell polarization enriched in lymphoid tissues. Immunology 116:164–171
Koelle MR, Horvitz HR (1996) EGL-10 regulates G protein signaling in the C. elegans nervous system and shares a conserved domain with many mammalian proteins. Cell 84:115–125
Krummel MF, Macara I (2006) Maintenance and modulation of T-cell polarity. Nat Immunol 7:1143–1149
Lammermann T, Bader BL, Monkley SJ et al (2008) Rapid leukocyte migration by integrin-independent flowing and squeezing. Nature 453:51–55
Ledgerwood LG, Lal G, Zhang N et al (2008) The sphingosine 1-phosphate receptor 1 causes tissue retention by inhibiting the entry of peripheral tissue T lymphocytes into afferent lymphatics. Nat Immunol 9:42–53
Lee JH, Katakai T, Hara T et al (2004) Roles of p-ERM and Rho-ROCK signaling in lymphocyte polarity and uropod formation. J Cell Biol 167:327–337
Li Z, Jiang H, Xie W et al (2000) Roles of PLC-beta2 and -beta3 and PI3Kgamma in chemoattractant-mediated signal transduction. Science 287:1046–1049

Manes S, Gomez-Mouton C, Lacalle RA et al (2005) Mastering time and space: immune cell polarization and chemotaxis. Semin Immunol 17:77–86

Matheu MP, Deane JA, Parker I et al (2007) Class IA phosphoinositide 3-kinase modulates basal lymphocyte motility in the lymph node. J Immunol 179:2261–2269

Matloubian M, Lo CG, Cinamon G et al (2004) Lymphocyte egress from thymus and peripheral lymphoid organs is dependent on S1P receptor 1. Nature 427:355–360

McVerry BJ, Garcia JG (2005) In vitro and in vivo modulation of vascular barrier integrity by sphingosine 1-phosphate: mechanistic insights. Cell Signal 17:131–139

Miller MJ, Wei SH, Cahalan MD, Parker I (2003) Autonomous T-cell trafficking examined in vivo with intravital two-photon microscopy. Proc Natl Acad Sci USA 100:2604–2609

Miller MJ, Wei SH, Parker I, Cahalan MD (2002) Two-photon imaging of lymphocyte motility and antigen response in intact lymph node. Science 296:1869–1873

Moratz C, Hayman JR, Gu H, Kehrl JH (2004) Abnormal B-cell responses to chemokines, disturbed plasma cell localization, and distorted immune tissue architecture in Rgs1$^{-/-}$ mice. Mol Cell Biol 24:5767–5775

Neer EJ (1995) Heterotrimeric G proteins: organizers of transmembrane signals. Cell 80(2):249–257

Neptune ER, Iiri T, Bourne HR (1999) Galphai is not required for chemotaxis mediated by Gi-coupled receptors. J Biol Chem 274:2824–2828

Nombela-Arrieta C, Lacalle RA, Montoya MC et al (2004) Differential requirements for DOCK2 and phosphoinositide-3-kinase gamma during T and B lymphocyte homing. Immunity 21:429–441

Nombela-Arrieta C, Mempel TR, Soriano SF et al (2007) A central role for DOCK2 during interstitial lymphocyte motility and sphingosine-1-phosphate-mediated egress. J Exp Med 204:497–510

Oak JS, Matheu MP, Parker I et al (2007) Lymphocyte cell motility: the twisting, turning tale of phosphoinositide 3-kinase. Biochem Soc Trans 35(Pt 5):1109–1113

Offermanns S, Simon MI (1996) Organization of transmembrane signalling by heterotrimeric G proteins. Cancer Surv 27:177–198

Okada T, Cyster JG (2007) CC chemokine receptor 7 contributes to Gi-dependent T-cell motility in the lymph node. J Immunol 178:2973–2978

Okada T, Miller MJ, Parker I et al (2005) Antigen-engaged B-cells undergo chemotaxis toward the T zone and form motile conjugates with helper T-cells. PLoS Biol 3:e150

Okada T, Ngo VM, Ekland EH et al (2002) Chemokine requirements for B-cell entry to lymph nodes and Peyer's patches. J Exp Med 196:65–75

Ortolano S, Hwang IY, Han SB, Kehrl JH (2006) Roles for phosphoinositide 3-kinases, Bruton's tyrosine kinase, and Jun kinases in B lymphocyte chemotaxis and homing. Eur J Immunol 36:1285–1295

Pappu R, Schwab SR, Cornelissen I et al (2007) Promotion of lymphocyte egress into blood and lymph by distinct sources of sphingosine-1-phosphate. Science 316:295–298

Pasvolsky R, Feigelson SW, Kilic SS et al (2007) A LAD-III syndrome is associated with defective expression of the Rap-1 activator CalDAG-GEFI in lymphocytes, neutrophils, and platelets. J Exp Med 204:1571–1582

Pham TH, Okada T, Matloubian M et al (2008) S1P1 receptor signaling overrides retention mediated by G alpha i-coupled receptors to promote T-cell egress. Immunity 28:122–133

Reif K, Okkenhaug K, Sasaki T et al (2004) Cutting edge: differential roles for phosphoinositide 3-kinases, p110gamma and p110delta, in lymphocyte chemotaxis and homing. J Immunol 173:2236–2240

Rosen H, Gonzalez-Cabrera P, Marsolais D et al (2008) Modulating tone: the overture of S1P receptor immunotherapeutics. Immunol Rev 223:221–235

Rudolph U, Finegold MJ, Rich SS et al (1995) Ulcerative colitis and adenocarcinoma of the colon in G alpha i2-deficient mice. Nat Genet 10:143–150

Sakata D, Taniguchi H, Yasuda S et al (2007) Impaired T lymphocyte trafficking in mice deficient in an actin-nucleating protein, mDia1. J Exp Med 204:2031–2038

Sanna MG, Wang SK, Gonzalez-Cabrera PJ et al (2006) Enhancement of capillary leakage and restoration of lymphocyte egress by a chiral S1P1 antagonist in vivo. Nat Chem Biol 2:434–441

Schwickert TA, Lindquist RL, Shakhar G et al (2007) In vivo imaging of germinal centres reveals a dynamic open structure. Nature 446:83–87
Shimonaka M, Katagiri K, Nakayama T et al (2003) Rap1 translates chemokine signals to integrin activation, cell polarization, and motility across vascular endothelium under flow. J Cell Biol 161:417–427
Snapper SB, Meelu P, Nguyen D et al (2005) WASP deficiency leads to global defects of directed leukocyte migration in vitro and in vivo. J Leukoc Biol 77:993–998
Stachowiak AN, Wang Y, Huang YC, Irvine DJ (2006) Homeostatic lymphoid chemokines synergize with adhesion ligands to trigger T and B lymphocyte chemokinesis. J Immunol 177:2340–2348
Sumen C, Mempel TR, Mazo IB, von Andrian UH (2004) Intravital microscopy: visualizing immunity in context. Immunity 21:315–329
Takesono A, Horai R, Mandai M et al (2004) Requirement for Tec kinases in chemokine-induced migration and activation of Cdc42 and Rac. Curr Biol 14:917–922
Thelen M, Stein JV (2008) How chemokines invite leukocytes to dance. Nat Immunol 9:953–959
Vicente-Manzanares M, Rey M, Perez-Martinez M et al (2003) The RhoA effector mDia is induced during T-cell activation and regulates actin polymerization and cell migration in T lymphocytes. J Immunol 171:1023–1034
von Andrian UH (1996) Intravital microscopy of the peripheral lymph node microcirculation in mice. Microcirculation 3:287–300
Wei SH, Parker I, Miller MJ, Cahalan MD (2003) A stochastic view of lymphocyte motility and trafficking within the lymph node. Immunol Rev 195:136–159
Wei SH, Rosen H, Matheu MP et al (2005) Sphingosine 1-phosphate type 1 receptor agonism inhibits transendothelial migration of medullary T-cells to lymphatic sinuses. Nat Immunol 6:1228–35
Woolf E, Grigorova I, Sagiv A et al (2007) Lymph node chemokines promote sustained T lymphocyte motility without triggering stable integrin adhesiveness in the absence of shear forces. Nat Immunol 8:1076–1085
Worbs T, Bernhardt G, Forster R (2008) Factors governing the intranodal migration behavior of T lymphocytes. Immunol Rev 221:44–63
Worbs T, Mempel TR, Bolter J et al (2007) CCR7 ligands stimulate the intranodal motility of T lymphocytes in vivo. J Exp Med 204:489–495

New Insights Into Leukocyte Recruitment by Intravital Microscopy

Alexander Zarbock and Klaus Ley

Contents

Abstract Leukocyte recruitment to sites of inflammation requires adhesion to and transmigration through the blood vessel wall. Recent progress in optical equipment and new genetic and molecular tools have revealed additional steps in the leukocyte adhesion cascade beyond rolling, adhesion, and transmigration. *In vivo* studies using intravital microscopy (IVM) were essential for the discovery of slow rolling, postadhesion strengthing, intraluminal crawling, and different routes of transmigration. IVM revealed unique features of leukocyte recruitment in different organs. This review focuses on insights into the leukocyte adhesion cascade gained by IVM.

A. Zarbock
Department of Anesthesiology and Intensive Care Medicine, University of Münster, Münster, Germany
e-mail: zarbock@uni-muenster.de

K. Ley
La Jolla Institute for Allergy and Immunology, La Jolla, CA, USA

D. McGavern and M. Dustin (eds.), *Visualizing Immunity,* Current Topics in Microbiology and Immunology 334,
DOI: 10.1007/978-3-540-93864-4_6,

1 Introduction

The immune system is composed of a complex network of cells and molecules that work together in order to protect the body from bacterial, fungal, parasitic, and viral infections and from tumor cells. One common feature of an immune response is the movement of immune cells from the blood to tissues in order to provide effector function. Leukocyte recruitment from the blood into lymph nodes or peripheral tissue is tightly regulated. A disturbance of this process leads to an altered immune response such as seen in leukocyte adhesion deficiency (LAD). This disease is characterized by a defective recruitment of neutrophils and is associated with recurrent bacterial infections due to an inappropriate inflammatory response to injury or infection (Anderson and Springer 1987). On the other end of the spectrum, overwhelming activation of leukocytes is also associated with tissue damage (Nathan 2006).

The study of translucent tissues of live animals by microscopy was a breakthrough, as this technique enabled the surveillance and investigation of the interaction of blood cells with the vessel wall in real time. Early studies in the nineteenth century described for the first time the interaction of blood cells with the endothelium in postcapillary venules and an increased frequency and intensity of this interaction during inflammation (Wagner 1839; Cohnheim 1889). Intravital microscopy (IVM) studies during the last 30 years have revealed molecular mechanisms of leukocyte recruitment and have developed a paradigm of this process (Butcher 1991; Springer 1994). The application of enhanced molecular and genetic tools and sophisticated optical and electronic equipment in IVM contributed to the further understanding of leukocyte recruitment and refined the recruitment cascade to include capturing, rolling, slow rolling, arrest, postadhesion events, and paracellular or transcellular transmigration (Ley et al. 2007). Interaction of selectins with their counter-receptors mediates the initial capturing and rolling of leukocytes on the endothelium. During the rolling of leukocytes along the endothelium, selectins and chemokines presented on inflamed endothelium regulate integrin activation. Binding of activated integrins to their counter-receptors leads, depending on the conformational state of the integrin, either to reduced rolling velocity or arrest. The integrins have, in addition to mediating slow rolling and firm arrest, the ability to transduce extracellular signals into the cell (Ginsberg et al. 2005). This process – called outside-in signaling – activates different signaling pathways that induce adhesion strengthening, transmigration, and respiratory burst.

Reductionist *in vitro* approaches helped discover the molecular mechanisms of leukocyte recruitment, but these systems were insufficient in identifying the physiologic importance of each molecular interaction. The use of IVM has helped to show that leukocyte recruitment in various tissues is regulated differently. This article will focus on mechanisms of neutrophil recruitment that were discovered or confirmed *in vivo* by IVM. In addition, we will give a perspective on further technical improvements that are needed to better understand leukocyte recruitment in different tissues.

2 Advantages and Disadvantages of IVM

One advantage of the conventional transillumination IVM is the high resolution and good image quality that allow to investigate the different steps of the leukocyte recruitment cascade *in vivo*. Furthermore, this technique makes it possible to scrutinize the different steps of the recruitment cascade under physiological conditions, and consequently it is more powerful than *in vitro* approaches.

However, the use of transillumination intravital microscopy is limited due to the restricted ability to penetrate deep tissues. Only tissues that are less than 100 μm thick like the mouse and rat cremaster muscle (Baez 1973), various other thin muscles

Table 1 Intravital microscopy techniques[a]

Technique	Used for	Characteristic features	Examples
Transillumination	Thin tissues like cremaster, mesentery, cheek pouch	Low cost, ease of use, excellent resolution, low penetration depth	Wagner (1839)
Transillumination fluorescence	Thin tissues; leukocyte trafficking; vascular permeability	Best results at low magnification and large fields of view	Bjork et al. (1982)
Epifluorescence	Thin and solid organs, leukocyte trafficking, permeability	Limited penetration depth, light toxicity	Slaaf et al. (1987)
Ratiometric epifluorescence	For calcium transients in endothelial cells and leukocytes, membrane potential	Fluo and Indo dyes work at single wavelength, but FURA ratiometric needed for quantification	Kunkel et al. (2000); Beach et al. (1996); Dubois et al. (2007); Hamilton and Sims (1987)
Stroboscopic epifluorescence	Thin and solid organs, leukocyte, platelet trafficking, micro-PIV	Low light toxicity, high temporal resolution if double strobe is used	Ley and Gaehtgens (1991); Smith et al. (2003); Tangelder et al. (1986)
Oblique reflected light illumination	Thin tissues for extravascular leukocytes	Yields pseudo-3D effect, enhances contrast for leukocytes in tissue	Mempel et al. (2003)
Multiphoton microscopy	Lymph nodes, skin, solid organs	Limited light toxicity, good penetration, best suited for cell migration	Auffray et al. (2007); Mempel et al. (2004)
Spinning disk confocal microscopy	Thin tissues and solid organs; used for leukocyte rolling and transmigration.	Reasonable light toxicity, useful for leukocyte subset identification; good time resolution	Norman et al. (2008); Chiang et al. (2007); Falati et al. (2002)
Laser scanning confocal	Thin tissues and solid organs	Rarely used because of excessive light toxicity and slow scan speed	Finkenauer et al. (1999); Lorenzl et al. (1993)

[a]Note that contrast-enhancing methods common in cell biology like phase contrast, differential interference contrast or Hoffman are not used in intravital microscopy

(Engelson et al. 1985; Lindbom et al. 1982), the rat, cat, or rabbit mesentery (Zweifach 1973), and the hamster cheek pouch (Duling 1973) can be investigated. Due to the availability of knockout and transgenic mice as well as excellent blocking antibodies for chemokines and adhesion molecules, mouse models are very popular.

Another disadvantage of this technique is the tissue trauma that occurs during the surgical preparation. The trauma induces the release of different proinflammatory mediators as well as upregulation of adhesion molecules on the endothelium that subsequently modulate leukocyte recruitment (Fiebig et al. 1991; Ley 1994; Thorlacius et al. 1994). The surgical trauma can be minimized and controlled, but rarely eliminated (bat wing).

New techniques like laser scanning confocal microscopy, multiphoton microscopy and spinning-disk confocal microscopy (Table 1) have contributed to the further progress in understanding the different steps of the recruitment cascade *in vivo*. Multiphoton microscopy achieves excellent penetration depth and is therefore suitable for following cell movements in complex and deep tissues like lymph nodes. Spinning disk confocal microscopy was recently deployed in order to track T cells and neutrophils (Norman et al. 2008). More technical details of IVM techniques can be found in (Ley et al. 2008).

3 Capturing

IVM studies were critical in discovering leukocyte capturing, also known as tethering, which means the first contact between leukocytes and endothelial cells. Almost all capturing events are mediated by the selectin class of adhesion molecules.

3.1 Selectins

P-selectin is stored in α-granules of platelets and in secretory granules called Weibel-Palade bodies of endothelial cells. Following stimulation with inflammatory mediators such as leukotrienes, histame, or thrombin, P-selectin can be rapidly, within minutes, mobilized to the cell surface where it is involved in leukocyte capturing and rolling. In addition, endothelial cells can transcriptionally regulate the surface expression of P-selectin (Gotsch et al. 1994), but the mechanism is differently regulated in humans and mice (Yao et al. 1999).

E-selectin expression on activated endothelium can be induced by several proinflammatory cytokines including tumor necrosis factor (TNF)-α and interleukin (IL)-1. Its expression is regulated by nuclear factor-κB (NF-κB) and translational mechanisms (Kraiss et al. 2003; Read et al. 1995). Surface expression on the endothelium can be found as early as 30 min after stimulation (Kansas 1996). Endothelium of the skin represents an exception, because these cells constitutively express E-selectin, mediating leukocyte rolling even under noninflammatory conditions (Hwang et al. 2004). Although E-selectin expression is delayed, it overlaps with P-selectin temporally, thus helping to enhance leukocyte recruitment during inflammation (Frenette et al. 1996).

L-selectin is constitutively expressed on most leukocytes. The localization of L-selectin on the top of microvilli is important for optimal leukocyte–endothelial cell interactions (von Andrian et al. 1995; Stein et al. 1999). L-selectin is involved in leukocyte recruitment to sites of inflammation (Ley et al. 2007) and lymphocyte homing (Rosen 2004). Following activation of leukocytes, L-selectin can be shed near the cell surface. A disintegrin and metallopeptidase (ADAM)-17 and at least one other enzyme are involved in activated and constitutive shedding of L-selectin, respectively (Smalley and Ley 2005).

3.2 Selectin Ligands

3.2.1 P-Selectin Glycoprotein Ligand (PSGL)-1

All selectins specifically bind to carbohydrates on selectin ligands in a calcium-dependent fashion. The most important selectin ligand on leukocytes is P-selectin glycoprotein ligand (PSGL)-1. This molecule is a homodimeric mucin-like glycoprotein, preferentially located in lipid rafts on the tip of microvilli (Abbal et al. 2006). PSGL-1 is comprised of an extracellular, transmembrane, and cytoplasmic domain (McEver and Cummings 1997). The cytoplasmic tail of human PSGL-1 consists of 63 amino acids and is, through the interaction with several cytoplasmic proteins (McEver and Cummings 1997), indispensable for downstream signaling following engagement (Miner et al. 2008). A juxtamembrane region of 18 amino acids constitutively interacts with Nef-associated factor 1 (Naf1), which is involved in P-selectin-induced signaling through PSGL-1 (Wang et al. 2007). The juxtamembrane region of the cytoplasmic tail of PSGL-1 can also directly interact with the immunoreceptor tyrosine-based activating motif (ITAM)-like sequence containing molecules of the ezrin-moesin-radixin (ERM) family (Urzainqui et al. 2002). These molecules link the juxtamembrane region of the cytoplasmic tail of PSGL-1 with the cytoskeleton in the uropod of migrating cells (Urzainqui et al. 2002; Alonso-Lebrero et al. 2000) and participate in the formation of protrusive membrane structures (Bretscher et al. 2000).

Under physiological shear flow conditions, PSGL-1 can bind P- (Mayadas et al. 1993), E- (Xia et al. 2002), and L-selectin (Sperandio et al. 2003). For optimal selectin binding, PSGL-1 must undergo posttranslational modifications (McEver and Cummings 1997). Several glycosyltransferases are involved in this process including core 2 *N*-acetylglucosaminyltransferase (Ellies et al. 1998), α1-3 fucosyltransferases FucT-VII and FucT-IV (Maly et al. 1996; Weninger et al. 2000), β1-4 galactosyltransferase-1 (Asano et al. 2003), and α2-3 sialyltransferase ST3GalIV (Ellies et al. 2002). Studies with mice lacking one or multiples of these glycosyltransferases demonstrated a broad array of leukocyte rolling defects ranging from a mild capturing defect on E-selectin *in vivo* seen in ST3GalIV-deficient mice

(Ellies et al. 2002) to an almost complete loss of leukocyte rolling in FucT-VII-deficient mice (Weninger et al. 2000; Homeister et al. 2001). In addition to these modifications, sulfation of tyrosines near the N-terminus of PSGL-1 by distinct tyrosylsulfotransferases contributes to the enhanced P-selectin binding to PSGL-1 (Ouyang and Moore 1998; Ouyang et al. 1998; Pouyani and Seed 1995). Eliminating the polypeptide N-acetylgalactosamine transferase-1 reduces the binding capacity of PSGL-1 to P- and E-selectin ex vivo and *in vivo* (unpublished observation) under flow (Tenno et al. 2007).

Blocking of PSGL-1 by antibodies that bind to the NH_2-terminal region inhibits rolling on L- and P-selectin (Sperandio et al. 2003; Norman et al. 1995), but not E-selectin. Recently, the E-selectin binding site on PSGL-1 was mapped to the mucin repeat region (Tauxe et al. 2008). Elimination of PSGL-1 by gene-targeting diminishes the number of leukocytes interacting with the endothelium more than antibody blockade (Xia et al. 2002; Yang et al. 1999), increases rolling velocity on P-selectin and E-selectin (Xia et al. 2002; Zarbock et al. 2007), and reduces neutrophil recruitment into the inflamed peritoneal cavity (Yang et al. 1999).

3.2.2 Other Selectin Ligands

PSGL-1 is the main selectin ligand on neutrophils. However, neutrophils express additional ligands that can bind E-selectin (Zarbock and Ley 2008) including E-selectin ligand (ESL)-1 (Hidalgo et al. 2007), CD44 (Katayama et al. 2005), macrophage antigen (Mac)-1 ($\alpha_M\beta_2$) (Crutchfield et al. 2000; Zen et al. 2007), and other unknown ligands (Ramos et al. 1997; Alon et al. 1995). Human, but not mouse, L-selectin also binds E-selectin (Zollner et al. 1997; Picker et al. 1991).

3.3 Selectin-Mediated Capturing In Vivo

Although all three selectins mediate capturing, the rolling behavior of leukocytes on the three selectins is different. By using P-selectin-deficient mice, Mayadas et al. convincingly demonstrated the key role of P-selectin in leukocyte interaction with the vessel wall (Mayadas et al. 1993). Leukocyte rolling in P-selectin-deficient mice is initially virtually absent in mesenteric venules (Mayadas et al. 1993). This rolling defect of leukocytes is accompanied by a delayed recruitment of leukocytes into inflamed tissue (Mayadas et al. 1993). In the acutely exteriorized mouse cremaster, leukocyte rolling is mediated by P-selectin (Ley et al. 1995) and the rolling velocity is 40 $\mu m\ s^{-1}$ (Jung et al. 1996).

IVM studies on inflamed postcapillary cremaster venules revealed that E-selectin-mediated rolling (3–7 $\mu m\ s^{-1}$) (Zarbock et al. 2007; Kunkel and Ley 1996) is slower than P-selectin-mediated rolling. Blocking or eliminating E-selectin does not influence neutrophil recruitment into the peritoneal cavity of C57BL/6 mice (Labow et al. 1994). However, in this model, neutrophil recruitment is almost completely dependent on E-selectin in the absence of Gα_i-signaling, suggesting that these two pathways have overlapping functions (Smith et al. 2004).

L-selectin-mediated rolling velocity is very fast (130 µm s^{-1}) (Jung et al. 1996). L-selectin-dependent rolling in the systemic circulation is mediated by the interaction between the L-selectin on free flowing leukocytes and the PSGL-1 presented on adherent leukocytes (Eriksson et al. 2001) or leukocyte-derived fragments (Sperandio et al. 2003).

3.4 *Selectin-Independent Capturing In Vivo*

Capturing of leukocytes to the endothelium is mainly selectin-dependent. Elimination of all three selectins almost abolishes leukocyte interaction with the vessel wall (Collins et al. 2001). The remaining interaction is mediated by α_4-integrin (Collins et al. 2001). The absence of leukocyte rolling in postcapillary venules of the cremaster is not accompanied by a complete abolishment of neutrophil recruitment into the inflamed peritoneal cavity (Collins et al. 2001). These data suggest that selectin-independent mechanisms can contribute to neutrophil recruitment to the site of inflammation.

4 Rolling

During the initial contact of the leukocytes with the endothelium, engagement of adhesion receptors and G-protein coupled receptors (GPCR) on leukocytes by selectins and chemokines presented on the enflamed endothelium can initiate intracellular signaling and subsequent activation of the leukocytes. Initiation of certain intracellular signaling pathways leads to activation of integrins (inside-out signaling) that is characterized by a profound conformational change. At least three integrin activation states (low-, intermediate-, and high-affinity state) are known (Shimaoka et al. 2003), but additional states may exist. Following conformational change, the integrins can bind to their counter-receptor and, depending on the affinity state, the integrins may regulate rolling velocity or induce firm arrest. The engaged integrins also have the ability to transmit signals from outside the cell to the inside the cell. This process, called 'outside-in signaling', activates intracellular signaling pathways which participate in postadhesion strengthening (Ley and Zarbock 2006), neutrophil migration (Lindbom and Werr 2002), respiratory burst (Nathan et al. 1989), phagocytosis (Mayadas and Cullere 2005), polarization, and is a linkage between the extracellular matrix and cytoskeleton (Berton and Lowell 1999).

Neutrophils express the β_2-integrins $\alpha_M\beta_2$ [macrophage antigen (Mac)-1], $\alpha_L\beta_2$ (lymphocyte function antigen (LFA)-1), $\alpha_x\beta_2$, and low levels of $\alpha_4\beta_1$. Under inflammatory conditions *in vivo*, the slow rolling velocity of approximately 8 µm s^{-1} on P- and E-selectin is regulated by engagement of the β_2-integrins Mac-1 and LFA-1 (Kunkel and Ley 1996; Jung et al. 1998; Hogg and Leitinger 2001), as blocking both integrins increased rolling velocity (Dunne and Ballantyne 2002). In order to unravel which selectin is responsible for the activation of the different integrins, a recent study used a blocking P-selectin antibody and pertussis toxin under inflammatory conditions to eliminate P-selectin- and $G\alpha_i$-coupled receptor-induced signaling (Zarbock et al. 2007). Under these circumstances, the rolling velocity on E-selectin was 4 $\mu m\ s^{-1}$ and only LFA-1 regulated the rolling velocity, as the additional blocking of Mac-1 did not further increase the rolling velocity (Zarbock et al. 2007). These data were confirmed in an autoperfused flow chamber assay on isolated adhesion molecules. PSGL-1 engagement by E-selectin activated LFA-1 in a spleen tyrosine kinase (Syk)-dependent manner (Zarbock et al. 2007). By using IVM in combination with *in vitro* experiments, it was demonstrated that an ITAM-dependent pathway involving the Src-family kinase Fgr and the ITAM-containing adapter proteins DAP12 and FcRγ is involved in E-selectin-mediated slow rolling (Zarbock et al. 2008). In order to show that the Src-family kinase Fgr and the ITAM-containing adaptor proteins DAP12 and FcRγ are indispensable for E-selectin-mediated slow rolling *in vivo*, IVM was performed on mixed chimeric mice that were generated by injecting bone marrow cells from LysM-GFP$^+$ mice (Faust et al. 2000) and gene-deficient mice into lethally irradiated wild type (wt) mice (Fig. 1). The advantage of this system is that gene-targeted leukocytes can be compared side-by-side with leukocytes from WT mice under the same hemodynamic conditions in the same venules. Flow chamber experiments with isolated bone-marrow-derived neutrophils demonstrated that LFA-1 regulates the rolling velocity on P-selectin and ICAM-1 (Miner et al. 2008). As $G\alpha_i$-coupled receptor-induced signaling was blocked under inflammatory conditions *in vivo* and proinflammatory cytokines were lacking in the flow chamber assay, one possible explanation would be that GPCRs are able to regulate the rolling velocity by Mac-1 activation. The rolling velocity on P-selectin and E-selectin under inflammatory conditions *in vivo* is also regulated by CD44 (Katayama et al. 2005).

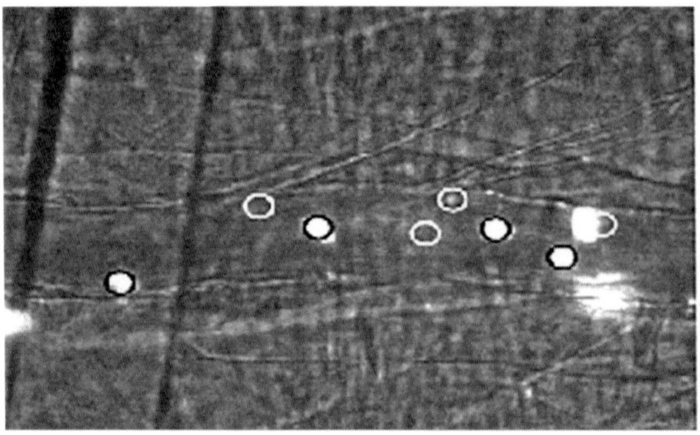

Fig. 1 Intravital microscopy of the cremaster muscle of mixed chimeric mice 6 weeks after bone marrow transplantation. The recipient mouse was lethally irradiated (2 × 600 rad) and reconstituted with 5 million unfractionated bone marrow cells mixed from two donor mice. Neutrophils from LysM-GFP$^+$ mice (*black circles*) and gene-deficient mice (GFP$^-$ cells, *white circles*) roll and adhere in this postcapillary venule. Their behavior can be compared in the same vessel under the exact same conditions of shear stress and flow

In the acutely exteriorized mouse cremaster muscle microcirculation, leukocyte rolling is known to be mediated by P-selectin (Ley et al. 1995). The rolling velocity of leukocytes in this model is 40–50 μm s^{-1} (Ley et al. 1995). P-selectin binding to PSGL-1 can prime neutrophils, but not fully activate integrins and induce arrest. These data are supported by flow chamber experiments that showed that the rolling velocity of neutrophils in whole blood is reduced on P-selectin and ICAM-1 compared to P-selectin alone (Zarbock et al. 2007; Chesnutt et al. 2006). P-selectin may act synergistically with other proinflammatory stimuli and contribute to full integrin activation (Wang et al. 2007).

In vitro and *in vivo* evidence demonstrates that L-selectin engagement can induce intracellular signaling and subsequent neutrophil activation. Altering the surface expression of L-selectin by a hydroxamic acid-based protease inhibitor amplified L-selectin-mediated signal input (Hafezi-Moghadam et al. 2001). The increased L-selectin expression on neutrophils was associated with an increased 'smoothness' of rolling (Hafezi-Moghadam et al. 2001), reduced rolling velocity on P-selectin (Hafezi-Moghadam and Ley 1999), enhanced adhesion, and transmigration (Hafezi-Moghadam et al. 2001). Based on *in vivo* data with a metalloprotease inhibitor, L-selectin-deficient mice have a substantial defect in neutrophil recruitment into inflammatory sites (Tedder et al. 1995). These data suggest that L-selectin has an important signaling role in neutrophil activation and recruitment.

5 Arrest

During rolling, neutrophils are exposed to different chemokines and other chemoattractants presented by inflamed endothelial cells. The activation of G-protein coupled receptors (GPCR) by these mediators rapidly triggers arrest that is mediated by the binding of neutrophil integrins to immunoglobulin superfamily members, such as ICAM-1 and VCAM-1, expressed on endothelial cells. During inflammation, endothelial cells are activated by different proinflammatory mediators and subsequently express adhesion molecules, as well as synthesize and release chemokines and lipid mediator that are presented on the cell surface.

Binding of chemokines to GPCRs on neutrophils induces activation of complex intracellular signaling networks which regulate integrin adhesiveness (Zarbock et al. 2007; Ley et al. 1993). Integrin adhesiveness can be modulated by changing the affinity and the avidity of the integrin (Kinashi and Katagiri 2004; Laudanna et al. 2002; Luo et al. 2007). Integrin affinity is regulated by conformational changes of individual integrins, which leads to an altered ligand-binding energy and ligand reduced dissociation rate. The avidity is influenced by the affinity and the valency of integrins, where valency refers to the density of integrins per cell surface area involved in cell adhesion. Valency is dependent on the surface expression of integrins and the lateral mobility. Signaling induced by GPCR leading to rapid activation of integrins is referred to as 'inside-out signaling'.

The specificity of leukocyte arrest is based on the differential expression of adhesion molecules, signaling molecules, integrins, and chemokine receptors. In addition, chemokine-induced signaling pathways can modulate different integrins in distinct leukocyte subsets, thus emphasizing the existence of cell-specific arrest pathways. As most studies have investigated GPCR signaling either in monocytes or lymphocytes, the intracellular signaling cascade that connects GPCRs with integrins in neutrophils is still incompletely understood.

Based on *in vivo* studies, the intracellular signaling cascade downstream of GPCRs can be subdivided into different steps. Following engagement of GPCR by chemokines, the $G\alpha_i$-subunit dissociates from the $G\beta\gamma$-complex. Neutrophils express three different $G\alpha_i$-subunits ($G\alpha_{i1-3}$), but these subunits are not redundant as the elimination of the $G\alpha_{i2}$-subunit significantly reduced chemokine-induced arrest *in vitro* and *in vivo* (Zarbock et al. 2007). Following dissociation, the $G\beta\gamma$-complex can interact and activate phospholipase C (PLC) (Camps et al. 1992). Activated PLC hydrolyzes phosphatidylinositol 4,5-biphosphate to inositol triphosphate and diacylglycerol. It was demonstrated that PLC is involved in chemokine-induced arrest of primary neutrophils ex vivo (Zarbock et al. 2007). An increase in the IP_3 concentration triggers Ca^{2+}-release from the endoplasmatic reticulum. Diacylglycerol activates some isoforms of protein kinase C. Calcium is a pleiotropic signaling molecule capable of binding to many proteins and is thought to be critically involved in chemokine-induced arrest. This assumption is supported by an *in vivo* study that demonstrated that intracellular Ca^{2+}-concentration increases in neutrophils under inflammatory conditions (Kunkel et al. 2000). However, this study did not address the question whether the increased intracellular Ca^{2+}-concentration was induced by GPCR signaling. An increased Ca^{2+}-concentration can also be induced by selectins (Schaff et al. 2008) or integrin outside-in signaling (Walzog et al. 1994). In addition to the interaction with PLC, the $G\beta\gamma$-complex can also activate other molecules including P-Rex-1 (Welch et al. 2002) and PI3Kγ (Hirsch et al. 2000). PI3Kγ-deficient mice have normal chemokine-induced arrest *in vivo*, but show a defect in postadhesion strengthening (Smith et al. 2006). A recently published study revealed that PI3Kγ in leukocytes is involved in early leukocyte migration in response to different chemokines, whereas leukocyte emigration after prolonged chemokine exposure is PI3Kγ-independent (Liu et al. 2007).

Different *in vitro* and *in vivo* studies in humans and mice demonstrated that the guanine nucleotide exchange factor (GEF) CalDAG-GEFI is involved in chemokine-induced neutrophil arrest (Bergmeier et al. 2007; Pasvolsky et al. 2007). However, it still remains unknown whether only calcium and diacylglycerol are required to fully activate CalDAG-GEFI. The small GTPase RAP1/2, which is involved in chemokine-induced arrest, is activated by CalDAG-GEFI (Shimonaka et al. 2003; Kinashi 2005). The molecules connecting Rap1 with the integrin in neutrophils remain to be elucidated. Studies on platelets and leukocytes demonstrated that talin1 can interact with the cytoplasmatic tail of the β-chain of integrins and modulate the conformational change of $\alpha_{IIb}\beta_3$ (Wegener et al. 2007; Kim et al. 2003;

Tadokoro et al. 2003). Elimination of talin1 in platelets leads to spontaneous hemorrhage, pathological bleeding, impaired $\alpha_{IIb}\beta_3$-mediated platelet aggregation, and β_1 integrin-mediated platelet adhesion (Petrich et al. 2007; Nieswandt et al. 2007). However, it is unknown whether talin1 is also involved in chemokine-induced arrest in neutrophils.

6 Postadhesion Events

Under static conditions, each of the two major β_2-integrins, Mac-1 and LFA-1, can mediate adhesion by binding to ICAM-1 (Petri and Bixel 2006). However, neutrophil adhesion under flow conditions *in vivo* is mainly mediated by the high affinity state β_2-integrin LFA-1 (Phillipson et al. 2006), whereas Mac-1 appears to be more important for emigration of neutrophils.

In addition to mediating adhesion, integrins transfer extracellular signals into the cell (outside-in signaling) in order to activate different signaling pathways that regulate various cellular functions, including proliferation, apoptosis, cell motility, phagocytosis, superoxide production, and degranulation (Luo et al. 2007). Conformational change and integrin clustering allow outside-in signaling that is required for the recruitment and activation of protein tyrosine kinases. In an *in vivo* study using IVM, Giagulli et al. (2006) demonstrated that elimination of two, Hck and Fgr, of the three Src family kinases in neutrophils is associated with impaired outside-in signaling and consequently with a decrease of sustained neutrophils adhesion to the vessel wall. However, the mechanism how integrins activate Src family kinase is still unknown. The activated Src kinases phosphorylate the ITAM-containing adaptor molecules DAP12 and FcRγ, which bind to and activate Syk that consequently initiates further downstream signaling including respiratory burst (Mocsai et al. 2006). Disruption of outside-in signaling at different levels, such as PI3kγ (Smith et al. 2006) and Vav1 und 3 (Gakidis et al. 2004), renders leukocyte adhesion unstable and leads to detachment. An *in vitro* study using neutrophils deficient in Wiskott-Aldrich Syndrome (WAS) protein (Zhang et al. 2006) showed that this defect is associated with a reduced adhesion and transendothelial migration under conditions of physiologic shear flow.

Before crossing the inflamed endothelium, neutrophils crawl along the endothelium (Phillipson et al. 2006). This process, which is Mac-1 and ICAM-1 dependent (Phillipson et al. 2006), is thought to be important to find the site of transmigration. Disabling crawling by eliminating Mac-1 changes the route of endothelial migration and delays neutrophils recruitment (Phillipson et al. 2006). A monocyte subpopulation expressing CXC3CR1 crawls continuously on unstimulated endothelium in mesenteric and dermal blood vessels (Auffray et al. 2007). In contrast to neutrophils, this 'patrolling' depends on the integrin LFA-1 and the chemokine receptor CX_3CR1 and is required for rapid tissue invasion at the site of inflammation.

7 Transmigration

The last step in the recruitment cascade is transmigration, when leukocytes leave the intravascular compartment and enter the tissue. During this process, transmigrating leukocytes encounter three different barriers: endothelium, basement membrane, and pericytes. It is known that the different leukocyte subsets require different molecules for transmigration and these differences have been reviewed elsewhere (Vestweber 2007; Engelhardt and Wolburg 2004).

Leukocyte transendothelial cell migration can occur either directly through individual endothelial cells (transcellular route) or between endothelial junctions (paracellular route).

7.1 *Paracellular Route*

The attachment of leukocytes to endothelial cells and the engagement of endothelial cell adhesion molecules may diminish interendothelial cell contacts and facilitate leukocyte migration through endothelial cell junctions. Attachment of activated neutrophils to endothelial cells induces a transient intracellular Ca^{2+} increase in endothelial cells that is required for opening intracellular junctions and subsequent migration of neutrophils (Huang et al. 1993). Rearrangement of junctional molecules during inflammation may further facilitate paracellular migration of leukocytes. Molecules that hamper transmigration, such as vascular endothelial-cadherin, are displaced from the junction (Shaw et al. 2001) and adhesion molecules that can directly interact with their counter receptors on leukocytes and facilitate cell–cell interaction, such as platelet/endothelial cell adhesion molecule-1 and junctional adhesion molecule A, are mobilized to the luminal surface (Muller 2003).

Using intravital microscopy *in vivo*, Nourshargh et al. demonstrated that the junctional adhesion molecules PECAM-1, JAM-A, and ICAM-2 mediate leukocyte transmigration in response to interleukin (IL)-1β (Nourshargh et al. 2006). However, leukocyte recruitment was independent of these molecules when inflammation was induced by TNF-α, leukotriene B4 (LTB4), or N-formyl-methionyl-leucyl-phenylalanine (fMLP) (Nourshargh et al. 2006). Elimination of endothelial cell-selective adhesion molecule (ESAM), which is also located in endothelial junctions, does not affect leukocyte rolling or adhesion in the microcirculation of the cremaster muscle, but reduced leukocyte extravasation by 50% (Wegmann et al. 2006). Bixel et al. (2007) demonstrated by using intravital and electron microscopy of cremaster venules that blocking CD99L2 inhibited leukocyte transmigration through the vessel wall at the level of the perivascular basement membrane. Phillipson et al. (2006) demonstrated that approximately 90% of neutrophils emigrate through the paracellular route in response to MIP-2. Blocking or eliminating Mac-1 did not influence the ability of neutrophils to adhere to the endothelium, but almost completely abolished crawling. Under these circumstances, neutrophil emigration was delayed and half the adherent neutrophils emigrated transcellularly, suggesting that neutrophils can emigrate transcellularly but normally use the paracellular route.

7.2 *Transcellular Route*

Although the majority of leukocytes take the paracellular route to transmigrate, some leukocytes transmigrate directly through endothelial cells (transcellular route).

Adherent leukocytes move laterally via interactions between β_2-integrins and ICAM-1. High levels of ICAM-1 on the surface of endothelial cells induced by TNF-α lead to an increase of neutrophil migration through the transcellular route. Elimination of the cytoplamatic tail of ICAM-1 reduces transcellular migration, suggesting that ICAM-1 engagement induces intracellular signaling, and hence facilitates transcellular migration (Yang et al. 2005).

During lateral movement, leukocytes form actin-dependent protrusive organelles on their ventral aspect ('podosomes') (Linder and Aepfelbacher 2003). These 'podosomes' force invaginations (termed 'podo-prints') into the surface of the endothelium that cause displacement of cytoplasm and other cytoplasmic structures (Carman et al. 2007). At the site of transcellular migration, podosomes become invasive and induce transcellular pore formation (Carman and Springer 2008). This process triggers SNARE-mediated vesicle fusion (Carman et al. 2007; Millan et al. 2006) and transcellular pore formation (Carman and Springer 2008). The binding of leukocyte β_2-integrins to endothelial cell adhesion molecules can induce the formation of "transmigratory cups" (Carman and Springer 2004). These transmigratory cups may be preceded by endothelial docking structures, projections rich in VCAM-1, ICAM-1, tetraspanins, and ERM (ezrin, radixin, moesin) proteins (Mayadas and Cullere 2005; Carman and Springer 2004; Barreiro et al. 2002).

An *in vivo* study by Feng et al. (1998) demonstrated that almost all neutrophils (approximately 90%) emigrate from venules of the skin through the transcellular route in response to the bacterial peptide fMLP. This study stands in contrast to many other studies which showed that the paracellular route is the dominant pathway. However, the data presented by Feng et al. (1998) are very convincing, as they used electron microscopy of serial tissue sections to study transmigration. One possible explanation for the observed differences is that in some tissues the paracellular route dominates, whereas in other tissues leukocytes favor the transcellular route. This could be caused by the different expression of adhesion molecules in various tissues, such as the constitutive expression of E-selectin on endothelial cells in the skin (Hwang et al. 2004). The type of stimulus to elicit leukocyte transmigration could also influence the decision which transmigration route the leukocytes takes, as different chemokines can activate several intracellular signaling pathways (Bokoch 1995; Heit et al. 2002). Therefore, further studies will be necessary to elucidate which factors determine the route of transmigration.

7.3 Endothelial Basement Membrane and Pericyte Sheath

After crossing the endothelium, leukocytes penetrate the endothelial basement membrane and, in most venules, the pericyte sheath. The endothelial basement membrane consists of collagen, different laminins, and other molecules (Hallmann et al. 2005).

By using immunofluorescence staining and confocal microscopy, Wang et al. (2006) have recently shown in unstimulated mouse cremasteric venules that the expression of certain basement membrane components are lower in some regions compared to other regions in the same vessel. Regions of low expression are closely associated with gaps between pericytes. In cytokine stimulated venules, neutrophils preferentially used the regions of low expression to migrate through the endothelial basement membrane (Wang et al. 2006).

8 Organ Specific Recruitment

In skeletal muscle, skin, and connective tissue, leukocyte recruitment follows the aforementioned cascade. However, studies investigating neutrophil recruitment in different organs show that some organs have unique features, and neutrophil recruitment into these organs do not follow the proposed paradigm.

8.1 Liver

Leukocyte recruitment in portal and central venules of the liver is very similar to that found in the venules of the systemic circulation (Lee and Kubes 2008). However, leukocyte trapping in the liver sinusoids is different from the systemic circulation due to distinctive structural and functional features of hepatic sinusoidal endothelium (Lee and Kubes 2008). In the systemic circulation, leukocyte adhesion is restricted to venules, whereas leukocyte adhesion in the liver is also seen in the hepatic capillaries (sinusoids) as well as postcapillary venules (Wong and Johnston 1997; Fox-Robichaud and Kubes 2000). *In vivo* studies employing intravital microscopy revealed that selectin-dependent rolling precedes adhesion in postcapillary venules of the liver, but adhesion was independent of selectins in sinusoids (Wong and Johnston 1997; Fox-Robichaud and Kubes 2000; Essani et al. 1998).

As the adhesion process in the sinusoids is selectin-independent, it was thought that, in addition to the narrow diameter and low shear forces in these vessels (Fox-Robichaud and Kubes 2000; Jaeschke and Smith 1997; Jaeschke et al. 1996), the reduced deformability of neutrophils following activation would modulate recruitment. However, a recently published study by Kubes and colleagues (McDonald et al. 2008) showed that hyaluronan is abundantly expressed in the liver sinusoids and that neutrophil CD44 was necessary for reversible neutrophil adhesion to hyaluronan within liver sinusoids *in vivo*. Disruption of the interaction of

CD44 with hyaluronan decreased liver pathophysiology and damage in response to bacterial endotoxin (McDonald et al. 2008).

8.2 Lung

The exact mechanism of neutrophil recruitment into the lung remains still unknown. Leukocyte recruitment into the lung is influenced by several factors including neutrophil deformability, adhesion molecules, and the unique capillary structure of the lung (Doerschuk 2001). On the way through the small pulmonary capillaries (Doerschuk et al. 1993), neutrophils have to stop several times in order to change their shape and subsequently squeeze through the small vessels. This process causes a prolonged transit time compared to red blood cells and leads to a significant accumulation of neutrophils in the lung ('marginated pool') (Doerschuk et al. 1987).

During inflammation, released proinflammatory mediators are able to activate neutrophils, which become stiff due to the polymerization of subcortical actin and sequester in the pulmonary microvasculature (Drost and MacNee 2002; Worthen et al. 1989). However, the role of adhesion molecules for neutrophil recruitment into the lung has not been fully elucidated. Elimination or blocking of L-selectin reduced neutrophil sequestration in the lung in response to the formyl peptide fMLP, but not C5a (Olson et al. 2002). A study using IVM of the rabbit lung demonstrated that L-selectin-mediated leukocyte–endothelial interaction is a necessary prerequisite for leukocyte sequestration in alveolar capillaries in response to endotoxemia (Kuebler et al. 2000). However, another study showed that blocking E-selectin and L-selectin by antibodies did not influence neutrophil sequestration in a model of sepsis-induced ALI (Carraway et al. 1998). IVM of the lung of gene-deficient mice would help to elucidate these contradictory data. However, IVM of the mouse lung is technically very challenging and not widely used. A recently published study showed convincingly that the employed surgical technique did not induce acute lung injury or edema formation and that medium-sized arterioles vasoconstricted in response to hypoxia or infusion of a thromboxane analog (Tabuchi et al. 2008).

These data suggest that neutrophil recruitment into the lung is different from other circulations, with different molecules required for neutrophil recruitment in response to different stimuli.

9 Perspectives

Although intravital microscopy has been employed to investigate the different steps of the leukocyte recruitment cascade in living animals for over 150 years, the studies were restricted to interaction at cellular level. However, the drastic improvement in optical equipment and the capability of manipulating the molecular structure and function of proteins and genes made it possible to investigate immune responses at the subcellular level under physiological conditions. The next step will be to use the sophisticated equipment in order to investigate and identify the unique features of leukocyte recruitment in different organs. Recent improvements in molecular techniques, microscopy, and imaging software now allow the direct visualization of molecular interactions in living animals. GFP fusion proteins and knockin mice have facilitated direct visualization of cell adhesion molecules and differential detection of leukocyte subsets, respectively. Novel microscopic techniques include multiphoton and confocal microscopy, stroboscopic epifluorescence microscopy, and oblique reflected light illumination. Improved algorithms for tracking cells and molecules along with sophisticated 3D reconstruction software vastly enhance the power of IVM. Understanding organ- and cell type-specific recruitment mechanisms will help guide target identification and the development of new therapeutic strategies.

Acknowledgments Supported by grants from Deutsche Forschungsgemeinschaft (AZ 428/2-1 and AZ 428/3-1 to A.Z.) and by grants from the National Institutes of Health to K.L.

References

Abbal C, Lambelet M, Bertaggia D, Gerbex C, Martinez M, Arcaro A, Schapira M, Spertini O (2006) Lipid raft adhesion receptors and Syk regulate selectin-dependent rolling under flow conditions. Blood 108:3352–3359

Alon R, Feizi T, Yuen CT, Fuhlbrigge RC, Springer TA (1995) Glycolipid ligands for selectins support leukocyte tethering and rolling under physiologic flow conditions. J Immunol 154:5356–5366

Alonso-Lebrero JL, Serrador JM, Dominguez-Jimenez C, Barreiro O, Luque A, del Pozo MA, Snapp K, Kansas G, Schwartz-Albiez R, Furthmayr H, Lozano F, Sanchez-Madrid F (2000) Polarization and interaction of adhesion molecules P-selectin glycoprotein ligand 1 and intercellular adhesion molecule 3 with moesin and ezrin in myeloid cells. Blood 95:2413–2419

Anderson DC, Springer TA (1987) Leukocyte adhesion deficiency: an inherited defect in the Mac-1, LFA-1, and p150,95 glycoproteins. Annu Rev Med 38:175–194

Asano M, Nakae S, Kotani N, Shirafuji N, Nambu A, Hashimoto N, Kawashima H, Hirose M, Miyasaka M, Takasaki S, Iwakura Y (2003) Impaired selectin-ligand biosynthesis and reduced inflammatory responses in beta-1,4-galactosyltransferase-I-deficient mice. Blood 102:1678–1685

Auffray C, Fogg D, Garfa M, Elain G, Join-Lambert O, Kayal S, Sarnacki S, Cumano A, Lauvau G, Geissmann F (2007) Monitoring of blood vessels and tissues by a population of monocytes with patrolling behavior. Science 317:666–670

Baez S (1973) An open cremaster muscle preparation for the study of blood vessels by in vivo microscopy. Microvasc Res 5:384–394

Barreiro O, Yanez-Mo M, Serrador JM, Montoya MC, Vicente-Manzanares M, Tejedor R, Furthmayr H, Sanchez-Madrid F (2002) Dynamic interaction of VCAM-1 and ICAM-1 with moesin and ezrin in a novel endothelial docking structure for adherent leukocytes. J Cell Biol 157:1233–1245

Beach JM, McGahren ED, Xia J, Duling BR (1996) Ratiometric measurement of endothelial depolarization in arterioles with a potential-sensitive dye. Am J Physiol 270:H2216–H2227

Bergmeier W, Goerge T, Wang HW, Crittenden JR, Baldwin AC, Cifuni SM, Housman DE, Graybiel AM, Wagner DD (2007) Mice lacking the signaling molecule CalDAG-GEFI represent a model for leukocyte adhesion deficiency type III. J Clin Invest 117:1699–1707

Berton G, Lowell CA (1999) Integrin signalling in neutrophils and macrophages. Cell Signal 11:621–635

Bixel MG, Petri B, Khandoga AG, Khandoga A, Wolburg-Buchholz K, Wolburg H, Marz S, Krombach F, Vestweber D (2007) A CD99-related antigen on endothelial cells mediates neutrophil but not lymphocyte extravasation in vivo. Blood 109:5327–5336
Bjork J, Hedqvist P, Arfors KE (1982) Increase in vascular permeability induced by leukotriene B4 and the role of polymorphonuclear leukocytes. Inflammation 6:189–200
Bokoch GM (1995) Chemoattractant signaling and leukocyte activation. Blood 86:1649–1660
Bretscher A, Chambers D, Nguyen R, Reczek D (2000) ERM-Merlin and EBP50 protein families in plasma membrane organization and function. Annu Rev Cell Dev Biol 16:113–143
Butcher EC (1991) Leukocyte-endothelial cell recognition: three (or more) steps to specificity and diversity. Cell 67:1033–1036
Camps M, Carozzi A, Schnabel P, Scheer A, Parker PJ, Gierschik P (1992) Isozyme-selective stimulation of phospholipase C-beta 2 by G protein beta gamma-subunits. Nature 360:684–686
Carman CV, Sage PT, Sciuto TE, de la Fuente MA, Geha RS, Ochs HD, Dvorak HF, Dvorak AM, Springer TA (2007) Transcellular diapedesis is initiated by invasive podosomes. Immunity 26:784–797
Carman CV, Springer TA (2004) A transmigratory cup in leukocyte diapedesis both through individual vascular endothelial cells and between them. J Cell Biol 167:377–388
Carman CV, Springer TA (2008) Trans-cellular migration: cell-cell contacts get intimate. Curr Opin Cell Biol 20:533–540
Carraway MS, Welty-Wolf KE, Kantrow SP, Huang YC, Simonson SG, Que LG, Kishimoto TK, Piantadosi CA (1998) Antibody to E- and L-selectin does not prevent lung injury or mortality in septic baboons. Am J Respir Crit Care Med 157:938–949
Chesnutt BC, Smith DF, Raffler NA, Smith ML, White EJ, Ley K (2006) Induction of LFA-1-dependent neutrophil rolling on ICAM-1 by engagement of E-selectin. Microcirculation 13:99–109
Chiang EY, Hidalgo A, Chang J, Frenette PS (2007) Imaging receptor microdomains on leukocyte subsets in live mice. Nat Methods 4:219–222
Cohnheim J (1889) Lectures on general pathology: a handbook for practitioners and students. The New Sydenham Society, London
Collins RG, Jung U, Ramirez M, Bullard DC, Hicks MJ, Smith CW, Ley K, Beaudet AL (2001) Dermal and pulmonary inflammatory disease in E-selectin and P-selectin double-null mice is reduced in triple-selectin-null mice. Blood 98:727–735
Crutchfield KL, Shinde Patil VR, Campbell CJ, Parkos CA, Allport JR, Goetz DJ (2000) CD11b/CD18-coated microspheres attach to E-selectin under flow. J Leukoc Biol 67:196–205
Doerschuk CM, Allard MF, Martin BA, MacKenzie A, Autor AP, Hogg JC (1987) Marginated pool of neutrophils in rabbit lungs. J Appl Physiol 63:1806–1815
Doerschuk CM, Beyers N, Coxson HO, Wiggs B, Hogg JC (1993) Comparison of neutrophil and capillary diameters and their relation to neutrophil sequestration in the lung. J Appl Physiol 74:3040–3045
Doerschuk CM (2001) Mechanisms of leukocyte sequestration in inflamed lungs. Microcirculation 8:71–88
Drost EM, MacNee W (2002) Potential role of IL-8, platelet-activating factor and TNF-alpha in the sequestration of neutrophils in the lung: effects on neutrophil deformability, adhesion receptor expression, and chemotaxis. Eur J Immunol 32:393–403
Dubois C, Panicot-Dubois L, Gainor JF, Furie BC, Furie B (2007) Thrombin-initiated platelet activation in vivo is vWF independent during thrombus formation in a laser injury model. J Clin Invest 117:953–960
Duling BR (1973) The preparation and use of the hamster cheek pouch for studies of the microcirculation. Microvasc Res 5:423–429
Dunne JL, Ballantyne CM, Beaudet AL, Ley K (2002) Control of leukocyte rolling velocity in TNF-alpha-induced inflammation by LFA-1 and Mac-1. Blood 99:336–341
Ellies LG, Sperandio M, Underhill GH, Yousif J, Smith M, Priatel JJ, Kansas GS, Ley K, Marth JD (2002) Sialyltransferase specificity in selectin ligand formation. Blood 100:3618–3625

Ellies LG, Tsuboi S, Petryniak B, Lowe JB, Fukuda M, Marth JD (1998) Core 2 oligosaccharide biosynthesis distinguishes between selectin ligands essential for leukocyte homing and inflammation. Immunity 9:881–890

Engelhardt B, Wolburg H (2004) Mini-review: Transendothelial migration of leukocytes: through the front door or around the side of the house. Eur J Immunol 34:2955–2963

Engelson ET, Skalak TC, Schmid-Schonbein GW (1985) The microvasculature in skeletal muscle. I. Arteriolar network in rat spinotrapezius muscle. Microvasc Res 30:29–44

Eriksson EE, Xie X, Werr J, Thoren P, Lindbom L (2001) Importance of primary capture and L-selectin-dependent secondary capture in leukocyte accumulation in inflammation and atherosclerosis in vivo. J Exp Med 194:205–218

Essani NA, Fisher MA, Simmons CA, Hoover JL, Farhood A, Jaeschke H (1998) Increased P-selectin gene expression in the liver vasculature and its role in the pathophysiology of neutrophil-induced liver injury in murine endotoxin shock. J Leukoc Biol 63:288–296

Falati S, Gross P, Merrill-Skoloff G, Furie BC, Furie B (2002) Real-time in vivo imaging of platelets, tissue factor and fibrin during arterial thrombus formation in the mouse. Nat Med 8:1175–1181

Faust N, Varas F, Kelly LM, Heck S, Graf T (2000) Insertion of enhanced green fluorescent protein into the lysozyme gene creates mice with green fluorescent granulocytes and macrophages. Blood 96:719–726

Feng D, Nagy JA, Pyne K, Dvorak HF, Dvorak AM (1998) Neutrophils emigrate from venules by a transendothelial cell pathway in response to FMLP. J Exp Med 187:903–915

Fiebig E, Ley K, Arfors KE (1991) Rapid leukocyte accumulation by "spontaneous" rolling and adhesion in the exteriorized rabbit mesentery. Int J Microcirc Clin Exp 10:127–144

Finkenauer V, Bissinger T, Funk RH, Karbowski A, Seiffge D (1999) Confocal laser scanning microscopy of leukocyte adhesion in the microcirculation of the inflamed rat knee joint capsule. Microcirculation 6:141–152

Fox-Robichaud A, Kubes P (2000) Molecular mechanisms of tumor necrosis factor alpha-stimulated leukocyte recruitment into the murine hepatic circulation. Hepatology 31:1123–1127

Frenette PS, Mayadas TN, Rayburn H, Hynes RO, Wagner DD (1996) Susceptibility to infection and altered hematopoiesis in mice deficient in both P- and E-selectins. Cell 84:563–574

Gakidis MA, Cullere X, Olson T, Wilsbacher JL, Zhang B, Moores SL, Ley K, Swat W, Mayadas T, Brugge JS (2004) Vav GEFs are required for beta2 integrin-dependent functions of neutrophils. J Cell Biol 166:273–282

Giagulli C, Ottoboni L, Caveggion E, Rossi B, Lowell C, Constantin G, Laudanna C, Berton G (2006) The Src family kinases Hck and Fgr are dispensable for inside-out, chemoattractant-induced signaling regulating beta2 integrin affinity and valency in neutrophils, but are required for beta2 integrin-mediated outside-in signaling involved in sustained adhesion. J Immunol 177:604–611

Ginsberg MH, Partridge A, Shattil SJ (2005) Integrin regulation. Curr Opin Cell Biol 17:509–516

Gotsch U, Jager U, Dominis M, Vestweber D (1994) Expression of P-selectin on endothelial cells is upregulated by LPS and TNF-alpha in vivo. Cell Adhes Commun 2:7–14

Hafezi-Moghadam A, Ley K (1999) Relevance of L-selectin shedding for leukocyte rolling in vivo. J Exp Med 189:939–948

Hafezi-Moghadam A, Thomas KL, Prorock AJ, Huo Y, Ley K (2001) L-selectin shedding regulates leukocyte recruitment. J Exp Med 193:863–872

Hallmann R, Horn N, Selg M, Wendler O, Pausch F, Sorokin LM (2005) Expression and function of laminins in the embryonic and mature vasculature. Physiol Rev 85:979–1000

Hamilton KK, Sims PJ (1987) Changes in cytosolic Ca2+ associated with von Willebrand factor release in human endothelial cells exposed to histamine. Study of microcarrier cell monolayers using the fluorescent probe indo-1. J Clin Invest 79:600–608

Heit B, Tavener S, Raharjo E, Kubes P (2002) An intracellular signaling hierarchy determines direction of migration in opposing chemotactic gradients. J Cell Biol 159:91–102

Hidalgo A, Peired AJ, Wild MK, Vestweber D, Frenette PS (2007) Complete identification of E-selectin ligands on neutrophils reveals distinct functions of PSGL-1, ESL-1, and CD44. Immunity 26:477–489

Hirsch E, Katanaev VL, Garlanda C, Azzolino O, Pirola L, Silengo L, Sozzani S, Mantovani A, Altruda F, Wymann MP (2000) Central role for G protein-coupled phosphoinositide 3-kinase gamma in inflammation. Science 287:1049–1053

Hogg N, Leitinger B (2001) Shape and shift changes related to the function of leukocyte integrins LFA-1 and Mac-1. J Leukoc Biol 69:893–898

Homeister JW, Thall AD, Petryniak B, Maly P, Rogers CE, Smith PL, Kelly RJ, Gersten KM, Askari SW, Cheng G, Smithson G, Marks RM, Misra AK, Hindsgaul O, von Andrian UH, Lowe JB (2001) The alpha(1,3)fucosyltransferases FucT-IV and FucT-VII exert collaborative control over selectin-dependent leukocyte recruitment and lymphocyte homing. Immunity 15:115–126

Huang AJ, Manning JE, Bandak TM, Ratau MC, Hanser KR, Silverstein SC (1993) Endothelial cell cytosolic free calcium regulates neutrophil migration across monolayers of endothelial cells. J Cell Biol 120:1371–1380

Hwang JM, Yamanouchi J, Santamaria P, Kubes P (2004) A critical temporal window for selectin-dependent CD4+ lymphocyte homing and initiation of late-phase inflammation in contact sensitivity. J Exp Med 199:1223–1234

Jaeschke H, Farhood A, Fisher MA, Smith CW (1996) Sequestration of neutrophils in the hepatic vasculature during endotoxemia is independent of beta 2 integrins and intercellular adhesion molecule-1. Shock 6:351–356

Jaeschke H, Smith CW (1997) Mechanisms of neutrophil-induced parenchymal cell injury. J Leukoc Biol 61:647–653

Jung U, Bullard DC, Tedder TF, Ley K (1996) Velocity differences between L- and P-selectin-dependent neutrophil rolling in venules of mouse cremaster muscle in vivo. Am J Physiol 271:H2740–H2747

Jung U, Norman KE, Scharffetter-Kochanek K, Beaudet AL, Ley K (1998) Transit time of leukocytes rolling through venules controls cytokine-induced inflammatory cell recruitment in vivo. J Clin Invest 102:1526–1533

Kansas GS (1996) Selectins and their ligands: current concepts and controversies. Blood 88:3259–3287

Katayama Y, Hidalgo A, Chang J, Peired A, Frenette PS (2005) CD44 is a physiological E-selectin ligand on neutrophils. J Exp Med 201:1183–1189

Kim M, Carman CV, Springer TA (2003) Bidirectional transmembrane signaling by cytoplasmic domain separation in integrins. Science 301:1720–1725

Kinashi T, Katagiri K (2004) Regulation of lymphocyte adhesion and migration by the small GTPase Rap1 and its effector molecule, RAPL. Immunol Lett 93:1–5

Kinashi T (2005) Intracellular signalling controlling integrin activation in lymphocytes. Nat Rev Immunol 5:546–559

Kraiss LW, Alto NM, Dixon DA, McIntyre TM, Weyrich AS, Zimmerman GA (2003) Fluid flow regulates E-selectin protein levels in human endothelial cells by inhibiting translation. J Vasc Surg 37:161–168

Kuebler WM, Borges J, Sckell A, Kuhnle GE, Bergh K, Messmer K, Goetz AE (2000) Role of L-selectin in leukocyte sequestration in lung capillaries in a rabbit model of endotoxemia. Am J Respir Crit Care Med 161:36–43

Kunkel EJ, Dunne JL, Ley K (2000) Leukocyte arrest during cytokine-dependent inflammation in vivo. J Immunol 164:3301–3308

Kunkel EJ, Ley K (1996) Distinct phenotype of E-selectin-deficient mice. E-selectin is required for slow leukocyte rolling in vivo. Circ Res 79:1196–1204

Labow MA, Norton CR, Rumberger JM, Lombard-Gillooly KM, Shuster DJ, Hubbard J, Bertko R, Knaack PA, Terry RW, Harbison ML, et al (1994) Characterization of E-selectin-deficient mice: demonstration of overlapping function of the endothelial selectins. Immunity 1:709–720

Laudanna C, Kim JY, Constantin G, Butcher E (2002) Rapid leukocyte integrin activation by chemokines. Immunol Rev 186:37–46

Lee WY, Kubes P (2008) Leukocyte adhesion in the liver: distinct adhesion paradigm from other organs. J Hepatol 48:504–512

Ley K, Baker JB, Cybulsky MI, Gimbrone MA, Jr., Luscinskas FW (1993) Intravenous interleukin-8 inhibits granulocyte emigration from rabbit mesenteric venules without altering L-selectin expression or leukocyte rolling. J Immunol 151:6347–6357

Ley K, Bullard DC, Arbones ML, Bosse R, Vestweber D, Tedder TF, Beaudet AL (1995) Sequential contribution of L- and P-selectin to leukocyte rolling in vivo. J Exp Med 181:669–675

Ley K, Gaehtgens P (1991) Endothelial, not hemodynamic, differences are responsible for preferential leukocyte rolling in rat mesenteric venules. Circ Res 69:1034–1041

Ley K, Laudanna C, Cybulsky MI, Nourshargh S (2007) Getting to the site of inflammation: the leukocyte adhesion cascade updated. Nat Rev Immunol 7:678–689

Ley K, Mestas J, Pospieszalska MK, Sundd P, Groisman A, Zarbock A (2008) Intravital microscopy investigation of leukocyte interactions with the blood vessel wall. Methods Enzymol

Ley K, Zarbock A (2006) Hold on to your endothelium: Postarrest steps of the leukocyte adhesion cascade. Immunity 25:185–187

Ley K (1994) Histamine can induce leukocyte rolling in rat mesenteric venules. Am J Physiol 267:H1017–H1023

Lindbom L, Tuma RF, Arfors KE (1982) Blood flow in the rabbit tenuissimus muscle. Influence of preparative procedures for intravital microscopic observation. Acta Physiol Scand 114:121–127

Lindbom L, Werr J (2002) Integrin-dependent neutrophil migration in extravascular tissue. Semin Immunol 14:115–121

Linder S, Aepfelbacher M (2003) Podosomes: adhesion hot-spots of invasive cells. Trends Cell Biol 13:376–385

Liu L, Puri KD, Penninger JM, Kubes P (2007) Leukocyte PI3Kgamma and PI3Kdelta have temporally distinct roles for leukocyte recruitment in vivo. Blood 110:1191–1198

Lorenzl S, Koedel U, Dirnagl U, Ruckdeschel G, Pfister HW (1993) Imaging of leukocyte-endothelium interaction using in vivo confocal laser scanning microscopy during the early phase of experimental pneumococcal meningitis. J Infect Dis 168:927–933

Luo BH, Carman CV, Springer TA (2007) Structural Basis of Integrin Regulation and Signaling. Annu Rev Immunol 25:619–647

Maly P, Thall A, Petryniak B, Rogers CE, Smith PL, Marks RM, Kelly RJ, Gersten KM, Cheng G, Saunders TL, Camper SA, Camphausen RT, Sullivan FX, Isogai Y, Hindsgaul O, von Andrian UH, Lowe JB (1996) The alpha(1,3)fucosyltransferase Fuc-TVII controls leukocyte trafficking through an essential role in L-, E-, and P-selectin ligand biosynthesis. Cell 86:643–653

Mayadas TN, Cullere X (2005) Neutrophil beta2 integrins: moderators of life or death decisions. Trends Immunol 26:388–395

Mayadas TN, Johnson RC, Rayburn H, Hynes RO, Wagner DD (1993) Leukocyte rolling and extravasation are severely compromised in P selectin-deficient mice. Cell 74:541–554

McDonald B, McAvoy EF, Lam F, Gill V, de la Motte C, Savani RC, Kubes P (2008) Interaction of CD44 and hyaluronan is the dominant mechanism for neutrophil sequestration in inflamed liver sinusoids. J Exp Med 205:915–927

McEver RP, Cummings RD (1997) Role of PSGL-1 binding to selectins in leukocyte recruitment. J Clin Invest 100:S97–S103

Mempel TR, Henrickson SE, Von Andrian UH (2004) T-cell priming by dendritic cells in lymph nodes occurs in three distinct phases. Nature 427:154–159

Mempel TR, Moser C, Hutter J, Kuebler WM, Krombach F (2003) Visualization of leukocyte transendothelial and interstitial migration using reflected light oblique transillumination in intravital video microscopy. J Vasc Res 40:435–441

Millan J, Hewlett L, Glyn M, Toomre D, Clark P, Ridley AJ (2006) Lymphocyte transcellular migration occurs through recruitment of endothelial ICAM-1 to caveola- and F-actin-rich domains. Nat Cell Biol 8:113–123

Miner JJ, Xia L, Yago T, Kappelmayer J, Liu Z, Klopocki AG, Shao B, McDaniel JM, Setiadi H, Schmidtke DW, McEver RP (2008) Separable requirements for cytoplasmic domain of PSGL-1 in leukocyte rolling and signaling under flow. Blood 112:2035–2045

Mocsai A, Abram CL, Jakus Z, Hu Y, Lanier LL, Lowell CA (2006) Integrin signaling in neutrophils and macrophages uses adaptors containing immunoreceptor tyrosine-based activation motifs. Nat Immunol 7:1326–1333

Muller WA (2003) Leukocyte-endothelial-cell interactions in leukocyte transmigration and the inflammatory response. Trends Immunol 24:327–334

Nathan C, Srimal S, Farber C, Sanchez E, Kabbash L, Asch A, Gailit J, Wright SD (1989) Cytokine-induced respiratory burst of human neutrophils: dependence on extracellular matrix proteins and CD11/CD18 integrins. J Cell Biol 109:1341–1349

Nathan C (2006) Neutrophils and immunity: challenges and opportunities. Nat Rev Immunol 6:173–182

Nieswandt B, Moser M, Pleines I, Varga-Szabo D, Monkley S, Critchley D, Fassler R (2007) Loss of talin1 in platelets abrogates integrin activation, platelet aggregation, and thrombus formation in vitro and in vivo. J Exp Med 204:3113–3118

Norman KE, Moore KL, McEver RP, Ley K (1995) Leukocyte rolling in vivo is mediated by P-selectin glycoprotein ligand-1. Blood 86:4417–4421

Norman MU, Hulliger S, Colarusso P, Kubes P (2008) Multichannel fluorescence spinning disk microscopy reveals early endogenous CD4 T cell recruitment in contact sensitivity via complement. J Immunol 180:510–521

Nourshargh S, Krombach F, Dejana E (2006) The role of JAM-A and PECAM-1 in modulating leukocyte infiltration in inflamed and ischemic tissues. J Leukoc Biol 80:714–718

Olson TS, Singbartl K, Ley K (2002) L-selectin is required for fMLP- but not C5a-induced margination of neutrophils in pulmonary circulation. Am J Physiol Regul Integr Comp Physiol 282:R1245–R1252

Ouyang Y, Lane WS, Moore KL (1998) Tyrosylprotein sulfotransferase: purification and molecular cloning of an enzyme that catalyzes tyrosine O-sulfation, a common posttranslational modification of eukaryotic proteins. Proc Natl Acad Sci USA 95:2896–2901

Ouyang YB, Moore KL (1998) Molecular cloning and expression of human and mouse tyrosylprotein sulfotransferase-2 and a tyrosylprotein sulfotransferase homologue in Caenorhabditis elegans. J Biol Chem 273:24770–24774

Pasvolsky R, Feigelson SW, Kilic SS, Simon AJ, Tal-Lapidot G, Grabovsky V, Crittenden JR, Amariglio N, Safran M, Graybiel AM, Rechavi G, Ben-Dor S, Etzioni A, Alon R (2007) A LAD-III syndrome is associated with defective expression of the Rap-1 activator CalDAG-GEFI in lymphocytes, neutrophils, and platelets. J Exp Med 204:1571–1582

Petri B, Bixel MG (2006) Molecular events during leukocyte diapedesis. FEBS J 273:4399–4407

Petrich BG, Marchese P, Ruggeri ZM, Spiess S, Weichert RA, Ye F, Tiedt R, Skoda RC, Monkley SJ, Critchley DR, Ginsberg MH (2007) Talin is required for integrin-mediated platelet function in hemostasis and thrombosis. J Exp Med 204:3103–3111

Phillipson M, Heit B, Colarusso P, Liu L, Ballantyne CM, Kubes P (2006) Intraluminal crawling of neutrophils to emigration sites: a molecularly distinct process from adhesion in the recruitment cascade. J Exp Med 203:2569–2575

Picker LJ, Warnock RA, Burns AR, Doerschuk CM, Berg EL, Butcher EC (1991) The neutrophil selectin LECAM-1 presents carbohydrate ligands to the vascular selectins ELAM-1 and GMP-140. Cell 66:921–933

Pouyani T, Seed B (1995) PSGL-1 recognition of P-selectin is controlled by a tyrosine sulfation consensus at the PSGL-1 amino terminus. Cell 83:333–343

Ramos CL, Kunkel EJ, Lawrence MB, Jung U, Vestweber D, Bosse R, McIntyre KW, Gillooly KM, Norton CR, Wolitzky BA, Ley K (1997) Differential effect of E-selectin antibodies on neutrophil rolling and recruitment to inflammatory sites. Blood 89:3009–3018

Read MA, Neish AS, Luscinskas FW, Palombella VJ, Maniatis T, Collins T (1995) The proteasome pathway is required for cytokine-induced endothelial-leukocyte adhesion molecule expression. Immunity 2:493–506

Rosen SD (2004) Ligands for L-selectin: homing, inflammation, and beyond. Annu Rev Immunol 22:129–156

Schaff UY, Yamayoshi I, Tse T, Griffin D, Kibathi L, Simon SI (2008) Calcium flux in neutrophils synchronizes beta2 integrin adhesive and signaling events that guide inflammatory recruitment. Ann Biomed Eng 36:632–646

Shaw SK, Bamba PS, Perkins BN, Luscinskas FW (2001) Real-time imaging of vascular endothelial-cadherin during leukocyte transmigration across endothelium. J Immunol 167:2323–2330

Shimaoka M, Xiao T, Liu JH, Yang Y, Dong Y, Jun CD, McCormack A, Zhang R, Joachimiak A, Takagi J, Wang JH, Springer TA (2003) Structures of the alpha L I domain and its complex with ICAM-1 reveal a shape-shifting pathway for integrin regulation. Cell 112:99–111

Shimonaka M, Katagiri K, Nakayama T, Fujita N, Tsuruo T, Yoshie O, Kinashi T (2003) Rap1 translates chemokine signals to integrin activation, cell polarization, and motility across vascular endothelium under flow. J Cell Biol 161:417–427

Slaaf DW, Tangelder GJ, Reneman RS, Jager K, Bollinger A (1987) A versatile incident illuminator for intravital microscopy. Int J Microcirc Clin Exp 6:391–397

Smalley DM, Ley K (2005) L-selectin: mechanisms and physiological significance of ectodomain cleavage. J Cell Mol Med 9:255–266

Smith DF, Deem TL, Bruce AC, Reutershan J, Wu D, Ley K (2006) Leukocyte phosphoinositide-3 kinase {gamma} is required for chemokine-induced, sustained adhesion under flow in vivo. J Leukoc Biol 80:1491–1499

Smith ML, Long DS, Damiano ER, Ley K (2003) Near-wall micro-PIV reveals a hydrodynamically relevant endothelial surface layer in venules in vivo. Biophys J 85:637–645

Smith ML, Olson TS, Ley K (2004) CXCR2- and E-selectin-induced neutrophil arrest during inflammation in vivo. J Exp Med 200:935–939

Sperandio M, Smith ML, Forlow SB, Olson TS, Xia L, McEver RP, Ley K (2003) P-selectin glycoprotein ligand-1 mediates L-selectin-dependent leukocyte rolling in venules. J Exp Med 197:1355–1363

Springer TA (1994) Traffic signals for lymphocyte recirculation and leukocyte emigration: the multistep paradigm. Cell 76:301–314

Stein JV, Cheng G, Stockton BM, Fors BP, Butcher EC, von Andrian UH (1999) L-selectin-mediated leukocyte adhesion in vivo: microvillous distribution determines tethering efficiency, but not rolling velocity. J Exp Med 189:37–50

Tabuchi A, Mertens M, Kuppe H, Pries AR, Kuebler WM (2008) Intravital microscopy of the murine pulmonary microcirculation. J Appl Physiol 104:338–346

Tadokoro S, Shattil SJ, Eto K, Tai V, Liddington RC, de Pereda JM, Ginsberg MH, Calderwood DA (2003) Talin binding to integrin beta tails: a final common step in integrin activation. Science 302:103–106

Tangelder GJ, Slaaf DW, Muijtjens AM, Arts T, oude Egbrink MG, Reneman RS (1986) Velocity profiles of blood platelets and red blood cells flowing in arterioles of the rabbit mesentery. Circ Res 59:505–514

Tauxe C, Xie X, Joffraud M, Martinez M, Schapira M, Spertini O (2008) P-selectin glycoprotein ligand-1 decameric repeats regulate selectin-dependent rolling under flow conditions. J Biol Chem PMID:18713749

Tedder TF, Steeber DA, Pizcueta P (1995) L-selectin-deficient mice have impaired leukocyte recruitment into inflammatory sites. J Exp Med 181:2259–2264

Tenno M, Ohtsubo K, Hagen FK, Ditto D, Zarbock A, Schaerli P, von Andrian UH, Ley K, Le D, Tabak LA, Marth JD (2007) Initiation of protein O glycosylation by the polypeptide GalNAcT-1 in vascular biology and humoral immunity. Mol Cell Biol 27:8783–8796

Thorlacius H, Raud J, Rosengren-Beezley S, Forrest MJ, Hedqvist P, Lindbom L (1994) Mast cell activation induces P-selectin-dependent leukocyte rolling and adhesion in postcapillary venules in vivo. Biochem Biophys Res Commun 203:1043–1049

Urzainqui A, Serrador JM, Viedma F, Yanez-Mo M, Rodriguez A, Corbi AL, Alonso-Lebrero JL, Luque A, Deckert M, Vazquez J, Sanchez-Madrid F (2002) ITAM-based interaction of ERM proteins with Syk mediates signaling by the leukocyte adhesion receptor PSGL-1. Immunity 17:401–412

Vestweber D (2007) Adhesion and signaling molecules controlling the transmigration of leukocytes through endothelium. Immunol Rev 218:178–96

von Andrian UH, Hasslen SR, Nelson RD, Erlandsen SL, Butcher EC (1995) A central role for microvillous receptor presentation in leukocyte adhesion under flow. Cell 82:989–999

Wagner R (1839) Erläuterungstafeln zur Physiologie und Entwicklungsgeschichte. Leopold Voss, Leipzig

Walzog B, Seifert R, Zakrzewicz A, Gaehtgens P, Ley K (1994) Cross-linking of CD18 in human neutrophils induces an increase of intracellular free Ca27, exocytosis of azurophilic granules, quantitative up-regulation of CD18, shedding of L-selectin, and actin polymerization. J Leukoc Biol 56:625–635

Wang HB, Wang JT, Zhang L, Geng ZH, Xu WL, Xu T, Huo Y, Zhu X, Plow EF, Chen M, Geng JG (2007) P-selectin primes leukocyte integrin activation during inflammation. Nat Immunol 8:882–892

Wang S, Voisin MB, Larbi KY, Dangerfield J, Scheiermann C, Tran M, Maxwell PH, Sorokin L, Nourshargh S (2006) Venular basement membranes contain specific matrix protein low expression regions that act as exit points for emigrating neutrophils. J Exp Med 203:1519–1532

Wegener KL, Partridge AW, Han J, Pickford AR, Liddington RC, Ginsberg MH, Campbell ID (2007) Structural basis of integrin activation by talin. Cell 128:171–182

Wegmann F, Petri B, Khandoga AG, Moser C, Khandoga A, Volkery S, Li H, Nasdala I, Brandau O, Fassler R, Butz S, Krombach F, Vestweber D (2006) ESAM supports neutrophil extravasation, activation of Rho, and VEGF-induced vascular permeability. J Exp Med 203:1671–1677

Welch HC, Coadwell WJ, Ellson CD, Ferguson GJ, Andrews SR, Erdjument-Bromage H, Tempst P, Hawkins PT, Stephens LR (2002) P-Rex1, a PtdIns(3,4,5)P3- and Gbetagamma-regulated guanine-nucleotide exchange factor for Rac. Cell 108:809–821

Weninger W, Ulfman LH, Cheng G, Souchkova N, Quackenbush EJ, Lowe JB, von Andrian UH (2000) Specialized contributions by alpha(1,3)-fucosyltransferase-IV and FucT-VII during leukocyte rolling in dermal microvessels. Immunity 12:665–676

Wong J, Johnston B, Lee SS, Bullard DC, Smith CW, Beaudet AL, Kubes P (1997) A minimal role for selectins in the recruitment of leukocytes into the inflamed liver microvasculature. J Clin Invest 99:2782–2790

Worthen GS, Schwab B, 3rd, Elson EL, Downey GP (1989) Mechanics of stimulated neutrophils: cell stiffening induces retention in capillaries. Science 245:183–186

Xia L, Sperandio M, Yago T, McDaniel JM, Cummings RD, Pearson-White S, Ley K, McEver RP (2002) P-selectin glycoprotein ligand-1-deficient mice have impaired leukocyte tethering to E-selectin under flow. J Clin Invest 109:939–950

Yang J, Hirata T, Croce K, Merrill-Skoloff G, Tchernychev B, Williams E, Flaumenhaft R, Furie BC, Furie B (1999) Targeted gene disruption demonstrates that P-selectin glycoprotein ligand 1 (PSGL-1) is required for P-selectin-mediated but not E-selectin-mediated neutrophil rolling and migration. J Exp Med 190:1769–1782

Yang L, Froio RM, Sciuto TE, Dvorak AM, Alon R, Luscinskas FW (2005) ICAM-1 regulates neutrophil adhesion and transcellular migration of TNF-alpha-activated vascular endothelium under flow. Blood 106:584–592

Yao L, Setiadi H, Xia L, Laszik Z, Taylor FB, McEver RP (1999) Divergent inducible expression of P-selectin and E-selectin in mice and primates. Blood 94:3820–3828

Zarbock A, Abram CL, Hundt M, Altman A, Lowell CA, Ley K (2008) PSGL-1 engagement by E-selectin signals through Src kinase Fgr and ITAM adapters DAP12 and FcRgamma to induce slow leukocyte rolling. J Exp Med 205(10):2339–2347

Zarbock A, Deem TL, Burcin TL, Ley K (2007) Galphai2 is required for chemokine-induced neutrophil arrest. Blood 110:3773–3779

Zarbock A, Ley K (2008) Mechanisms and consequences of neutrophil interaction with the endothelium. Am J Pathol 172:1–7

Zarbock A, Lowell CA, Ley K (2007) Spleen tyrosine kinase Syk is necessary for E-selectin-induced alpha(L)beta(2) integrin-mediated rolling on intercellular adhesion molecule-1. Immunity 26:773–783

Zen K, Cui LB, Zhang CY, Liu Y (2007) Critical role of mac-1 sialyl lewis x moieties in regulating neutrophil degranulation and transmigration. J Mol Biol 374:54–63
Zhang H, Schaff UY, Green CE, Chen H, Sarantos MR, Hu Y, Wara D, Simon SI, Lowell CA (2006) Impaired integrin-dependent function in Wikott-Aldrich Syndrome Protein-deficient murine and human neutrophils. Immunity 25:285–295
Zollner O, Lenter MC, Blanks JE, Borges E, Steegmaier M, Zerwes HG, Vestweber D (1997) L-selectin from human, but not from mouse neutrophils binds directly to E-selectin. J Cell Biol 136:707–716
Zweifach BW (1973) The microcirculation in the intestinal mesentery. Microvasc Res 5:363–367

Visualizing the Molecular and Cellular Events Underlying the Initiation of B-Cell Activation

Naomi E. Harwood and Facundo D. Batista

Contents

Abstract The appropriate activation of B cells is critical for the development of effective immune responses. B cell activation is initiated following the engagement of the B cell receptor (BCR) with specific antigen. The spatiotemporal characterization of the ensuing molecular and cellular events has been the subject of recent high-resolution imaging investigations. In this review we highlight information gathered thus far concerning the initial processes underlying the activation of B cells.

N.E. Harwood and F.D. Batista (✉)
Lymphocyte Interaction Laboratory, Cancer Research UK London Research Institute, Lincoln's Inn Fields Laboratories, 44 Lincoln's Inn Fields, London, WC2A 3PX, UK,
e-mail: facundo.batista@cancer.org.uk

D. McGavern and M. Dustin (eds.), *Visualizing Immunity,* Current Topics in Microbiology and Immunology 334,
DOI: 10.1007/978-3-540-93864-4_7,

First, we consider studies that have offered new insights into the early molecular events that occur within the B cell prior to formation of the immunological synapse. As such, BCR-microclusters formed on engagement with antigen have been identified as the sites of active signaling and assembly of "microsignalosomes." Furthermore, signaling through these "microsignalosomes" is propagated and enhanced through B cell spreading in response to membrane-antigen in a CD19-dependent manner. Finally, we discuss a number of multiphoton microscopy studies that have enabled dynamic characterization of the initial encounters between B cells and antigen *in vivo*. These investigations visualize the presentation of larger antigens to B cells via cell-mediated strategies, involving macrophages in the subcapsular sinus and dendritic cells in the paracortex.

1 Introduction

Components of the immune system operate in a highly coordinated manner, providing protection from a wide variety of potentially damaging situations. The adaptive arm of the immune system ensures that the responses raised during such situations are both specific and long lasting. These essential characteristics are conferred through the concerted action of B and T lymphocytes.

B cells are activated following recognition of specific antigen by the B cell receptor (BCR). The BCR is a cell surface receptor comprised of a membrane immunoglobulin (Ig), together with the Igα/β complex, allowing the B cell to both sense and respond to extracellular antigen (Reth 1989). Antigen engagement results in rapid phosphorylation of the BCR on tyrosine residues within immunoreceptor tyrosine activation motifs (ITAMs) of the Igα/β complex (Dal Porto et al. 2004; DeFranco 1997; Kurosaki 1999; Reth and Wienands 1997). In turn, this phosphorylation initiates the formation of the signalosome, an assembly of signaling molecules, including Vav, Bruton's tyrosine kinase (Btk), phosphoinositide 3-kinase (PI3K), and phospholipase Cγ2 (PLCγ2) (Dal Porto et al. 2004; Kurosaki and Kurosaki 1997; Scharenberg et al. 2007), and adaptor molecules, such as B cell linker (Blnk) (Fu et al. 1998; Goitsuka et al. 1998; Wienands et al. 1998). The signalosome allows extensive coordination of cellular responses, and depending on its precise composition can result in alteration of cellular activities including gene expression and cell morphology. One consequence of these responses is the extraction of tethered antigen from the presenting surface, prior to its BCR-mediated internalization (Batista and Neuberger 2000). The BCR accelerates the delivery of antigen to endosomes where they are processed and then bound to newly synthesized major histocompatibility complex (MHC) molecules (Aluvihare et al. 1997; Amigorena et al. 1994). Subsequent presentation of loaded MHC molecules on the B cell surface enables the recruitment of antigen-specific CD4$^+$ B cell help, allowing optimal B-cell activation (Lanzavecchia 1985; Rock et al. 1984).

Following their activation, B cells can contribute to the adaptive immune response through the production of two distinct "waves" of antibodies. The first wave of antibodies is generally of low-affinity and is secreted by plasma B cells located in the extrafollicular region of secondary lymphoid organs (SLOs) (MacLennan et al. 2003). In contrast, plasma cells generated within germinal centers are responsible for producing the second wave of antibodies, and can be observed around 5 days after initial stimulation (MacLennan 1994; Moller 1987). These antibodies are secreted by B cells that have undergone affinity maturation and are generally of very high affinity. Secreted antibodies can utilize a variety of strategies, such as virus neutralization and complement activation, to mediate their protective effects.

Over the past decade, the molecular and cellular processes underlying B cell activation have been visualized through a variety of high-resolution imaging techniques. These techniques have enabled a shift from classic static characterizations to more comprehensive spatiotemporal descriptions of the processes involved in B cell activation. Here, we will describe briefly both the imaging techniques and the specialized tools that have been employed in the investigation of B cell activation. We will then move on to discuss the insights gained from these studies in terms of the molecular requirements and interactions within the B cell, before concluding with a description of a number of recent investigations into the dynamics of the initiation of B cell activation in real time and *in vivo*.

2 Tools for Visualizing B Cell Activation Events

2.1 *B Cells*

B cell activation has been investigated in a wide range of B-cell lines developed over the past 40 years. These cell lines enable the production large quantities of B cells with relative ease for the investigation of processes underlying B cell activation induced by antibodies to the BCR. However, as this type of widespread cross-linking may not truly replicate the process of antigen-induced activation, cell lines have been transformed to express BCR specific for particular antigens. In spite of this development, observations collected from B cell lines must be assessed and verified in primary B cells obtained from animal models. The repertoire of primary B cells within an individual must be comprised of a huge diversity of possible BCR antigen specificities in order to maintain efficacy. However, monitoring antigen-specific stimulation in such repertoires would entail extraordinarily sensitive analytical methods, the like of which are not routinely available at present. To address these concerns, immunologists have utilized genetic manipulation strategies to influence the generation of primary B cells in animal models (reviewed in Brink et al. 2008). As such, through modification of the genes encoding the chains of the BCR, transgenic animals can be generated that express a bias in the antigen-specificity of the primary B cell repertoire (reviewed in Goodnow 1992). These transgenic models have proved extremely useful in the investigation of primary B cell activation in response to antigens such as hen egg lysozyme (Goodnow et al. 1988) and H-2K^K (Russell et al. 1991). More recently the development of "knockin" technology has enabled the physiological expression of transgenes in terms of their expression levels and tissue-specificity (Pelanda et al. 1997). This technology has been applied to allow the expression of antigen-specific class-switched BCRs on the B cell surface (Phan et al. 2003), allowing a more accurate representation of antigen-induced BCR stimulation under physiological conditions. Nevertheless, as the expressed transgenic BCRs are often already of extremely high affinity for antigen, it should be borne in mind that in these cases they might not absolutely mimic the encounter of the primary B cell repertoire with antigen.

2.2 Antigen

Through the BCR, B cells can recognize and respond to a wide-variety of potential antigens. In contrast to T cells whose activation occurs exclusively through recognition of antigenic fragments associated with MHCs on the surface of antigen-presenting cells (APCs), B cells recognize exposed epitopes of native antigen. This type of recognition dictates that B cells are able to mount responses to both soluble and membrane-associated antigen, though most early investigations into B cell activation employed soluble antigen for stimulation. However, it is becoming evident that the majority of naive B cell priming involves stimulation with membrane-associated antigen under physiological conditions (Carrasco and Batista 2006a). As a result, tools that were originally developed in order to characterize T cells have more recently been employed and tailored to enable visualization of B cell activation by membrane-antigen (Fig. 1). One such method involved the settling of T cells on glass coverslips coated with antibodies capable of cross-linking the TCR (Bunnell et al. 2001). Though this simple approach allows analysis of events initiated following TCR engagement, it is limited by the restricted mobility of the TCR ligand. An alternative strategy involves the assembly of planar lipid bilayers from liposomes (Grakoui et al. 1999). These bilayers allow the incorporation of particular antigen at a defined density, together with associated accessory molecules of choice, simulating the activity of an APC (Carrasco et al. 2004). However, though the bilayer enables lateral movement within and across the plane, it cannot be used to replicate or assess the contribution of the cytoskeleton within APC. Thus, it is necessary to ensure that observations from these studies involving planar bilayers are established as important in actual cell–cell interactions.

2.3 Imaging

Historically, various microscopy techniques have been hindered by the increased levels of sample damage associated with improved resolution. Accordingly, approaches such as confocal microscopy and total internal reflection microscopy (TIRFM) were introduced to improve sectioning capability as a means of improving

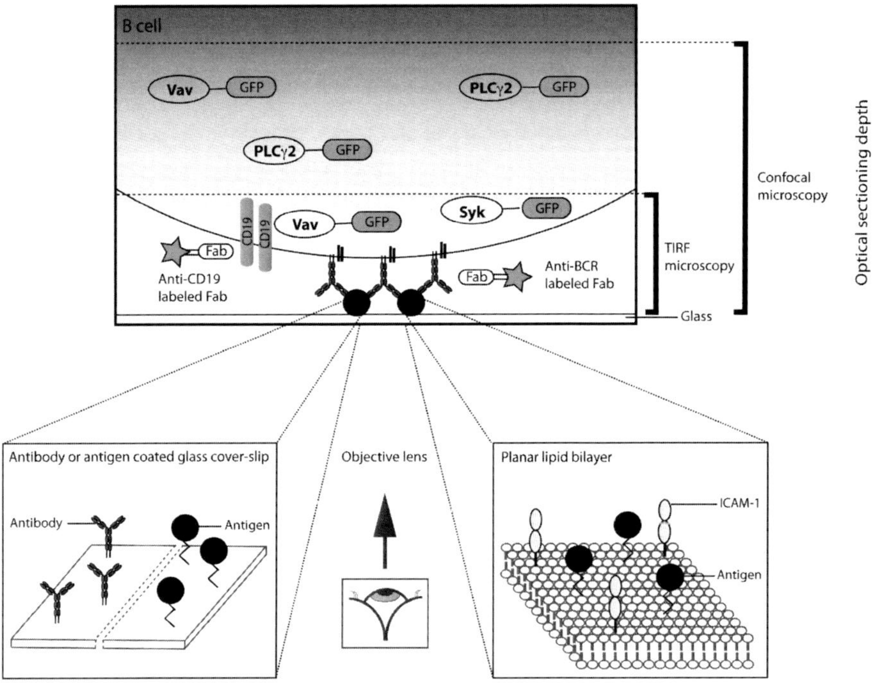

Fig. 1 Tools for visualizing B-cell activation. *Top panel*: The spatiotemporal dynamics of molecules within the B-cell following settling on an antigen-containing surface can be investigated using confocal microscopy and TIRFM. These two methods allow, through the objective lens, visualization of molecular events at different optical sectioning depths into the B cell. Two different strategies for visualizing molecules by these methods are shown – the first involves direct conjugation with GFP, and the second involves staining with specific antibody fragments (Fab) labeled with a fluorescent probe. The antigen for these investigations can be either coated onto the surface of a glass coverslip (*left box*), or embedded within a planar lipid bilayer, which can also contain accessory molecules such as the integrin ligand ICAM-1 (*right box*).

effective resolution (Fig. 1). Confocal microscopy involves illumination of a fine, restricted plane of the sample followed by detection that eliminates unwanted information from outside the plane of interest. This approach has been further improved to allow more rapid visualization of events through the introduction of spinning disc confocal microscopy (Bunnell et al. 2006). The parallel development of TIRFM has proved extremely useful for the investigation of interactions in close proximity to a surface at high resolution (Schneckenburger 2005). This technique, developed in the 1980s, allows visualization of molecules in contact with a surface to a depth of around 200 nm. This process is dependent on total internal reflection of incident light to generate evanescent waves capable of exciting flurophores in close proximity to the sample interface. Thus, TIRFM has been widely applied to the investigation of molecular events that occur in or around the plasma membrane. Although TIRFM and confocal microscopy have yielded much information as to the molecular interactions underlying cell events, they necessitate *in vitro* investigation of samples. One alternative approach, using multiphoton microscopy (MPM), allows the visualization of cellular dynamics within the correct physiological environment and in real time (Cahalan and Parker ; Germain et al. 2006). MPM uses two-photon excitation to minimize photodamage, and thereby images cellular interactions in three dimensions and to a depth of around 200 μm. This technique is currently yielding novel insights into to the dynamics of lymphocyte interactions *in vivo*.

The advent of high-resolution imaging techniques has required the concurrent development of approaches, such as fluorescent labeling (Lichtman and Conchello 2005), to identify and discriminate between selected groups of molecules. Early studies regularly employed fluorescent-labeled antibodies for staining of fixed sections. However, more recently, the spatiotemporal dynamics of specific molecules have been characterized extrinsically through binding of fluorescently labeled antibody fragments or intrinsically through expression of molecules conjugated with green fluorescent protein (GFP), or a similar derivative. These types of molecules can, for example, be utilized to gain information as to organization and mobility within membranes through techniques such as fluorescence recovery after photobleaching (FRAP) and single-particle tracking (SPT). Alternatively, fluorescence resonance energy transfer (FRET) and fluorescence lifetime imaging (FLIM) can be used to visualize the dynamics of intra- and intermolecular interactions between different fluorescent labels (reviewed in Treanor and Batista 2007).

3 Visualizing Molecular Events During the Initiation of B Cell Activation

The earliest classic descriptions of B cell behavior were derived from biochemical analyses following stimulation with soluble antigens. These early studies in essence represent static snapshots across whole populations and therefore offer little insight into the dynamics of molecular processes that occur on activation of individual B cells. However, more recently various combinations of the technologies described above have been used to dissect the molecular and cellular changes that occur within the B cell during activation (Fig. 2). In this section, we discuss recent investigations that will form the foundation for a complete spatiotemporal description of the initial events within B cells following antigenic stimulation.

3.1 The Immunological Synapse

Early investigations into the initiation of activation of lymphocytes, such as T, B, and NK cells, visualized a widespread reorganization of molecules within membranes in response to specific recognition of antigen (Monks et al. 1998; Grakoui et al. 1999; Batista et al. 2001; Davis et al. 1999; Stinchcombe et al. 2001). This molecular reorganization is now established as a common feature of antigen recognition by

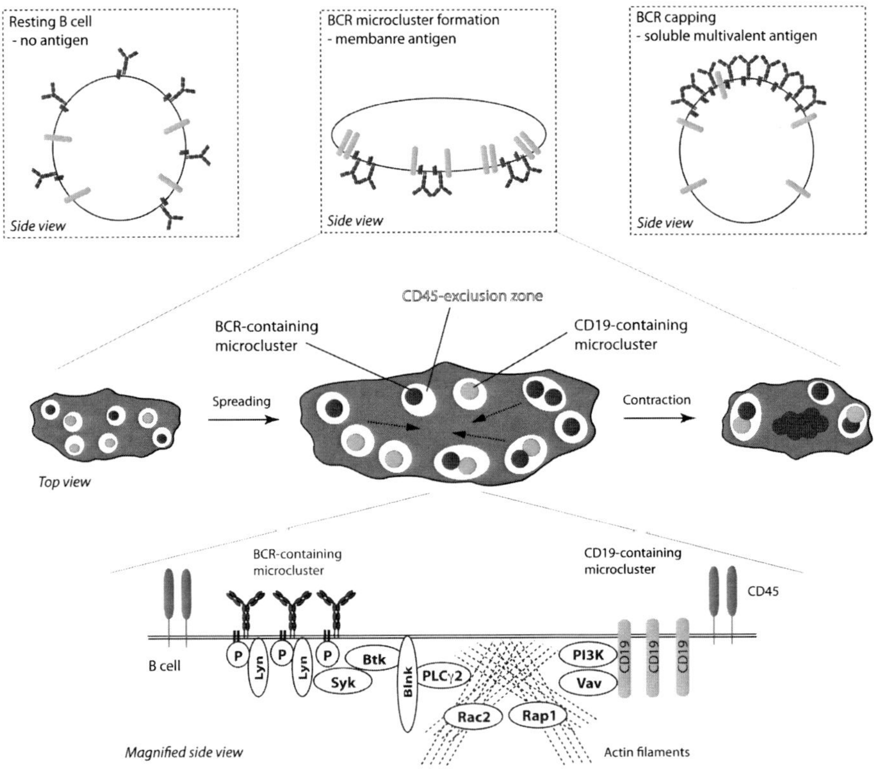

Fig. 2 Molecular processes underlying antigen-induced B cell activation. *Left box*: In the resting B cell membrane the BCR is thought to exist either as monomers or as higher-order oligomers. *Middle box*: Stimulation with membrane-bound antigen induces the formation of BCR-microclusters throughout the contact surface. *Middle panel*: A view from the antigenic surface (with the antigen-containing surface not shown for clarity) of this process over time is shown. After the initial contact, BCR microclusters, together with microclusters containing other regulatory molecules such as CD19, form throughout the contact area in sites that exclude inhibitory phosphatases such as CD45. The microclusters are propagated by the B cell spreading response. Following achievement of the maximum contact area, B cells undergo a slower contraction phase, gathering BCR microclusters into the cSMAC of the IS for antigen internalization. *Bottom panel*: A magnified view of the molecular constituents of the BCR-microsignalosome following the dynamic recruitment of a CD19-containing microcluster is shown. *Right box*: Stimulation with multivalent soluble antigen induces the formation of a "cap" of BCRs and allowing "cocapping" with other molecules such as CD19.

specific immunoreceptors and results in the formation of an immunological synapse (IS). The structure of the mature IS is characterized by an accumulation of immunoreceptors and antigen in a central supramolecular activation cluster (cSMAC) surrounded by a peripheral SMAC (pSMAC) containing integrins such as LFA-1 (Monks et al. 1998). Though the mature IS was first visualized in lymphocytes more than a decade ago the precise function of the cSMAC has remained controversial.

In view of the accumulation of immunoreceptors and ligands in the cSMAC, it was originally proposed that the cSMAC formed the site of active signaling within the IS (Grakoui et al. 1999). However, it was subsequently suggested that the cSMAC might function as a platform for internalization of antigen, as signaling was observed to occur primarily within the periphery of the IS (Lee et al. 2002). In line with this, more recent studies have shown that signaling in the periphery is necessary for "sustained" signaling and associated T cell activation (Bunnell et al. 2002; Campi et al. 2005; Yokosuka et al. 2005). Interestingly, a very recent study has demonstrated signaling within the T cell cSMAC can also occur, and this may play an important role in the initiation of activation in response to less potent antigens (Cemerski et al. 2008). In B cells, the mature IS is assembled over a period of approximately 10 min following BCR engagement (Carrasco et al. 2004), while intracellular signaling, as revealed by changes in intracellular calcium concentration, is initiated considerably more rapidly (Fleire et al. 2006). Furthermore, it has been shown that B cells extract antigen from cSMAC (Batista and Neuberger 2000), a process of particular importance as it allows the recruitment of T cell help and maximal B cell activation (Lanzavecchia 1985; Rock et al. 1984). Regardless of precise functional segregation within the IS, it remains clear that it is absolutely necessary to examine the molecular events prior to IS formation in order to fully characterize the mechanisms underlying the initiation of B cell activation.

3.2 BCR-Microclusters are Formed on Antigen Engagement

Given the ability of multivalent, but not monovalent, antigen to trigger B cell activation, it was originally expected that the dimerization of BCR monomers triggered the initiation of signaling. However, while FRET revealed the BCR to exist as monomers in the B cell membrane prior to stimulation (Tolar et al. 2005), biochemical characterizations suggested that BCRs may exist as multimers (Schamel and Reth 2000). It therefore currently remains unclear as to the state of the BCR in the "resting" cell membrane. In spite of this, it has been well established that that BCRs redistribute and form a "cap" within the cell membrane following engagement with multivalent soluble antigen (Fig. 2) (Schreiner et al. 1976; Stackpole et al. 1974; Unanue et al. 1972). Furthermore, it has been demonstrated that the capping of BCRs can induce cocapping with other components residing in the B cell membrane such as CD19 (Pesando et al. 1989; Phee et al. 2001). In contrast, engagement with membrane-tethered antigen induces the rapid formation of higher-order oligomers, seen by confocal microscopy and known as BCR-microclusters (Fig. 2) (Depoil et al. 2008). These microclusters are comprised of between 100 and 500 BCRs, and they actively exclude the inhibitory phosphatase CD45. As BCR-microclusters have been observed following antigenic stimulation of B cells expressing a BCR unable to transmit intracellular signaling, it is evident that their formation precedes and is not dependent on signaling through the BCR per se (Depoil et al. 2008). In contrast, the formation of these microclusters is dependent on reorganization of the actin cytoskeleton. Interestingly these microclusters are a critical feature associated with successful B cell stimulation by membrane-bound antigen, alluding to their significance in the process of B cell activation. Moreover, as similar structures have been visualized in T cells (Bunnell et al. 2002; Campi et al. 2005; Yokosuka et al. 2005), microclusters may function as common signaling units among lymphocytes. Though the molecular mechanism underlying the formation of these microclusters has not yet been determined, this might occur as a result of diffusion trapping following exclusion of inhibitory phosphatases, such as CD45 and CD148. The organization of immunoreceptors into microclusters allows exquisite and dynamic flexibility for regulation of their signaling activity. Unlike cap formation where all immunoreceptors are cross-linked at one site in the cell membrane and therefore inaccessible, individual microclusters can allow dynamic recruitment of other clusters containing regulators that might attenuate or enhance immunoreceptor signaling (Fig. 2). In line with this proposal, molecular segregation within membranes forming distinct functional domains has been visualized by transmission electron microscopy (Lillemeier et al. 2006).

3.3 BCR-Microclusters as "Microsignalosomes"

The intracellular molecular events initiated following the formation of BCR-microclusters in the plasma membrane have been investigated recently using TIRFM in both the chicken DT40 B cell line and in primary B cells taken from transgenic knockout mice (Weber et al. 2008). The DT40 cell line is particularly useful due to the ease of genetic recombination in this system, allowing the generation of knockout mutants that may be otherwise embryonically lethal in mice (Winding and Berchtold 2001). Consequently, a comprehensive panel of knockouts have been generated in this cell line (Shinohara and Kurosaki 2006), allowing the contribution of various signaling molecules to B cell activation to be assessed (Fig. 2). It was observed that DT40 B cells deficient in Lyn and the Src-family kinase Syk were unable able to trigger BCR-mediated signaling following antigenic stimulation (Weber et al. 2008). Reconstitution of these cells with fluorescently-labeled protein demonstrated the requirement for the sequential recruitment of Lyn and Syk kinases to the BCR-microcluster. Furthermore, it was observed that these microclusters become sites for the recruitment of additional intracellular signaling molecules, such as Vav and PLCγ2, and adaptor molecules, such as Blnk, thus demonstrating that the BCR-microclusters form the site of active signaling (Depoil et al. 2008), and redefining microclusters at this stage as "microsignalosomes" (Weber et al. 2008). Interestingly, it was also found that the key components, Vav and PLCγ2, influence the recruitment and retention of one another, suggesting cooperation within the microsignalosome. This cooperation is analogous to that described previously between components of signaling microclusters in T cells (Bunnell et al. 2006). Surprisingly DT40 B cells deficient in PI3K, in contrast to primary mouse B cells, did not exhibit any significant impairment in B cell activation (Depoil, Weber and Batista, unpublished data). These findings demonstrate that, though the principles of microsignalosome formation and function are widely applicable within B cells, slight differences in the precise molecular requirements for B cell activation may occur among cells of different origin.

3.4 B Cell Spreading Propagates Signaling Through "Microsignalosomes"

Given the significance of signaling microclusters in mediating activation, B cells employ a cellular response to enhance the generation of microsignalosomes. This response, visualized using scanning electron microscopy and confocal microscopy, involves spreading of the B cell across the antigen-containing membrane (Fleire et al. 2006) and is similar manner to that observed previously in T cells (Bunnell et al. 2001, 2002; Negulescu et al. 1996; Parsey and Lewis 1993). As B cell spreading did not occur in signaling incompetent T cells or in the presence of actin polymerization inhibitors, this cellular response absolutely requires BCR-mediated signaling and cytoskeletal reorganization (Fleire et al. 2006). Following initial contact of the B cell with antigen-containing membrane, a small number of microsignalosomes were formed throughout the contact area (Fig. 2). It is thought that these microsignalosomes then initiate a series of intracellular signaling events resulting in localized cytoskeleton rearrangements, stimulating an alteration in B cell morphology and resulting in spreading across the antigen-containing surface (Depoil et al. 2008; Weber et al. 2008). In concert, spreading increases the amount of antigen that can be engaged by the B cell and the number of microclusters formed. The maximum area of contact between B cell and antigen-presenting membrane during spreading generally reaches a maximum around 3–5 min after antigenic stimulation (Fleire et al. 2006). Following this response, the B cell undergoes a more prolonged contraction phase, resulting in collection of antigen in a central aggregate and the formation of the mature IS. The extent of the spreading response can be quantified by interference reflection microscopy and is dependent on the affinity and density of antigen present in the antigen-containing membrane (Fleire et al. 2006). Thus, increasing the affinity or density of antigen in the presenting membrane enlarges the maximal contact area achieved by spreading and enables the accumulation of increased amounts of antigen collected in the cSMAC. This spreading response provides a mechanism for propagating BCR-mediated signaling, facilitating B cell activation and allowing affinity discrimination between different antigenic ligands (Fleire et al. 2006). It would be expected that enhancing adhesion of B cells to the antigen-containing surface, and facilitating spreading would require lower thresholds of antigen to induce B cell activation. Indeed, this has been shown to be the case for the integrins LFA-1 and VLA-4 (Carrasco et al. 2004; Carrasco and Batista 2006b). These B cell surface molecules mediate adhesion to membranes containing integrin ligands including ICAM-1, VCAM, and fibronectin, enabling augmented B cell activation by membrane-bound antigen. In the case of LFA-1, it has been shown that enhanced spreading following BCR engagement was dependent on key cytoskeletal regulators, including Rac2 and Rap1 (Arana et al. 2008; Lin et al. 2008).

3.5 CD19 is Critical for the Initiation of B Cell Spreading

Although it has been observed that microcluster formation and intracellular signaling are required, these processes alone are not sufficient to initiate B cell spreading. During a detailed molecular dissection of the requirements for spreading, a novel and essential role for the B cell coreceptor CD19 was identified (Depoil et al. 2008). It was observed that CD19-deficient B cells could not efficiently initiate intracellular calcium signaling in response to membrane-antigen, rendering them less likely to become activated. Until this stage, CD19 was most commonly known as part of a complex containing CD21-CD81-TAPA2 that is responsible for enhanced B cell activation in response to complement-coated antigen (Fearon and Carter 1995). Crucially, however, the newly identified role for CD19 does not require either CD21 or complement-coated antigen. Interestingly, the dynamics of CD19 within the B cell membrane following stimulation by membrane-antigen were observed by TIRFM. It was found that fluorescently-labeled CD19 becomes transiently associated with signaling BCR-microclusters (Fig. 2) (Depoil et al. 2008). Indeed, this visualization offers a molecular explanation for the function of CD19 in facilitating the recognition of membrane-bound antigen. As the cytoplasmic region of CD19 contains binding sites for a number of key intracellular signaling molecules such as Vav and PI3K (Li et al. 1997; Tuveson et al. 1993), its transient association with BCR-microclusters stimulates the recruitment of additional signaling molecules and thus enhances signaling through the BCR (Sato et al. 1997). Such a mechanism is further supported by the observation of cooperation between components of microsignalosomes. Interestingly, this role for CD19 provides a satisfactory explanation for previous anomalies within the literature (Fearon and Carroll 2000). It was noted that CD19-deficient and wild type B cells were equally responsive to stimulation with soluble antigen (Fujimoto et al. 1999; Sato et al. 1997). However, CD19-deficient mice exhibit impaired B cell responses more severe than the phenotype displayed by CD21-deficient mice (Ahearn et al. 1996; Engel et al. 1995; Rickert et al. 1995). This demonstrates not only the critical role of CD19 but also the predominance of membrane-bound antigen recognition during the development of B cell responses (Depoil et al. 2008). Furthermore, it appears that B cells have evolved specialized mechanisms for the recognition of membrane-bound antigen.

3.6 Questions Remaining in the Initiation of B Cell Activation

Though the studies detailed above have proved extremely informative of the molecular processes occurring early in B cell activation, a number of important issues remain to be clarified. Of particular importance is establishing the molecular mechanism underlying the initiation of signaling within the B cell following antigen engagement of the BCR. It has been suggested that the equivalent process in T cells occurs as a result of "kinetic segregation" within the cell membrane (Choudhuri et al. 2005; Davis and van der Merwe 1996, 2006). The "kinetic segregation" model states that following recognition of antigen bulky phosphatases such as CD45 are excluded from sites of close contact between the T cell and APC. As such, these close contact sites become enriched in smaller molecules including TCRs and Lyn, and exhibit an altered phosphorylation balance rendering them more likely to be activated. We have suggested, and discussed in more detail, that a similar process might underlie the initiation of B cell signaling following recognition of antigen presented on a constrained surface (Harwood and Batista 2008). However, the connection between the initiation of B cell signaling and the actual formation of BCR-microclusters is yet to be established. Investigations addressing these issues might also offer an explanation as to the observed differences between B and T cells in terms of their sensitivity to antigen. Intriguingly, while full activation of T cells only requires engagement of between 4 and 10 TCRs (Irvine et al. 2002), B cells have been demonstrated to exhibit a higher threshold of BCR engagement to induce activation (Batista et al. 2007; Fleire et al. 2006).

Other interesting issues to be addressed in the future include the elucidation of downstream cellular effectors of the microsignalosome, including the cytoskeletal effectors responsible for mediating the morphological changes underlying the spreading response, and the identification of the molecular mechanism of B cell contraction and antigen internalization.

4 Visualizing Intercellular Events During the Initiation of B Cell Activation *In Vivo*

Detailed and intricate anatomical descriptions of immune system components have been derived from a wealth of historical electron microscopy investigations incorporating immunofluorescence labeling (reviewed in Bajénoff and Germain 2007). This type of investigation, in a manner analogous to classic biochemical analyses, involves the generation of static snapshots. However, the advent of MPM has enabled for the first time visualization of the intercellular interactions following antigenic stimulation, during the course of B cell activation, both in real time and *in vivo*. Indeed, the intravital application of this technology has heralded the arrival of a new era of dynamics, based on the foundation laid by previous static characterizations (Bajénoff and Germain 2007). The majority of MPM investigations thus far have been concerned with the spatiotemporal characterization of cellular dynamics within secondary lymphoid organs (SLOs), such as lymph nodes and the spleen. This type of study is technically less challenging in lymph nodes as compared with the deeply embedded and blood-rich spleen. As this is the case, we will include a basic description of the microarchitecture of the lymph node, before moving on to discuss in more detail the recent insights into the dynamics of cellular interactions associated with the initiation of B cell activation.

4.1 *Architecture of the Lymph Node*

Lymph nodes are surrounded by a protective outer collagenous capsule; with the interior divided into three discrete chemokine-defined regions – the cortex, paracortex, and medulla (Fig. 3). Directly beneath the subcapsular sinus (SCS), a macrophage-rich sheet surrounds the cortex, or B cell zone. B cells in this region represent the largest population of $IgM^{int}IgD^{high}CD21^{int}CD23^{high}$ cells in the body and are organized into follicles. Follicles also contain many radiation-resistant follicular dendritic cells (FDCs) that express high levels of complement and Fc receptors, and adhesion molecules, such as VCAM and ICAM-1 (Carroll 1998; Klaus et al. 1980; Mandel et al. 1980; Qin et al. 2000). The precise mechanism of FDC development is currently not well understood, but it is known that they originate from a mesenchymal source (reviewed in Cyster et al. 2000). Around 5 days after antigen administration, specialized structures, known as germinal centers, may be observed within the cortex. These specialized structures are comprised of rapidly proliferating B cells in association with an organized network of FDCs. Germinal centers form the sites for the affinity maturation process and are thus important during the development of humoral immune responses to T cell dependent antigen. Lymphocytes are supplied to the lymph node and carried back to the periphery by specialized capillaries known as high endothelial venules (HEVs) that are located within the paracortex, or T cell zone. These HEVs can also mediate the recruitment of other cells such as APCs that have accumulated antigen in the periphery. The paracortex also contains a highly organized network of collagenous conduit fibers specialized to allow the passage of lymph fluid from the SCS to HEVs and the medulla (reviewed in Gretz et al. 1997). The medulla, the innermost area of the node, contains both B and T cells organized into medullary cords, and is also rich in dendritic cells (DCs) and macrophages.

Lymph nodes are strategically positioned at branches of the lymphatic system throughout the body enabling extensive antigenic sampling of lymphatic fluid (Catron et al. 2004). Lymphatic fluid, containing various chemokines and antigens collected from throughout the body, is brought into the lymph node through the afferent vessel and is subsequently "filtered" by the lymph node as it is moved out towards the efferent vessel. A tightly packed network of macrophages and DCs beneath the SCS restricts the movement of lymphatic fluid into the lymph node interior. As a result lymphatic fluid can only access the interior through the intricate meshwork of fibroblastic reticular cells (FRC)-defined conduits extensively associated with DCs (Gretz et al. 2000; Sixt et al. 2005). These conduit structures exclude larger components of lymphatic fluid, but allow diffusion of components below approximately 70 kDa into the lymph node interior. Following its passage through the node towards the medulla, the lymph fluid selectively enriched for lymphocytes is drained through the efferent lymphatic vessel.

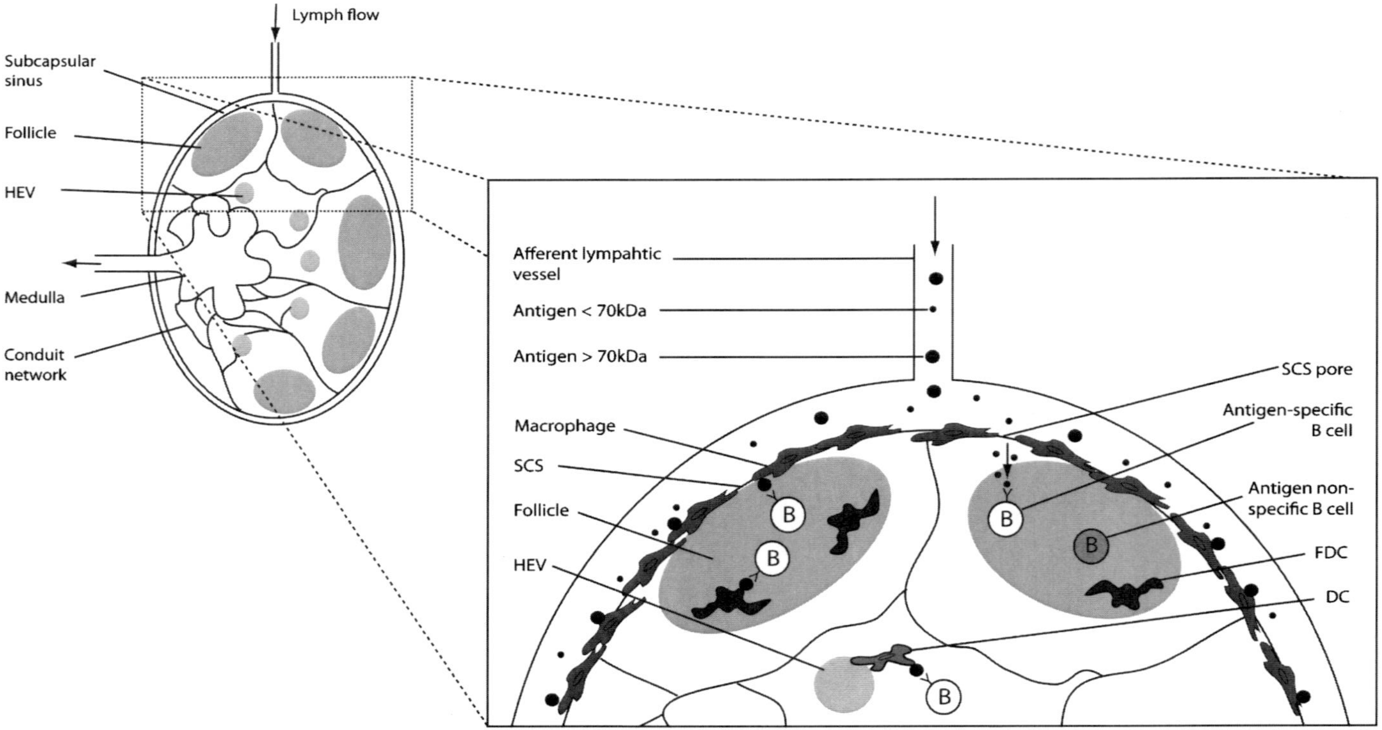

Fig. 3 Sites of antigen presentation to B cells in the lymph node *in vivo*. *Left*: A schematic representation of the lymph node is shown with the key anatomical features highlighted. *Right*: A magnified view of the lymph node is shown, indicating an area amenable to investigation by MPM. B cells within the follicle can encounter small soluble antigen, such as toxins, by direct diffusion through small pores in the subcapsular sinus (SCS). In contrast, larger antigens are presented to B cells on the surface of presenting cells, such as macrophages within the SCS, FDCs within the follicles and DCs in the proximity of the HEVs in the paracortex.

4.2 Lymphocyte Dynamics Within the "Resting" Lymph Node

The earliest MPM investigations, carried out in explanted or isolated lymph nodes, demonstrated the highly motile nature of lymphocytes within the resting lymph node (Bousso and Robey 2003; Miller et al. 2002; Stoll et al. 2002). As such highly polarized T and B cells, within their defined regions, were found to move along apparently random trajectories with average velocities of around 12 and 6 $\mu m\ min^{-1}$, respectively. However, the specific visualization of networks of FRCs and FDCs in chimeric GFP mice demonstrated a degree of order within the migration of labeled lymphocytes across these networks (Bajénoff et al. 2006). The rapid movement of lymphocytes across "networks of order" within the lymph node ensures that B and T cells are continually scanning their environment and ready to respond to antigen as soon as it is encountered. Interestingly, similar investigations have also offered insight into the behavior of CD11c$^+$ DCs within the paracortex (Lindquist et al. 2004). In contrast to lymphocytes, these DC were found to be relatively sessile moving at speeds of less than 2 $\mu m\ min^{-1}$, presumably as a result of their role in structural support of the conduit network. However, this restriction in motion did not prevent the rapid movement of dendritic processes, thus indicating that these DCs retain the capacity to sample and respond to environmental signals.

4.3 Antigen Encounter with B Cells

Antigen is supplied to the lymph node through the afferent lymph vessel extremely rapidly; indeed within minutes of subcutaneous administration antigen arrives at the SCS (Nossal et al. 1968). The encounter of B cells with antigen initiates a series of molecular events, detailed above, resulting in B cell activation. As B cells are capable of recognizing antigen in a number of different forms, it would be expected that these encounters might occur via a number of different mechanisms. Given the importance of this initial step in the generation of humoral immune responses, it has received considerable attention through MPM investigations.

4.3.1 Soluble Antigen Encounter

One mechanism by which antigen can enter the cortex of the lymph node involves direct diffusion through pores (0.1–1 μm in diameter), initially identified within the SCS by electron microscopy (Clark 1962; Farr et al. 1980; van Ewijk et al. 1988). The presence of these pores in the SCS allows small, soluble antigen, such as low molecular weight toxins, that enter the node to directly pass into the cortex and gain access to B cell follicles (Fig. 3) (Pape et al. 2007). Indeed, immunohistochemistry of fixed lymph node sections showed that this rapid diffusion of antigen was both independent of DCs and did not require movement of B cells from their follicular location. These observations are complementary to those of an earlier investigation into the dynamics of B cell migration in response to immunization with the same soluble antigen (Okada et al. 2005). In this case, stimulation with antigen resulted in a substantial decrease in the motility of antigen-specific B cells. Furthermore within 24 h, these B cells were able to present antigen in association with MHC-II on their surface, and had undergone directed migration towards the boundary between the cortex and the paracortex. Interestingly, it has been shown that direct diffusion from the SCS may also allow DCs resident in the cortex to accumulate antigen, representing one means by which DCs might present antigen to T cells in the lymph node (Itano et al. 2003).

4.3.2 Macrophage-Mediated Antigen Encounter

Whilst diffusion from the SCS provides a mechanism for exposing follicular B cells to small soluble antigens, the dimensions of these pores limits the access of larger antigens to B cells within the cortex. It was therefore postulated that specialized cells, possibly located within the SCS, might transport such antigens into the B cell-containing follicles (Farr et al. 1980; Fossum 1980; Szakal et al. 1983, 1989). Furthermore, the macrophages in the SCS were observed to form a distinct population from those within the medulla, displaying limited phagocytic activity and thereby enabling surface presentation of intact antigen (Fossum 1980; Szakal et al. 1983). Three independent MPM investigations have recently established a critical and novel role for these macrophages in the presentation of antigen to follicular B cells at the SCS (Fig. 3) (Carrasco and Batista 2007; Junt et al. 2007; Phan et al. 2007). These studies collectively identified that, following antigen administration, macrophages within the SCS rapidly accumulate a variety of larger antigens such as that found within immune complexes (containing antigen in association with antibody and/or complement fragments), bacteria, and viruses. In support of this, treatment with clodronate liposomes to deplete macrophages abrogates capture and retention of vesicular stomatitis virus at the SCS and results in systemic dissemination of the virus (Junt et al. 2007). Antigen accumulated at the SCS was subsequently presented intact by the macrophages, allowing recognition by follicular B cells in close proximity. Accordingly it was shown that antigen-specific B cells exhibited reduced migration after antigen administration and made stable contacts with the macrophage-rich SCS (Carrasco and Batista 2007; Junt et al. 2007; Phan et al. 2007). These contacts at the SCS allowed specific B cells to acquire and internalize antigen through their BCR, prior to their migration to the B–T cell boundary (Carrasco and Batista 2007). Such repositioning renders B cells in a region optimal for receiving T cell help, in line with previous observations (Garside et al. 1998; Okada et al. 2005). Together, these studies have provided the first clear demonstration of a role for macrophages in the initiation of follicular B cell responses.

4.3.3 DC-Mediated Antigen Encounter

As described above, lymphocytes enter the lymph node through HEVs in the paracortex. Furthermore, this region not only contains resident DCs in close association with FRCs lining the conduit system, but also recent immigrant DCs that have accumulated antigen in peripheral tissues. Thus, this region would seem an ideal site for immediate presentation of antigen to lymphocytes. Indeed, DCs are known to be extremely efficient in the MHC-mediated presentation of antigen to T cells (Delamarre et al. 2005; Dudziak et al. 2007; Itano et al. 2003). In contrast, antigen presentation to B cells would necessitate the stable accumulation of antigen on the cell surface and/or the recycling of antigen to the cell surface from degradation-resistant endosomes. Interestingly, a population of DCs in the paracortex has been identified that have the capacity of mediate presentation of intact antigen to B cells (Colino et al. 2002; Qi et al. 2006; Wykes et al. 1998). A detailed intravital MPM characterization demonstrated that this population of DCs are concentrated around HEVs such that recent B cell immigrants from the periphery can survey their antigenic contents (Fig. 3) (Qi et al. 2006). Following administration of antigen-loaded DCs, specific B cells exhibit a reduction in their speed of migration and a three-fold increase their residency time on DCs. The DC-mediated activation of B cells in the paracortex leaves B cells in an unfamiliar environment but perfectly placed to receive the necessary T cell help for their maximal activation (Qi et al. 2006). In support of this suggestion, "trios" of DCs, B, and T cells have been visualized within this region of the lymph node (Lindquist et al. 2004). This DC-mediated activation of B cells may yield extrafollicular plasma cells capable of early-antibody responses to antigen. Alternatively, fully activated B cells may migrate to the follicle and mediate germinal center formation.

4.3.4 FDC-Mediated Antigen Encounter

As FDCs retain antigen in the form of immune complexes, they have been classically considered as important sites of antigen presentation to B cells within SLOs (Chen et al. 1978; Klaus and Humphrey 1977; Mandel et al. 1980; Tew et al. 1979, 1980). In view of the ability of FDC to present native antigen it would seem likely that FDCs may play an important role during the initiation of early B cell responses within primary follicles. However, the relative importance and underlying mechanism of this FDC-mediated presentation of antigen to naive B cells has not yet been firmly established by high-resolution *in vivo* dynamic studies (Fig. 3). In contrast, three elegant MPM studies have recently investigated the spatiotemporal dynamics of B cells within germinal centers 6 days after administration of antigen (Allen et al. 2007; Hauser et al. 2007; Schwickert et al. 2007). Interestingly, these studies suggest that by this stage B cells may employ a different mechanism for recognition of membrane-associated antigen than during the initial phase of B cell activation.

4.4 The Role of B Cells in Antigen Transport and Presentation

MPM investigations visualizing the spatiotemporal dynamics of B-lymphocytes have also offered new insights into additional roles for these cells early in the initiation of B cell activation, in the transport and presentation of antigen.

4.4.1 B cells as Antigen Transporters

Originally a role in the transport of antigen to FDCs was proposed for marginal zone B cells (Brown et al. 1970; Gray et al. 1984). Marginal zone B cells are a population of nonrecirculating cells in the spleen, distinct from follicular B cells, and comprised of a large proportion of polyreactive clones. The location of B cells in the marginal zone makes them ideally positioned to sample and mount rapid responses to blood-borne antigen. Recently, the underlying mechanism of antigen transport by marginal zone B cells was visualized through an elegant study involving the treatment of chimeric mice with the sphingosine-1-phosphate receptor ($S1P_1$) antagonist FTY720 (Cinamon et al. 2007). In this study, it was observed that marginal zone B cells rapidly shuttled to and from follicles dependent on the expression of CXCR5 and $S1P_1$, respectively. This shuttling was observed even in the absence of antigen and represents a mechanism whereby immune complexes containing antigen might be effectively delivered from the blood to FDCs. However, at this stage it is not known if marginal zone B cells are capable of passing antigen directly to follicular B cells. In a manner similar to marginal zone B cells, follicular B cells themselves play a role in "antigen transport" to FDCs in the lymph node. As the SCS and the conduit system restrict the diffusion of antigens in the form of immune complexes from the lymphatic fluid, a cell-mediated mechanism is necessary to transport this type of antigen to the surface of FDCs. The transport of antigen by follicular B cells is independent of antigen-specific BCR recognition. In line with this, it was found that in the absence of specific B cells, antigen could be acquired on the surface of other follicular B cells (Carrasco and Batista 2007; Phan et al. 2007). Furthermore, this acquisition of antigen was shown to be dependent on the expression of complement receptors on the B cell surface (Phan et al. 2007). As FDCs express higher levels of complement receptors on their surface, it would be expected that they could compete effectively with B cells for antigen binding (Carroll 1998). Potentially, other receptors of the innate immune system, such as CD23, might also play a role in the transport of antigen by B cells in SLOs (Hjelm et al. 2008).

4.4.2 B Cells as Antigen Presenters

As B cells are professional APCs, capable of mediating MHC-II presentation of antigen to $CD4^+$ T helper cells, they occupy an exceptional position in the coordination of adaptive immune responses. Initial studies indicated that, compared with DCs,

B cells are relatively poor APCs, though the antigen-presenting potential of B cells improves following specific recognition of antigen (Sallusto and Lanzavecchia 1994). As a result, the importance of B cells in antigen presentation has remained controversial. Initial immunological tracking of antigen-specific lymphocytes in lymph nodes demonstrated that after antigen encounter, B cells migrate towards the B–T cell boundary at the edge of the follicles (Garside et al. 1998). The dynamics of this B cell antigen presentation have since been characterized by intravital MPM in response to stimulation with soluble antigen (Okada et al. 2005). Following accumulation of antigen, B cells were shown to undergo CCR7-mediated migration towards the B–T cell boundary. Upon reaching this boundary, B cells engaged antigen-specific T cells for up to an hour. Surprisingly, at this stage it was observed that B cell motility actually increased, leading to the transit of B–T cell complexes through the node. A similar B cell migration to the B–T cell boundary was observed in response to stimulation with particulate antigen or immune complexes following their accumulation at the SCS (Carrasco and Batista 2007; Phan et al. 2007). Thus, it appears that this mechanism of B cell antigen presentation to T cells contributes *in vivo* to the initiation of T cell responses. However, it should be borne in mind that the number of antigen-specific B cells in the primary repertoire *in vivo* is likely to be very low.

4.5 *Future Developments to Improve In Vivo Studies*

MPM affords the opportunity to visualize the dynamics of intercellular interactions underlying immune responses for the first time in living tissue. Such very recent studies have demonstrated the variety in mechanisms of antigen presentation to B cells, and have identified insights into additional roles of B cells in the transport and presentation of antigen. Further development of MPM technology to enable imaging for more extended periods of time, and its application to other secondary lymphoid organs, will be invaluable in the derivation of a more detailed description of the events underlying B cell activation *in vivo*.

5 Conclusions

Advances in high-resolution microscopy in combination with molecular biology have enabled the molecular and cellular events that underlie B cell activation to be visualized. Thus, over the past decade, much information has been gathered as to the dynamics of processes involved in the initiation of B cell activation both *in vitro* and *in vivo*. As such antigen-induced BCR-microclusters have been identified as the sites responsible for the initiation of B cell signaling and subsequent assembly of numerous microsignalosomes. Furthermore, B cells utilize a spreading response, dependent on the coreceptor CD19, to propagate signaling through these microsignalosomes that leads to enhanced B-cell activation in response to membrane-antigen. B cells can respond to antigen in a variety of different forms, and thus a number of different mechanisms of encountering antigen *in vivo* have been uncovered by MPM. The simplest of these is only possible for small soluble antigens and involves direct diffusion from pores within the SCS. Alternatively, B cells recognize intact antigen presented to them on the surface of macrophages in the SCS and on immigrant DCs in the paracortex. In addition, at later stages of the immune response, antigen presentation by FDCs in the germinal center plays an important role in affinity maturation. Though the progress described in this review has involved the collective effort of numerous investigators, challenges remain to fully understand the mechanisms underlying the initiation of B cell activation *in vivo*.

Acknowledgments We would like to thank members of the Lymphocyte Interaction Laboratory for helpful discussions and critical reading of the manuscript.

References

Ahearn J, Fischer M, Croix D, Goerg S, Ma M, Xia J, Zhou X, Howard R, Rothstein T, Carroll M (1996) Disruption of the Cr2 locus results in a reduction in B-1a cells and in an impaired B-cell response to T-dependent antigen. Immunity 4:251–262

Allen C, Okada T, Tang H, Cyster J (2007) Imaging of germinal center selection events during affinity maturation. Science 315:528–531

Aluvihare VR, Khamlichi AA, Williams GT, Adorini L, Neuberger MS (1997) Acceleration of intracellular targeting of antigen by the B-cell antigen receptor: importance depends on the nature of the antigen-antibody interaction. EMBO J 16:3553–3562

Amigorena S, Drake JR, Webster P, Mellman I (1994) Transient accumulation of new class II MHC molecules in a novel endocytic compartment in B lymphocytes. Nature 369:113–120

Arana E, Vehlow A, Harwood N, Vigorito E, Henderson R, Turner M, Tybulewicz V, Batista F (2008) Activation of the small GTPase Rac2 via the B-cell receptor regulates B-cell adhesion and immunological synpase formation. Immunity 28:88–99

Bajénoff M, Germain R (2007) Seeing is believing: a focus on the contribution of microscopic imaging to our understanding of immune system function. Eur J Immunol 37(Suppl 1):S18–S33

Bajénoff M, Egen J, Koo L, Laugier J, Brau F, Glaichenhaus N, Germain R (2006) Stromal cell networks regulate lymphocyte entry, migration, and territoriality in lymph nodes. Immunity 25:989–1001

Batista F, Neuberger M (2000) B-cells extract and present immobilized antigen: implications for affinity discrimination. EMBO J 19:513–20

Batista F, Iber D, Neuberger M (2001) B-cells acquire antigen from targeT-cells after synapse formation. Nature 411:489–494

Batista FD, Arana E, Barral P, Carrasco YR, Depoil D, Eckl-Dorna J, Fleire S, Howe K, Vehlow A, Weber M, Treanor B (2007). The role of intergrins and coreceptors in refining thresholds for B-cell responses. Immunol Rev 218:197–213

Bousso P, Robey E (2003) Dynamics of CD8+ T-cell priming by dendritic cells in intact lymph nodes. Nat Immunol 4:579–585

Brink R, Phan TG, Paus D, Chan TD (2008) Visualizing the effects of antigen affinity on T-dependent B-cell differentiation. Immunol Cell Biol 86:31–39

Brown J, De Jesus D, Holborow E, Harris G (1970) Lymphocyte-mediated transport of aggregated human gamma-globulin into germinal center areas of normal mouse spleen. Nature 228:367–369

Bunnell S, Kapoor V, Trible R, Zhang W, Samelson L (2001) Dynamic actin polymerization drives T-cell receptor-induced spreading: a role for the signal transduction adaptor LAT. Immunity 14:315–329

Bunnell S, Hong D, Kardon J, Yamazaki T, McGlade C, Barr V, Samelson L (2002) T-cell receptor ligation induces the formation of dynamically regulated signaling assemblies. J Cell Biol 158:1263–1275

Bunnell S, Singer A, Hong D, Jacque B, Jordan M, Seminario M, Barr V, Koretzky G, Samelson L (2006) Persistence of cooperatively stabilized signaling clusters drives T-cell activation. Mol Cell Biol 26:7155–7166

Cahalan M, Parker I (2008) Choreography of Cell Motility and Interaction Dynamics Imaged by Two-Photon Microscopy in Lymphoid Organs. Annu Rev Immunol 26:585–626

Campi G, Varma R, Dustin M (2005) Actin and agonist MHC-peptide complex-dependent T-cell receptor microclusters as scaffolds for signaling. J Exp Med 202:1031–1036

Carrasco YR, Batista FD (2006a) B-cell recognition of membrane-bound antigen: an exquisite way of sensing ligands. Curr Opin Immunol 18:286–291

Carrasco YR, Batista FD (2006b) B-cell activation by membrane-bound antigens is facilitated by the interaction of VLA-4 with VCAM-1. EMBO J 25:889–899

Carrasco Y, Batista F (2007) B-cells acquire particulate antigen in a macrophage-rich area at the boundary between the follicle and the subcapsular sinus of the lymph node. Immunity 27:160–171

Carrasco Y, Fleire S, Cameron T, Dustin M, Batista F (2004) LFA-1/ICAM-1 interaction lowers the threshold of B-cell activation by facilitating B-cell adhesion and synapse formation. Immunity 20:589–599

Carroll M (1998) The role of complement and complement receptors in induction and regulation of immunity. Annu Rev Immunol 16:545–568

Catron D, Itano A, Pape K, Mueller D, Jenkins M (2004) Visualizing the first 50 hr of the primary immune response to a soluble antigen. Immunity 21:341–347

Cemerski S, Das J, Giurisato E, Markiewicz MA, Allen PM, Chakraborty AK, Shaw AS (2008) The balance between T-cell receptor signaling and degradation at the center of the immunological synapse is determined by antigen quality. Immunity 29:414–422

Chen L, Frank A, Adams J, Steinman R (1978) Distribution of horseradish peroxidase (HRP)-anti-HRP immune complexes in mouse spleen with special reference to follicular dendritic cells. J Cell Biol 79:184–199

Choudhuri K, Wiseman D, Brown M, Gould K, van der Merwe P (2005) T-cell receptor triggering is critically dependent on the dimensions of its peptide-MHC ligand. Nature 436:578–582

Cinamon G, Zachariah M, Lam O, Foss F, Cyster J (2007) Follicular shuttling of marginal zone B-cells facilitates antigen transport. Nat Immunol 9:54–62

Clark S (1962) The reticulum of lymph nodes in mice studied with the electron microscope. Am J Anat 110:217–257

Colino J, Shen Y, Snapper C (2002) Dendritic cells pulsed with intact Streptococcus pneumoniae elicit both protein- and polysaccharide-specific immunoglobulin isotype responses in vivo through distinct mechanisms. J Exp Med 195:1–13

Cyster J, Ansel K, Reif K, Ekland E, Hyman P, Tang H, Luther S, Ngo V (2000) Follicular stromal cells and lymphocyte homing to follicles. Immunol Rev 176:181–193

Dal Porto J, Gauld S, Merrell K, Mills D, Pugh-Bernard A, Cambier J (2004) B-cell antigen receptor signaling 101. Mol Immunol 41:599–613

Davis S, van der Merwe P (1996) The structure and ligand interactions of CD2: implications for T-cell function. Immunol Today 17:177–187

Davis S, van der Merwe P (2006) The kinetic-segregation model: TCR triggering and beyond. Nat Immunol 7:803–809

Davis D, Chiu I, Fassett M, Cohen G, Mandelboim O, Strominger J (1999) The human natural killer cell immune synapse. Proc Natl Acad Sci USA 96:15062–15067

DeFranco A (1997) The complexity of signaling pathways activated by the BCR. Curr Opin Immunol 9:296–308

Delamarre L, Pack M, Chang H, Mellman I, Trombetta E (2005) Differential lysosomal proteolysis in antigen-presenting cells determines antigen fate. Science 307:1630–1634

Depoil D, Fleire S, Treanor BL, Weber M, Harwood NE, Marchbank KL, Tybulewicz VL, Batista FD (2008) CD19 is essential for B-cell activation by promoting B-cell receptor-antigen microcluster formation in response to membrane-bound ligand. Nat Immunol 9:63–72

Dudziak D, Kamphorst A, Heidkamp G, Buchholz V, Trumpfheller C, Yamazaki S, Cheong C, Liu K, Lee H, Park C, Steinman R, Nussenzweig M (2007) Differential antigen processing by dendritic cell subsets in vivo. Science 315:107–111

Engel P, Zhou L, Ord D, Sato S, Koller B, Tedder T (1995) Abnormal B lymphocyte development, activation, and differentiation in mice that lack or overexpress the CD19 signal transduction molecule. Immunity 3:39–50

Farr A, Cho Y, De Bruyn P (1980) The structure of the sinus wall of the lymph node relative to its endocytic properties and transmural cell passage. Am J Anat 157:265–284

Fearon D, Carroll M (2000) Regulation of B lymphocyte responses to foreign and self-antigens by the CD19/CD21 complex. Annu Rev Immunol 18:393–422

Fearon D, Carter R (1995) The CD19/CR2/TAPA-1 complex of B lymphocytes: linking natural to acquired immunity. Annu Rev Immunol 13:127–149

Fleire S, Goldman J, Carrasco Y, Weber M, Bray D, Batista F (2006) B-cell ligand discrimination through a spreading and contraction response. Science 312:738–741

Fossum S (1980) The architecture of rat lymph nodes. IV. Distribution of ferritin and colloidal carbon in the draining lymph nodes after foot-pad injection. Scand J Immunol 12:433–441

Fu C, Turck C, Kurosaki T, Chan A (1998) BLNK: a central linker protein in B-cell activation. Immunity 9:93–103

Fujimoto M, Bradney A, Poe J, Steeber D, Tedder T (1999) Modulation of B lymphocyte antigen receptor signal transduction by a CD19/CD22 regulatory loop. Immunity 11:191–200

Garside P, Ingulli E, Merica R, Johnson J, Noelle R, Jenkins M (1998) Visualization of specific B and T lymphocyte interactions in the lymph node. Science 281:96–99

Germain R, Miller M, Dustin M, Nussenzweig M (2006) Dynamic imaging of the immune system: progress, pitfalls and promise. Nat Rev Immunol 6:497–507

Goitsuka R, Fujimura Y, Mamada H, Umeda A, Morimura T, Uetsuka K, Doi K, Tsuji S, Kitamura D (1998) BASH, a novel signaling molecule preferentially expressed in B-cells of the bursa of Fabricius. J Immunol 161:5804–5808

Goodnow CC (1992) Transgenic mice and analysis of B-cell tolerance. Annu Rev Immunol 10:489–518

Goodnow C, Crosbie J, Adelstein S, Lavoie T, Smith-Gill S, Brink R, Pritchard-Briscoe H, Wotherspoon J, Loblay R, Raphael K (1988) Altered immunoglobulin expression and functional silencing of self-reactive B lymphocytes in transgenic mice. Nature 334:676–682

Grakoui A, Bromley S, Sumen C, Davis M, Shaw A, Allen P, Dustin M (1999) The immunological synapse: a molecular machine controlling T-cell activation. Science 285:221–227

Gray D, Kumararatne D, Lortan J, Khan M, MacLennan I (1984) Relation of intra-splenic migration of marginal zone B-cells to antigen localization on follicular dendritic cells. Immunology 52:659–669

Gretz J, Anderson A, Shaw S (1997) Cords, channels, corridors and conduits: critical architectural elements facilitating cell interactions in the lymph node cortex. Immunol Rev 156:11–24

Gretz J, Norbury C, Anderson A, Proudfoot A, Shaw S (2000) Lymph-borne chemokines and other low molecular weight molecules reach high endothelial venules via specialized conduits while a functional barrier limits access to the lymphocyte microenvironments in lymph node cortex. J Exp Med 192:1425–1440

Harwood NE, Batista FD (2008) New insights into the early molecular events underlying B-cell activation. Immunity 28:609–619

Hauser A, Junt T, Mempel T, Sneddon M, Kleinstein S, Henrickson S, von Andrian U, Shlomchik M, Haberman A (2007) Definition of germinal-center B-cell migration in vivo reveals predominant intrazonal circulation patterns. Immunity 26:655–667

Hjelm F, Karlsson M, Heyman B (2008) A novel B-cell-mediated transport of IgE-immune complexes to the follicle of the spleen. J Immunol 180:6604–6610

Irvine DJ, Purbhoo MA, Krogsgaard M, Davis MM (2002). Driect observation of ligand recognition by T cells. Nature 419:845–849

Itano A, McSorley S, Reinhardt R, Ehst B, Ingulli E, Rudensky A, Jenkins M (2003) Distinct dendritic cell populations sequentially present antigen to CD4 T-cells and stimulate different aspects of cell-mediated immunity. Immunity 19:47–57

Junt T, Moseman E, Iannacone M, Massberg S, Lang P, Boes M, Fink K, Henrickson S, Shayakhmetov D, Di Paolo N, van Rooijen N, Mempel T, Whelan S, von Andrian U (2007) Subcapsular sinus macrophages in lymph nodes clear lymph-borne viruses and present them to antiviral B-cells. Nature 450:110–114

Klaus G, Humphrey J (1977) The generation of memory cells. I. The role of C3 in the generation of B memory cells. Immunology 33:31–40

Klaus G, Humphrey J, Kunkl A, Dongworth D (1980) The follicular dendritic cell: its role in antigen presentation in the generation of immunological memory. Immunol Rev 53:3–28

Kurosaki T (1999) Genetic analysis of B-cell antigen receptor signaling. Annu Rev Immunol 17:555–592

Kurosaki T, Kurosaki M (1997) Transphosphorylation of Bruton's tyrosine kinase on tyrosine 551 is critical for B-cell antigen receptor function. J Biol Chem 272:15595–15598

Lanzavecchia A (1985) Antigen-specific interaction between T and B-cells. Nature 314:537–539

Lee K, Holdorf A, Dustin M, Chan A, Allen P, Shaw A (2002) T-cell receptor signaling precedes immunological synapse formation. Science 295:1539–1542

Li X, Sandoval D, Freeberg L, Carter R (1997) Role of CD19 tyrosine 391 in synergistic activation of B lymphocytes by coligation of CD19 and membrane Ig. J Immunol 158:5649–5657

Lichtman J, Conchello J (2005) Fluorescence microscopy. Nat Methods 2:910–919

Lillemeier BF, Pfeiffer JR, Surviladze Z, Wilson BS, Davis MM (2006) Plasma membrane-associated proteins are clustered into islands attached to the cytoskeleton. Proc Natl Acad Sci U S A 103:18992–18997

Lin K, Freeman S, Zabetian S, Brugger H, Weber M, Lei V, Dang-Lawson M, Tse K, Santamaria R, Batista F, Gold M (2008) The Rap GTPases regulate B-cell morphology, immune synpase formation and signaling by particulate B-cell receptor ligands. Immunity 28:75–87

Lindquist R, Shakhar G, Dudziak D, Wardemann H, Eisenreich T, Dustin M, Nussenzweig M (2004) Visualizing dendritic cell networks in vivo. Nat Immunol 5:1243–1250

MacLennan I (1994) Germinal centers. Annu Rev Immunol 12:117–139

MacLennan I, Toellner K, Cunningham A, Serre K, Sze D, Zuniga E, Cook M, Vinuesa C (2003) Extrafollicular antibody responses. Immunol Rev 194:8–18

Mandel T, Phipps R, Abbot A, Tew J (1980) The follicular dendritic cell: long term antigen retention during immunity. Immunol Rev 53:29–59

Miller M, Wei S, Parker I, Cahalan M (2002) Two-photon imaging of lymphocyte motility and antigen response in intact lymph node. Science 296:1869–1873

Moller G (1987) Role of somatic mutation in the generation of lymphocyte diversity. Immunol Rev 96:1–162

Monks C, Freiberg B, Kupfer H, Sciaky N, Kupfer A (1998) Three-dimensional segregation of supramolecular activation clusters in T-cells. Nature 395:82–86

Negulescu PA, Krasieva TB, Khan A, Kerschbaum HH, Cahalan MD (1996) Polarity of T-cell shape, motility, and sensitivity to antigen. Immunity 4:421–430

Nossal G, Abbot A, Mitchell J, Lummus Z (1968) Antigens in immunity. XV. Ultrastructural features of antigen capture in primary and secondary lymphoid follicles. J Exp Med 127:277–290

Okada T, Miller M, Parker I, Krummel M, Neighbors M, Hartley S, O'Garra A, Cahalan M, Cyster J (2005) Antigen-engaged B-cells undergo chemotaxis toward the T zone and form motile conjugates with helper T-cells. PLoS Biol 3:e150

Pape K, Catron D, Itano A, Jenkins M (2007) The humoral immune response is initiated in lymph nodes by B-cells that acquire soluble antigen directly in the follicles. Immunity 26: 491–502

Parsey M, Lewis G (1993) Actin polymerization and pseudopod reorganization accompany anti-CD3-induced growth arrest in Jurkat T-cells. J Immunol 151:1881–1893

Pelanda R, Schwers S, Sonoda E, Torres RM, Nemazee D, Rajewsky K (1997) Receptor editing in a transgenic mouse model: site, efficiency, and role in B-cell tolerance and antibody diversification. Immunity 7:765–775

Pesando JM, Bouchard LS, McMaster BE (1989) CD19 is functionally and physically associated with surface immunoglobulin. J Exp Med 170:2159–2164

Phan TG, Amesbury M, Gardam S, Crosbie J, Hasbold J, Hodgkin PD, Basten A, Brink R (2003) B-cell receptor-independent stimuli trigger immunoglobulin (Ig) class switch recombination and production of IgG autoantibodies by anergic self-reactive B-cells. J Exp Med 197:845–860

Phan T, Grigorova I, Okada T, Cyster J (2007) Subcapsular encounter and complement-dependent transport of immune complexes by lymph node B-cells. Nat Immunol 8:992–1000

Phee H, Rodgers W, Coggeshall KM (2001) Visualization of negative signaling in B-cells by quantitative confocal microscopy. Mol Cell Biol 21:8615–8625

Qi H, Egen JG, Huang AY, Germain RN (2006) Extrafollicular activation of lymph node B-cells by antigen-bearing dendritic cells. Science 312:1672–1676

Qin D, Wu J, Vora K, Ravetch J, Szakal A, Manser T, Tew J (2000) Fc gamma receptor IIB on follicular dendritic cells regulates the B-cell recall response. J Immunol 164:6268–6275

Reth M (1989) Antigen receptor tail clue. Nature 338:383–384

Reth M, Wienands J (1997) Initiation and processing of signals from the B-cell antigen receptor. Annu Rev Immunol 15:453–479

Rickert R, Rajewsky K, Roes J (1995) Impairment of T-cell-dependent B-cell responses and B-1 cell development in CD19-deficient mice. Nature 376:352–355

Rock K, Benacerraf B, Abbas A (1984) Antigen presentation by hapten-specific B lymphocytes. I. Role of surface immunoglobulin receptors. J Exp Med 160:1102–1113

Russell DM, Dembi Z, Morahan G, Miller JF, Bürki K, Nemazee D (1991) Peripheral deletion of self-reactive B-cells. Nature 354:308–311

Sallusto F, Lanzavecchia A (1994) Efficient presentation of soluble antigen by cultured human dendritic cells is maintained by granulocyte/macrophage colony-stimulating factor plus interleukin 4 and downregulated by tumor necrosis factor alpha. J Exp Med 179:1109–1118

Sato S, Miller A, Howard M, Tedder T (1997) Regulation of B lymphocyte development and activation by the CD19/CD21/CD81/Leu 13 complex requires the cytoplasmic domain of CD19. J Immunol 159:3278–3287

Schamel W, Reth M (2000) Monomeric and oligomeric complexes of the B-cell antigen receptor. Immunity 13:5–14

Scharenberg A, Humphries L, Rawlings D (2007) Calcium signalling and cell-fate choice in B-cells. Nat Rev Immunol 7:778–789

Schneckenburger H (2005) Total internal reflection fluorescence microscopy: technical innovations and novel applications. Curr Opin Biotechnol 16:13–18

Schreiner GF, Braun J, Unanue ER (1976) Spontaneous redistribution of surface immunoglobulin in the motile B lymphocyte. J Exp Med 144:1683–1688

Schwickert T, Lindquist R, Shakhar G, Livshits G, Skokos D, Kosco-Vilbois M, Dustin M, Nussenzweig M (2007) In vivo imaging of germinal centers reveals a dynamic open structure. Nature 446:83–87

Shinohara H, Kurosaki T (2006) Genetic analysis of B-cell signaling. Subcell Biochem 40:145–187

Sixt M, Kanazawa N, Selg M, Samson T, Roos G, Reinhardt D, Pabst R, Lutz M, Sorokin L (2005) The conduit system transports soluble antigens from the afferent lymph to resident dendritic cells in the T-cell area of the lymph node. Immunity 22:19–29

Stackpole CW, Jacobson JB, Lardis MP (1974) Two distinct types of capping of surface receptors on mouse lymphoid cells. Nature 248:232–234

Stinchcombe J, Bossi G, Booth S, Griffiths G (2001) The immunological synapse of CTL contains a secretory domain and membrane bridges. Immunity 15:751–761

Stoll S, Delon J, Brotz T, Germain R (2002) Dynamic imaging of T-cell-dendritic cell interactions in lymph nodes. Science 296:1873–1876

Szakal A, Holmes K, Tew J (1983) Transport of immune complexes from the subcapsular sinus to lymph node follicles on the surface of nonphagocytic cells, including cells with dendritic morphology. J Immunol 131:1714–1727

Szakal A, Kosco M, Tew J (1989) Microanatomy of lymphoid tissue during humoral immune responses: structure function relationships. Annu Rev Immunol 7:91–109

Tew J, Mandel T, Burgess A, Hicks J (1979) The antigen binding dendritic cell of the lymphoid follicles: evidence indicating its role in the maintenance and regulation of serum antibody levels. Adv Exp Med Biol 114:407–410

Tew J, Phipps R, Mandel T (1980) The maintenance and regulation of the humoral immune response: persisting antigen and the role of follicular antigen-binding dendritic cells as accessory cells. Immunol Rev 53:175–201

Tolar P, Sohn H, Pierce S (2005) The initiation of antigen-induced B-cell antigen receptor signaling viewed in living cells by fluorescence resonance energy transfer. Nat Immunol 6:1168–1176

Treanor B, Batista F (2007) Mechanistic insight into lymphocyte activation through quantitative imaging and theoretical modelling. Curr Opin Immunol 19:476–483

Tuveson D, Carter R, Soltoff S, Fearon D (1993) CD19 of B-cells as a surrogate kinase insert region to bind phosphatidylinositol 3-kinase. Science 260:986–989

Unanue ER, Perkins WD, Karnovsky MJ (1972) Ligand-induced movement of lymphocyte membrane macromolecules. I. Analysis by immunofluorescence and ultrastructural radioautography. J Exp Med 136:885–906

van Ewijk W, Brekelmans P, Jacobs R, Wisse E (1988) Lymphoid microenvironments in the thymus and lymph node. Scann Microsc 2:2129–2140

Weber M, Treanor B, Depoil D, Shinohara H, Harwood NE, Hikida M, Kurosaki T, Batista FD (2008) Phospholipase C-gamma2 and Vav cooperate within signaling microclusters to propagate B-cell spreading in response to membrane-bound antigen. J Exp Med 205:853–868

Wienands J, Schweikert J, Wollscheid B, Jumaa H, Nielsen P, Reth M (1998) SLP-65: a new signaling component in B lymphocytes which requires expression of the antigen receptor for phosphorylation. J Exp Med 188:791–795

Winding P, Berchtold MW (2001) The chicken B-cell line DT40: a novel tool for gene disruption experiments. J Immunol Methods 249:1–16

Wykes M, Pombo A, Jenkins C, MacPherson G (1998) Dendritic cells interact directly with naive B lymphocytes to transfer antigen and initiate class switching in a primary T-dependent response. J Immunol 161:1313–1319

Yokosuka T, Sakata-Sogawa K, Kobayashi W, Hiroshima M, Hashimoto-Tane A, Tokunaga M, Dustin M, Saito T (2005) Newly generated T-cell receptor microclusters initiate and sustain T-cell activation by recruitment of Zap70 and SLP-76. Nat Immunol 6:1253–1262

Tracking the Dynamics of *Salmonella* Specific T Cell Responses

James J. Moon and Stephen J. McSorley

Contents

Abstract Over the last decade, significant advances have been made in the methodology for studying immune responses *in vivo*. It is now possible to follow almost every aspect of pathogen-specific immunity using *in vivo* models that incorporate physiological infectious doses and natural routes of infection.

J.J. Moon
Department of Microbiology, University of Minnesota Medical School, Minneapolis, MN, 55455, USA
Center for Immunology, University of Minnesota Medical School, Minneapolis, MN, 55455, USA

S.J. McSorley (✉)
Department of Medicine, University of Minnesota Medical School, Minneapolis, MN, 55455, USA
Center for Immunology, University of Minnesota Medical School, Minneapolis, MN, 55455, USA
Center for Infectious Diseases and Microbiology Translational Research, University of Minnesota Medical School, Minneapolis, MN, 55455, USA
e-mail: mcsor002@umn.edu

D. McGavern and M. Dustin (eds.), *Visualizing Immunity,* Current Topics in Microbiology and Immunology 334,
DOI: 10.1007/978-3-540-93864-4_8,

This new ability to study immunity in a relevant physiological context will greatly expand our understanding of the dynamic interplay between host and pathogen. Visualizing the resolution of primary infection and the development of long-term immunological memory should also aid the development of new vaccines and therapeutics for infectious diseases. In this review, we will describe the application of *in vivo* visualization technology to *Salmonella* infection, describe our current understanding of *Salmonella*-specific immunity, and discuss some unanswered questions that remain in this model.

1 Introduction

During the twentieth century, increased understanding of immunity to infectious disease led to the development of successful vaccines and therapeutics for viral and bacterial infections which plagued previous generations (Amanna and Slifka 2005; Brines 1996). Despite this success, there still remains a significant need for vaccine and therapeutic development. The emergence of new pathogens (Woolhouse et al. 2005), the acquired resistance of pathogens to currently effective therapeutics (Martinez-Cajas and Wainberg 2007; Okeke et al. 2007), a lack of research and development in the diseases of the developing world (Walker 2005), and the deliberate transmission of infectious agents for ideological purposes (Moran et al. 2008), all generate considerable demand for new vaccines and therapeutics.

If these new vaccines and therapeutics are to be developed in a rational manner, the interplay between microbes and the host immune response must be understood in considerable detail. Fortunately, over the last 10 years or so, the development of new imaging methodologies now allow direct study of the immune response in a physiological context (Germain and Jenkins 2004; Henrickson and von Andrian 2007; Huang et al. 2004; Jenkins et al. 2001; Negrin and Contag 2006). So far, the application of these visualization technologies has been limited to a select number of infectious disease models. However, visualization of immune responses to infection is now becoming more widely used and therefore generates significant potential for discovery in many infectious disease models. In this review, we will focus on the development and use of visualization approaches for studying immunity to *Salmonella* infection. We will describe how imaging tools have confirmed and extended our knowledge of host immunity to *Salmonella*, discuss current limitations in our knowledge, and speculate on the potential for future advances in studying this disease.

2 Studying Immunity to *Salmonella*

Salmonella are gram-negative enterobacteria that cause significant human and veterinary disease in the US and elsewhere (Grassl and Finlay 2008; Parry et al. 2002; Rabsch et al. 2001). The immune response to *Salmonella* has been studied in infected humans and in animal models of disease.

2.1 Human Disease Caused by Salmonella

After several nomenclature revisions (Su and Chiu 2007), the *Salmonella* genus now contains three species, *S. enterica, S. bongori*, and *S. subterranean*, but almost all significant human disease is caused by a single subspecies, *S. enterica* subsp. *enterica*. This subspecies contains both the typhi and paratyphi Serovars that cause typhoid and paratyphoid fever in developing countries and over 2,000 different serovars that cause *Salmonella* gastroenteritis in humans and animals (Grassl and Finlay 2008; Parry et al. 2002; Rabsch et al. 2001). The global incidence of typhoid was recently estimated at 21,650,974 infections and 216,510 deaths per year (Crump et al. 2004; Jones and Falkow 1996). Numerous closely related *Salmonella* serovars cause nontyphoidal Salmonellosis, an increasingly important food-borne infection in the US (Mead et al. 1999). Both diseases have been recognized as significant biothreats to the US food and water supply (Jones and Falkow 1996; Sobel et al. 2002). During a typhoid epidemic in Tajikistan, more than 90% of clinical isolates were multidrug resistant and 82% of these were resistant to ciprofloxacin (Tarr et al. 1999), the antibiotic of choice for treating typhoid in developed nations. Recent analysis of the bacterial proteome indicates that all major targets of *Salmonella* metabolism have already been targeted by antibiotic development strategies (Becker et al. 2006); thus the potential for generating new antibiotics is not encouraging. Understanding the generation of an adaptive immune response to *Salmonella* is therefore of considerable medical and economic importance.

2.2 Animal Models of Salmonella Infection

Although *S. enterica* Serovar typhi causes typhoid fever in humans, it does not cause typhoid in other mammals (Fierer and Guiney 2001). Therefore, Serovar typhi is usually studied *in vitro* or using *in vivo* models that require unusual routes of administration (Bueno et al. 2008; Pasetti et al. 2002; Subramanian and Qadri 2006). In contrast, oral infection of mice with *S. enterica* Serovar typhimurium causes a systemic typhoid-like disease involving penetration of intestinal Peyer's patches and rapid dissemination to the liver, spleen, and bone marrow (Srinivasan and McSorley 2006). Therefore, murine infection with *S. enterica* Serovar typhimurium bears a striking similarity to human typhoid and is the most commonly studied laboratory model of this disease (Mastroeni and Sheppard 2004; Santos et al. 2001). However, despite the utility of this mouse *Salmonella* model, it does have some limitations as a model of human disease. First, not all bacterial virulence factors that are important for murine disease are required for the pathogenesis of human typhoid. Therefore, as with most animal models, data from the murine typhoid model should not be overinterpreted and needs to be considered alongside clinical studies of typhoid in humans to develop a complete picture of typhoid immunity (Pasetti et al. 2003). Second, *Salmonella* infection of mice does not cause diarrhea, yet diarrhea is a prominent feature of *Salmonella* enterocolitis in humans.

Therefore, the mouse model is much more useful for studying typhoid than as a model of *Salmonella* gastroenteritis. A better model of *Salmonella* gastroenteritis in humans is the bovine Salmonellosis model where intestinal disease and diarrhea develop rapidly in a similar manner to human disease (Santos et al. 2001).

3 Visualizing *Salmonella In Vivo*

Using the mouse model of typhoid, a number of studies have visualized the colonization and growth of *Salmonella in vivo*. Our laboratory performed fluorescence microscopy on midline sections of whole infected mice to simultaneously visualize *Salmonella* in all organs (McSorley and Jenkins, unpublished). These studies revealed that the spleen and liver are the major sites of *Salmonella* replication (Fig. 1). More detailed studies using confocal microscopy demonstrated that bacteria in the liver are initially associated with infiltrating neutrophils but subsequently are found exclusively within macrophages (Richter-Dahlfors et al. 1997). Similar studies, some using a *Salmonella* strain expressing green fluorescent protein (GFP), demonstrated that *Salmonella* in the spleen are also found within macrophages (Matsui et al. 2000; Salcedo et al. 2001). Thus, visualization studies in mice confirm that *Salmonella* is an intracellular pathogen that replicates extensively within macrophages found in the spleen and liver. A more recent study of *Salmonella* in the liver found that the bacteria replicate intracellularly in discrete foci, each of which is likely to arise from the clonal expansion of a single bacterium (Sheppard et al. 2003). Thus, individual foci of *Salmonella* appear to grow independently until reaching a certain threshold, after which the bacteria redistribute to new foci.

4 Adaptive Immunity in the Murine Typhoid Model

The mouse model of *Salmonella* infection has long been of interest to scientists trying to understand bacterial pathogenesis and immunity to infection (Blanden et al. 1966; Gowen and Calhoun 1943; Zinkernagel 1976). As *Salmonella* preferentially infects and resides within macrophages, the activation of these cells by IFN-γ produced by Th1 cells plays a prominent role in bacterial killing (Mastroeni 2002; Mittrucker and Kaufmann 2000; Wick 2003). Indeed, nude mice (Sinha et al. 1997), mice lacking TCR β (Hess et al. 1996; Weintraub et al. 1997), CD4 (Hess et al. 1996), CD28 (McSorley and Jenkins 2000; Mittrucker 1999), MHC class-II (Hess et al. 1996), IFN-γ (VanCott et al. 1998), IFN-γR (Hess et al. 1996), or the Th1 transcription factor T-bet (Ravindran et al. 2005) all fail to resolve primary infection with a live vaccine strain (LVS) of *Salmonella*. Furthermore, depletion of IFN-γ (Mastroeni et al. 1992; Nauciel and Espinasse-Maes 1992), IL-12 (Mastroeni et al. 1996), or TNF-α (Mastroeni et al. 1992; Nauciel and Espinasse-Maes 1992)

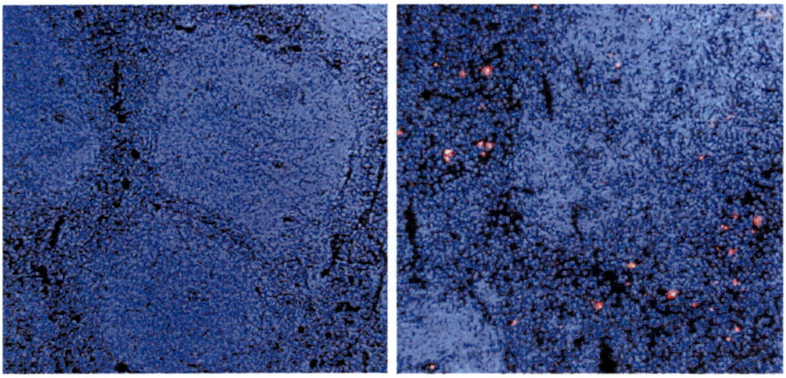

Fig. 1 *Salmonella* in the splenic red pulp of infected mice. C57BL/6 mice were infected orally with 5×10^9 attenuated *Salmonella*, SL3261 and whole body sections were taken 3 days later. Sections were blocked and stained with an antibody specific to *Salmonella* LPS and the signal amplified using biotinyl tyramide and Streptavidin-Cy5 (*red*). Tissue was counterstained with DAPI (*blue*) to stain nuclei. Images show staining of spleens from control or *Salmonella*-infected mice

abrogates or reduces the protective immunity conferred by LVS-immunization. The relevance of these findings in the murine model are strengthened by clinical reports of increased susceptibility to *Salmonella* in patients with primary immune deficiencies in IL-12 or IFN-γ receptor signaling (Cleary et al. 2003; Jouanguy et al. 1999). Thus, a considerable amount of data in mouse and human typhoid models indicates that CD4 Th1 cells play an important protective role.

In a somewhat surprising finding for an intracellular pathogen, our laboratory and others have demonstrated that *Salmonella*-specific antibody participates in immunity to typhoid (Mastroeni et al. 2000; McSorley and Jenkins 2000; Mittrucker et al. 2000). Although B cell-deficient mice survive vaccination with LVS-*Salmonella*, they do not acquire protective immunity to secondary typhoid, indicating that antibody is absolutely required for effective immunity (Mastroeni et al. 2000; McSorley and Jenkins 2000; Mittrucker et al. 2000). It is currently unclear how antibody is protective against an intracellular pathogen. One possibility is that serum antibody prevents cell–cell transmission of *Salmonella* following macrophage apoptosis (Ravindran and McSorley 2005). Alternatively, an antibody response could alter the processing of *Salmonella* antigens to enhance antigen presentation to CD4 T cells and subsequent cellular immunity (Bueno et al. 2007; Mastroeni et al. 2001; Ugrinovic et al. 2003). Another possibility is that *Salmonella*-specific mucosal IgA secreted into the intestine can inhibit initial bacterial penetration of epithelial cells or simply reduce the overall load of bacteria in the gut lumen (Wijburg et al. 2006). Whatever the mechanism, effective immunity against typhoid appears to require the combined activity of *Salmonella*-specific Th1 cells and antibody.

The role of CD8 T cells in adaptive immunity to *Salmonella* is less clear. While CD8 CTL responses to *Salmonella* have been characterized (Lo et al. 1999; Mastroeni et al. 1992; Nauciel 1990), mice lacking CD8 T cells display only a mild reduction in their ability to resolve infections with attenuated *Salmonella* (Hess et al. 1996; Lo et al. 1999).

5 Tracking Antigen-Specific Responses in *Salmonella* Infection

The importance of adaptive immunity in mediating host resistance to *Salmonella* has fueled extensive research into *Salmonella*-specific T cell responses (Ravindran and McSorley 2005; Bueno et al. 2007). As with other models of infectious disease, these studies have benefited greatly from the development of techniques to track antigen-specific T cells in mice (Jenkins et al. 2001). The application of these techniques to *in vivo* models of *Salmonella* pathogenesis has revealed valuable information about where, when, and how *Salmonella*-specific T cells behave during the course of infection (Ravindran and McSorley 2005). However, with this knowledge comes a host of new questions, providing the impetus for further development of immunological tools to improve studies of *Salmonella in vivo*.

5.1 *Target Antigen and Epitope Identification*

The relationship between *Salmonella*-specific T and B cells and the antigens they recognize constitutes a critical parameter determining the outcome of an infection. Thus, one of the most relevant issues concerning the study of adaptive immune responses to *Salmonella* is the precise identification of antigenic epitopes recognized by host lymphocytes. Explicit knowledge of these epitopes is essential for detailed characterization of T and B cell responses to *Salmonella* and may also provide potential candidates for a future subunit typhoid vaccine.

The challenge of *Salmonella* epitope-discovery is a daunting one, as the genome of Serovars typhi and typhimurium contain around 4,300 open reading frames (McClelland et al. 2001; Parkhill et al. 2001), each with the potential to encode multiple peptide epitopes recognized by T cells. Moreover, the timing and localization of bacterial protein expression during different stages of the disease adds a further layer of complexity to the availability of antigenic targets for T cells (Becker et al. 2006; Eriksson et al. 2003; Rollenhagen and Bumann 2006). Nevertheless, a handful of defined CD8 and CD4 T cell epitopes recognized by *Salmonella*-specific T cells have been successfully identified (Table 1).

Early studies demonstrated a role for CD8 T cells in immunity to *Salmonella* and led to a search for MHC class-I restricted peptide epitopes (Lo et al. 1999;

Table 1 Identified *Salmonella* epitopes

Peptide	MHC restriction	Reference
GroEL 192–200	Qa-1^b	Lo et al. (2000)
OmpC 73–80	K^b	Diaz-Quinonez et al. (2004)
OmpC 132–139	K^b	Diaz-Quinonez et al. (2004)
FliC 80–94	I-A^k	Bergman et al. (2005)
FliC 339–350	I-A^k	Cookson and Bevan (1997)
FliC 427–441	I-A^b	McSorley et al. (2000)
FliC 455–469	I-A^b	Bergman et al. (2005)
SipC 381–394	I-A^d	Musson et al. (2002)

Mastroeni et al. 1992; Nauciel 1990). Lo and colleagues found that a high proportion of CD8 CTL responses in mice infected with attenuated *Salmonella* recognized epitopes presented by nonclassical MHC class Ib molecules, particularly Qa-1^b (Lo et al. 1999). MALDI mass spectroscopy of peptides eluted from Qa-1^b molecules on *Salmonella*-infected cells identified the 192–200 peptide of the GroEL protein as an immunodominant epitope for CTL responses (Lo et al. 2000). In a different study, specialized software was used to predict K^b-binding peptides within the sequence of the immunogenic *Salmonella* outer membrane protein C (OmpC) porin, resulting in the identification of residues 73–80 and 132–139 as two immunodominant peptide epitopes (Diaz-Quinonez et al. 2004). However, our laboratory has been unable to detect CD8 T cell responses to either of these epitopes in *Salmonella*-infected mice (McSorley et al., unpublished observations), although it should be noted that the methodology used was different from that of the original report. Thus, despite the identification of a few class-I *Salmonella* epitopes, class-I tetramer reagents for tracking *Salmonella*-specific CD8 T cells have not yet been developed. This is unfortunate, since the development of ex vivo staining reagents that would allow identification of endogenous *Salmonella*-specific CD8 T cells should enable detailed analysis of the contribution of CD8 T cells to *Salmonella* immunity. Hopefully, this is an area that can be addressed in the future in both murine and human typhoid research.

Early work on CD4 T cells demonstrated significant protection after the transfer of *Salmonella*-specific T cell lines, or after immunization with *Salmonella* porins or other cell surface proteins (Misfeldt and Johnson 1979; Paul et al. 1985, 1988; Tabaraie et al. 1994); however, the epitopes recognized by CD4 T cells remained undefined. Cookson and Bevan were the first to identify an epitope of *Salmonella* flagellin recognized by *Salmonella*-specific CD4 T cells from C3H/HeJ mice in the context of I-A^k (Cookson and Bevan 1997). Flagellin is the major structural protein of bacterial flagella, is produced in large quantities by cultured bacteria, and is a known ligand for the innate immune receptor TLR5 (Salazar-Gonzalez and McSorley 2005; Hayashi, Aderem et al. 2001). Thus, it is perhaps not surprising that it is also a target of the adaptive response. The targeting of flagellin by *Salmonella*-specific CD4 T cells was subsequently confirmed in C57BL/6 mice, and currently four different MHC class-II flagellin epitopes have been reported (Bergman et al. 2005; McSorley et al. 2000). These initial studies also demonstrated the presence of long-lived memory T cell responses to flagellin epitopes, suggesting a role for these antigen-specific responses in protective immunity (Cookson and Bevan 1997; McSorley et al. 2000). Consistent with this observation, immunization of susceptible mice with flagellin conferred resistance to low dose challenge with virulent *Salmonella* (McSorley et al. 2000; Strindelius et al. 2002). The only other defined epitope recognized by *Salmonella*-specific CD4 T cells is the I-A^d restricted peptide 381–394 of *Salmonella* invasion protein C (SipC) (Musson et al. 2002). Interestingly, expression of both flagellin and SipC are expressed by cultured bacteria, but is tightly regulated during the transition to intramacrophage growth (Eriksson et al. 2003). Whether this is relevant for the targeting of these antigens by CD4 T cell responses is currently unclear.

In a more comprehensive screen for *Salmonella* target antigens, Rollenhagen and colleagues created a library of *Salmonella* promoter elements that were used to drive expression of a fluorescent reporter gene (Rollenhagen et al. 2004). Their logic was that identifying the *Salmonella* proteins that are highly expressed *in vivo* would include antigens that were most likely to be dominant T cell targets. Using this methodology, a list of candidate *Salmonella* proteins was generated that may be an important source of epitopes during an immune response. Within this list of highly expressed proteins, Mig-14 and SseB provided protective immunity when used as subunit vaccines. Efforts are now underway by our colleagues to fine map peptide epitopes from these proteins by screening T cells from immunized mice with libraries of overlapping peptide sequences.

As should be clear from the limited list of candidate antigens discussed above, epitope discovery still remains a major goal in the study of immunity to *Salmonella*. However, this situation is little different from numerous other infectious disease models that have been somewhat neglected by immunologists. Fortunately, epitope identification in a number of infectious disease models is now receiving greater attention (Peters and Sette 2007). The identification of new epitopes opens up the possibility for development of antigen-specific reagents such as T cell clones, TCR transgenic mice, and peptide:MHC tetramers that can facilitate visualization of *Salmonella*-specific T cells *in vivo*.

5.2 *Visualizing Salmonella-Specific Responses Using Surrogate Antigens*

Due to the scarcity of defined T cell epitopes in the *Salmonella* model, surrogate antigens have been used to track *Salmonella*-specific T cell responses. Proof of principle for this approach was established by early studies exploring the use of attenuated strains of *Salmonella* expressing model antigens, either as full-length proteins or short peptide sequences embedded within a native *Salmonella* protein (Maskell et al. 1987; Newton et al. 1989). B cell, cytotoxic T cell, and helper T cell responses to a wide variety of model antigens were elicited in this manner, thereby demonstrating potential roles for each of these cell populations in host adaptive immunity to *Salmonella* (Aggarwal et al. 1990; Khan et al. 1994; McSorley et al. 1997). In most of these experiments, freshly isolated T cells from immunized mice were restimulated *in vitro* with model antigens and a common read-out such as cell proliferation or specific lysis of target cells was recorded. These indirect functional assays can indicate the presence or absence of epitope-specific T cells in an immunized or infected animal, but they reveal very little about the biology of epitope-specific T cells *in vivo*.

The development of TCR transgenic mice that contain a peripheral CD4 population with monoclonal antigen specificity allowed investigators to move beyond this low level of resolution. TCR transgenic mice were initially infected with *Salmonella* expressing the model antigen recognized by peripheral T cells in the host (Verma et al. 1995), but a TCR transgenic host provides a highly artificial environment in which to model an immune response. The true power of TCR transgenic technology was realized through the use of an adoptive transfer approach where antigen specific T cells are seeded at relatively low frequency among polyclonal endogenous CD4 T cells in a wild type mouse (Kearney et al. 1994; Pape et al. 1997). Chen and Jenkins were the first to use this approach to examine the CD4 T cell response to *Salmonella* (Chen and Jenkins 1999). In this study, naive OVA-specific T cells from DO11.10 TCR transgenic mice were adoptively transferred into MHC-compatible BALB/c mice. The transferred cells were distinguished from endogenous T cells using an antibody specific for the DO11.10 TCR (Haskins et al. 1983). After infection with a transgenic strain of *Salmonella* expressing OVA, DO11.10 T cells proliferated extensively in the draining lymph nodes and spleen of both resistant and susceptible strains of BALB/c host mice, demonstrating that susceptibility was unrelated to the frequency of *Salmonella*-specific T cells but rather was dependent on the ability of these cells to secrete IFN-γ (Chen and Jenkins 1999).

The coordinated use of pathogen-expressed surrogate antigens and TCR transgenic T cell adoptive transfer has now become a standard protocol for investigating T cell responses to pathogens (Bertholet et al. 2006; Chen and Jenkins 1999; Pope et al. 2001). Although this approach certainly provides *in vivo* antigen-specific information, it also has some limitations. First, the complex regulation of pathogen proteins in response to the host environment is very poorly modeled by the overexpression of a transgene. Indeed, the selection of a promoter for antigen expression in *Salmonella* can have profound effects on the induction of an adaptive immune response (McSorley et al. 1997). This issue is evident in recent adoptive transfer studies using OVA-specific OT-I TCR transgenic CD8 T cells and *Salmonella* expressing OVA. A delayed CD8 T cell response to OVA-expressing *Salmonella* was reported in one study, but was not detected in a second study using a different antigen expression system (Jones-Carson et al. 2007; Luu et al. 2006). Thus, transgenic overexpression studies are heavily dependent on the strength and location of promoter activity *in vivo*. Second, the high level of model antigen expression in many of these systems can have deleterious effects on the viability of host bacteria (Bumann 2001b; Wick and Pfeifer 1996). Forcing a pathogen to devote a large percentage of protein expression machinery to the production of an irrelevant protein is likely to disrupt the natural life cycle of the pathogen and give rise to results that may be difficult to interpret. Third, in many overexpression studies, the visualized CD4 or CD8 T cell response to pathogen-expressed model antigen does not contribute to protective immunity and may therefore provide very little information about the nature of endogenous protective responses to pathogen proteins. However, when dealing with infectious disease models where epitope identification is still in its infancy, the overexpression of model antigens can still be a useful strategy to uncover antigen-specific responses *in vivo*.

In an attempt to examine both antigen expression and T cell responses simultaneously, an attenuated strain of *Salmonella* was generated expressing green fluorescent protein (GFP) fused to a minimal peptide sequence containing the immunodominant CD4 T cell epitope of OVA (Bumann 2001a). In contrast to the earlier study (Chen and Jenkins 1999), the number of OVA-specific T cells correlated precisely with the presence of antigen *in vivo*, which led the author to conclude that the transient nature of the T cell response in the former study was due to the enormous dose of OVA-expressing *Salmonella* initially administered (Bumann 2003).

5.3 Salmonella-Specific TCR Transgenic Mice

The artificial nature of surrogate antigen expression in visualizing *Salmonella*-specific immunity provided the motivation for the development of an experimental system to examine T cell responses to endogenous *Salmonella* epitopes. However, due to the limited number of defined epitopes noted above, almost all these efforts have focused on a single antigen, flagellin.

To provide a source of T cells with specificity to a bona fide *Salmonella* antigen, a transgenic mouse called SM1 was generated using TCR α and β genes from a CD4 T cell clone specific for residues 427–441 of flagellin in the context of I-A^b (McSorley et al. 2002, 2000). SM1 mice were backcrossed onto a *RAG-2*-deficient background to eliminate endogenous TCR rearrangement, which occurs at high frequency in this particular TCR transgenic mouse (McSorley, unpublished observations). SM1 mice have also been crossed to CD90.1 and CD45.1 congenic backgrounds, allowing identification of SM1 cells in adoptively transferred recipients using antibodies specific for these allelic markers (Srinivasan et al. 2004b; Srinivasan and McSorley 2007). The advantage of this system over the use of an OVA-specific system (Bumann 2001a; Chen and Jenkins 1999) is that SM1 cells allow examination of a CD4 T cell response to an endogenously expressed *Salmonella* protein. Similar adoptive transfer systems that allow visualization of T cell responses to endogenous bacterial antigens have recently been developed in models of Mycobacterial infection (Reiley et al. 2008; Wolf et al. 2008).

Oral infection of SM1 T cell-transferred mice with a virulent strain of *Salmonella* initiated rapid activation and proliferation of SM1 cells in the intestinal Peyer's patches and mesenteric lymph nodes (McSorley et al. 2002). A similar, highly localized response to antigen produced in the draining lymph nodes was noted in the Mycobacterial model and may therefore be a common feature of mucosal infection (Reiley et al. 2008; Wolf et al. 2008). Interestingly, in the SM1 model, the flagellin-specific response remained localized to the Peyer's patch and mesenteric lymph nodes despite systemic spread of the bacteria to the spleen and liver (McSorley et al. 2002). The explanation for this finding is still not clear but several subsequent experiments have addressed some of the potential issues.

A follow-up study indicated that strong competition from endogenous *Salmonella*-specific T cells contributes to poor persistence of SM1 cells in response to live, but not heat-killed, bacteria (Srinivasan et al. 2004a). An intriguingly similar scenario was recently reported in which a CD4 TCR transgenic T cell clone initially proliferated but quickly died off during infection with recombinant *Listeria monocytogenes* (Williams et al. 2008). In this study, the authors argued that the TCR transgenic clones faced competition from endogenous T cell clones specific for the same antigen. However, flagellin only constitutes a fraction of the overall CD4 T cell response to *Salmonella* (Srinivasan et al. 2004a), and the frequency of naive endogenous T cells specific for the flagellin I-A^b epitope in C57BL/6 mice is very low (Moon et al. 2007). Thus, competition may not be specifically for flagellin:I-A^b ligands but rather access to resources such as essential cytokines or antigen-presenting cells (APC).

Another possibility is that the expression of flagellin itself is simply not maintained following initial infection. Indeed, Cookson and colleagues have reported differential expression of flagellin in the spleen versus intestinal lymphoid tissue (Alaniz et al. 2006; Cummings et al. 2006). However, the transfer of *Salmonella*-infected splenocytes into naive mice causes activation of SM1 cells, arguing that flagellin is available in the spleen, albeit this activation was relatively weak suggesting that antigen may indeed be limiting (Srinivasan et al. 2004b). One possibility is that flagellin is expressed in infected spleens in a sequestered environment away from SM1 cells, but this has not been demonstrated experimentally. Whatever the mechanism for the lack of a systemic SM1 response, flagellin is one of the most abundant proteins expressed by *Salmonella*, and the evasion of a flagellin-specific T cell response in the spleen may thus represent an important immune evasion tactic by *Salmonella*

5.4 *Salmonella*-Specific Tetramers

Despite the advantages of using the SM1 system, there are also drawbacks to the use of TCR transgenic adoptive transfer systems. One inherent limitation of this approach is that the epitope-specific T cell population under study is comprised of a single clonotype, raising the possibility that the behavior of the monoclonal population may not accurately reflect the full polyclonal repertoire of endogenous T cells responding to the given epitope. Also, it is now known that modification of the naive precursor frequency by the transfer of TCR transgenic T cells can alter the half-life of

T cells, affect the kinetics of T cell activation, and have deleterious effects on the development of immune memory (Badovinac et al. 2007; Ford et al. 2007; Foulds and Shen 2006; Hataye et al. 2006; Marzo et al. 2005). These issues raise the possibility that an endogenous epitope-specific T cell population may differ substantially from a relatively high frequency TCR transgenic population.

Direct visualization of endogenous T cell responses in unmanipulated mice would avoid these limitations, but the low frequency of epitope-specific endogenous cells makes this approach technically difficult, and indeed necessitated the development of the TCR transgenic adoptive transfer approach in the first place. However, recent developments in the construction of peptide:MHC tetramers combined with new techniques for detecting low frequency T cell populations now makes endogenous T cell tracking possible (Hataye et al. 2006; Moon et al. 2007). We have recently generated a flagellin$_{427–441}$:I-A^b tetramer and have used it to examine endogenous flagellin-specific T cell responses *in vivo* (Moon et al. 2007). Our initial studies have demonstrated that naïve flagellin$_{427–441}$:I-A^b-specific T cells are found at very low frequency in C57BL/6 mice (around 20 cells per mouse) but expand several hundred-fold to a systemic dose of flagellin peptide and LPS. We are currently studying the development of endogenous flagellin-specific T cell responses in *Salmonella* infected mice using this tracking methodology. With this new technology, the labor-intensive generation of T cell lines and TCR transgenic mice will no longer be needed to study T cell responses to newly identified epitopes in pathogen organisms.

6 Development and Function of Th1 Cells During *Salmonella* Infection

As noted above, numerous studies have demonstrated a requirement for Th1 cells producing IFN-γ for the resolution of *Salmonella* infection. However, the development of *in vivo* tracking approaches have allowed more detailed insight into the development of *Salmonella*-specific Th1 cells and how these cells mediate their effector function *in vivo*. These studies raise important questions about exactly when Th1 cells contribute to *Salmonella* immunity and the mechanism by which Th1 cells activate infected macrophages.

6.1 *When do Th1 Cells Activate Salmonella-Infected Macrophages?*

Initial studies tracking SM1 T cells in response to live infection demonstrated that these cells are activated extremely rapidly in intestinal lymphoid tissues of the gut (McSorley et al. 2002). We now know that this early activation is dependent upon the rapid migration of CCR6$^+$ dendritic cells in response to *Salmonella* penetration of the Peyer's patch epithelium (Salazar-Gonzalez et al. 2006).

Thus, *Salmonella*-specific T cells are activated within a few hours of oral infection. These data are somewhat at odds with earlier studies using gene-deficient mice which suggest that *Salmonella*-specific T cells only contribute in the late stage of disease resolution. Mice with deficiencies in Th1 cells or the development of Th1 cells all display a profound deficiency in bacterial clearance from the spleen and/or liver (Hess et al. 1996; Ravindran et al. 2005; VanCott et al. 1998; Weintraub et al. 1997), but this was only evident several weeks after infection. Thus, while recent visualization studies indicate activation of T cells within hours, bacterial colonization studies suggest T cells do not affect bacterial growth until several weeks later. The development of Th1 cells and their contribution to *Salmonella* immunity therefore requires more detailed study. One attractive possibility is that *Salmonella* actively inhibit the function of Th1 cells *in vivo*.

The ability of *Salmonella* to inhibit T cells has been examined in some depth and has generally focused on the ability of bacteria to regulate antigen presentation to avoid T cell activation (Cheminay et al. 2005; Qimron et al. 2004; Svensson et al. 2000; Tobar et al. 2004, 2006 ; van der Velden et al. 2003; Yrlid et al. 2000). However, much of this work has been carried out *in vitro* and the *in vivo* significance is not readily apparent, especially given the fact that efficient activation of SM1 T cells occurs in response to oral or intravenous infection (McSorley et al. 2002; Salazar-Gonzalez et al. 2006; Srinivasan et al. 2004b). Considerably less attention has been given to the alternative possibility, that *Salmonella* can inhibit the survival or function of *Salmonella*-specific T cells after initial expansion and activation have already occurred. Indeed, we have already noted that SM1 T cells expand but fail to survive long-term in mice exposed to live bacteria (Srinivasan et al. 2004a). Our more recent data suggest that these SM1 cells die by apoptosis in a process dependent on bacterial virulence factors (Srinivasan and McSorley, unpublished data). Thus, *Salmonella* can directly inhibit the function of *Salmonella*-specific Th1 cells that are activated early in the response. Understanding the nature of this inhibitory effect may lead to the development of more effective therapeutics or more immunogenic vaccine vectors for typhoid.

6.2 How do Th1 Cells Activate Salmonella-Infected Macrophages?

Naive CD4 T cells transit through the blood and secondary lymphoid tissues until they are activated via the TCR by peptides presented by dendritic cells in the context of MHC class-II (Jenkins et al. 2001). After an initial round of expansion, these activated T cells acquire effector functions and the ability to enter nonlymphoid tissues (Lefrancois 2006; Swain et al. 2006). It is generally assumed that an effector Th1 cell trafficking to the liver will produce IFN-γ following recognition of peptide:MHC presented by a *Salmonella*-infected macrophage. Therefore, this theoretical model requires two independent recognition events via the TCR: the first in secondary lymphoid tissues, and the second at the infected site. However, recent data raise questions about whether this second recognition event is actually required.

Given the current lack of knowledge about T cell epitopes in *Salmonella* infection, our laboratory developed a crude means of visualizing endogenous *Salmonella*-specific T cells *in vivo* (Srinivasan et al. 2004a). This simply involved the injection of infected mice with a bacterial lysate and examining IFN-γ and TNF-α production directly ex vivo. A sizable population of CD4 and CD8 T cells was found producing these effector cytokines in infected mice but not in naive mice (Srinivasan et al. 2004a), suggesting that this method allows visualization of polyclonal endogenous CD4 T cells responding to *Salmonella*. Surprisingly, CD4 and CD8 T cells were also activated to produce IFN-γ when *Salmonella*-infected mice were injected with ultrapure LPS (Srinivasan et al. 2007), demonstrating that T cell activation in this assay occurred by innate stimulation. Furthermore, previously activated *Salmonella*-specific T cells could respond and produce IFN-γ even after transfer and stimulation in MHC class-II-deficient mice (Srinivasan et al. 2007). This "innate stimulation" of CD4 T cells was only evident during active infection and was partially dependent on IL-18 signaling (Srinivasan et al. 2007). Very similar results have been reported for virus-specific CD8 T cells in infected mice (Beadling and Slifka 2005; Berg and Forman 2006; Kambayashi et al. 2003), suggesting that innate stimulation of previously activated T cells is a fairly common phenomenon in infectious disease models. Thus, although TCR ligation is required for initial expansion of *Salmonella*-specific T cells in secondary lymphoid tissues, it may not be required for elaboration of IFN-γ production in the infected liver. The ability of effector T cells to respond to innate stimuli such as LPS or inflammatory cytokines may be important for amplifying the antibacterial effector response in the face of rapid bacterial replication. However, the relative importance of innate activation of CD4 or CD8 T cells versus TCR ligation has not been examined directly in any infectious disease model.

7 Conclusion

Visualization of pathogen-specific immune responses *in vivo* allows a more detailed understanding of immunity to infectious disease. The development of TCR transgenic adoptive transfer systems and new tetramer approaches will surely expand what we currently know about adaptive immune responses from traditional methodologies. This new knowledge is likely to lead to an in-depth understanding of T cell activation, effector function, and memory development, and holds considerable promise for the generation of novel vaccines and therapeutics to treat infectious disease.

References

Aggarwal A, Kumar S, Jaffe R, Hone D, Gross M, Sadoff J (1990) Oral *Salmonella*: malaria circumsporozoite recombinants induce specific CD8+ cytotoxic T cells. J Exp Med 172: 1083–1090

Alaniz RC, Cummings LA, Bergman MA, Rassoulian-Barrett SL, Cookson BT (2006) *Salmonella* typhimurium coordinately regulates FliC location and reduces dendritic cell activation and antigen presentation to CD4+ T cells. J Immunol 177:3983–3993

Amanna I, Slifka MK (2005) Public fear of vaccination: separating fact from fiction. Viral Immunol 18:307–315

Badovinac VP, Haring JS, Harty JT (2007) Initial T cell receptor transgenic cell precursor frequency dictates critical aspects of the CD8(+) T cell response to infection. Immunity 26:827–841

Beadling C, Slifka MK (2005) Differential regulation of virus-specific T-cell effector functions following activation by peptide or innate cytokines. Blood 105:1179–1186

Becker D, Selbach M, Rollenhagen C, Ballmaier M, Meyer TF, Mann M, Bumann D (2006) Robust *Salmonella* metabolism limits possibilities for new antimicrobials. Nature 440:303–307

Berg RE, Forman J (2006) The role of CD8 T cells in innate immunity and in antigen non-specific protection. Curr Opin Immunol 18:338–343

Bergman MA, Cummings LA, Alaniz RC, Mayeda L, Fellnerova I, Cookson BT (2005) CD4+-T-cell responses generated during murine *Salmonella* enterica serovar Typhimurium infection are directed towards multiple epitopes within the natural antigen FliC. Infect Immun 73:7226–7235

Bertholet S, Goldszmid R, Morrot A, Debrabant A, Afrın F, Collazo-Custodio C, Houde M, Desjardins M, Sher A, Sacks D (2006) Leishmania antigens are presented to CD8+ T cells by a transporter associated with antigen processing-independent pathway *in vitro* and *in vivo*. J Immunol 177:3525–3533

Blanden RV, Mackaness GB, Collins FM (1966) Mechanisms of acquired resistance in mouse typhoid. J Exp Med 124:585–600

Brines R (1996) Two hundred years on: Jenner and the discovery of vaccination. Immunol Today 17:203–204

Bueno SM, Gonzalez PA, Carreno LJ, Tobar JA, Mora GC, Pereda CJ, Salazar-Onfray F, Kalergis AM (2008) The capacity of *Salmonella* to survive inside dendritic cells and prevent antigen presentation to T cells is host specific. Immunology

Bueno SM, Gonzalez PA, Schwebach JR, Kalergis AM (2007) T cell immunity evasion by virulent *Salmonella* enterica. Immunol Lett 111:14–20

Bumann D (2001a) *In vivo* visualization of bacterial colonization antigen expression and specific T-cell induction following oral administration of live recombinant *Salmonella* enterica serovar typhimurium. Infect Immun 69:4618–4626

Bumann D (2001b) Regulated antigen expression in live recombinant *Salmonella* enterica serovar Typhimurium strongly affects colonization capabilities and specific CD4(+)-T-cell responses. Infect Immun 69:7493–7500

Bumann D (2003) T cell receptor-transgenic mouse models for studying cellular immune responses to *Salmonella in vivo*. FEMS Immunol Med Microbiol 37:105–109

Cheminay C, Mohlenbrink A, Hensel M (2005) Intracellular *Salmonella* inhibit antigen presentation by dendritic cells. J Immunol 174:2892–2899

Chen ZM, Jenkins MK (1999) Clonal expansion of antigen-specific CD4 T cells following infection with *Salmonella* typhimurium is similar in susceptible (Itys) and resistant (Ityr) BALB/c mice. Infect Immun 67:2025–2029

Cleary AM, Tu W, Enright A, Giffon T, Dewaal-Malefyt R, Gutierrez K, Lewis DB (2003) Impaired accumulation and function of memory CD4 T cells in human IL-12 receptor beta 1 deficiency. J Immunol 170:597–603

Cookson BT, Bevan MJ (1997) Identification of a natural T cell epitope presented by *Salmonella*-infected macrophages and recognised by T cells from orally immunised mice. J Immunol 158:4310–4319

Crump JA, Luby SP, Mintz ED (2004) The global burden of typhoid fever. Bull World Health Organ 82:346–353

Cummings LA, Wilkerson WD, Bergsbaken T, Cookson BT (2006) *In vivo* fliC expression by *Salmonella* enterica serovar Typhimurium is heterogeneous regulated by ClpX, and anatomically restricted. Mol Microbiol 61:795–809

Diaz-Quinonez A, Martin-Orozco N, Isibasi A, Ortiz-Navarrete V (2004) Two *Salmonella* OmpC K(b)-restricted epitopes for CD8+-T-cell recognition. Infect Immun 72:3059–3062

Eriksson S, Lucchini S, Thompson A, Rhen M, Hinton JC (2003) Unravelling the biology of macrophage infection by gene expression profiling of intracellular *Salmonella* enterica. Mol Microbiol 47:103–118

Fierer J, Guiney DG (2001) Diverse virulence traits underlying different clinical outcomes of *Salmonella* infection. J Clin Invest 107:775–780

Ford ML, Koehn BH, Wagener ME, Jiang W, Gangappa S, Pearson TC, Larsen CP (2007) Antigen-specific precursor frequency impacts T cell proliferation differentiation and requirement for costimulation. J Exp Med 204:299–309

Foulds KE, Shen H (2006) Clonal competition inhibits the proliferation and differentiation of adoptively transferred TCR transgenic CD4 T cells in response to infection. J Immunol 176:3037–3043

Germain RN, Jenkins MK (2004) *In vivo* antigen presentation. Curr Opin Immunol 16:120–125

Gowen JW, Calhoun ML (1943) On the Physical Basis for Genetic Resistance to Mouse Typhoid *Salmonella* Typhimurium. Proc Natl Acad Sci U S A 29:144–149

Grassl GA, Finlay BB (2008) Pathogenesis of enteric *Salmonella* infections. Curr Opin Gastroenterol 24:22–26

Haskins K, Kubo R, White J, Pigeon M, Kappler J, Marrack P (1983) The major histocompatibility complex-restricted antigen receptor on T cells. IIsolation with a monoclonal antibody. J Exp Med 157:1149–1169

Hataye J, Moon JJ, Khoruts A, Reilly C, Jenkins MK (2006) Naive and memory CD4+ T cell survival controlled by clonal abundance. Science 312:114–116

Henrickson SE, von Andrian UH (2007) Single-cell dynamics of T-cell priming. Curr Opin Immunol 19:249–258

Hess J, Ladel C, Miko D, Kaufmann SH (1996) *Salmonella* typhimurium aroA- infection in gene-targeted immunodeficient mice: major role of CD4+ TCR-alpha beta cells and IFN-gamma in bacterial clearance independent of intracellular location. J Immunol 156:3321–3326

Huang AY, Qi H, Germain RN (2004) Illuminating the landscape of *in vivo* immunity: insights from dynamic in situ imaging of secondary lymphoid tissues. Immunity 21:331–339

Jenkins MK, Khoruts A, Ingulli E, Mueller DL, McSorley SJ, Reinhardt RL, Itano A, Pape KA (2001) *In vivo* activation of antigen-specific CD4 T cells. Annu Rev Immunol 19:23–45

Jones BD, Falkow S (1996) Salmonellosis: host immune responses and bacterial virulence determinants. Annu Rev Immunol 14:533–561

Jones-Carson J, McCollister BD, Clambey ET, Vazquez-Torres A (2007) Systemic CD8 T-cell memory response to a *Salmonella* pathogenicity island 2 effector is restricted to *Salmonella* enterica encountered in the gastrointestinal mucosa. Infect Immun 75:2708–2716

Jouanguy E, Doffinger R, Dupuis S, Pallier A, Altare F, Casanova JL (1999) IL-12 and IFN-gamma in host defense against mycobacteria and *salmonella* in mice and men. Curr Opin Immunol 11:346–351

Kambayashi T, Assarsson E, Lukacher AE, Ljunggren HG, Jensen PE (2003) Memory CD8+ T cells provide an early source of IFN-gamma. J Immunol 170:2399–2408

Kearney ER, Pape KA, Loh DY, Jenkins MK (1994) Visualization of peptide-specific T cell immunity and peripheral tolerance induction *in vivo*. Immunity 1:327–339

Khan CM, Villarreal-Ramos B, Pierce RJ, Riveau G, Demarco de Hormaeche R, McNeill H, Ali T, Fairweather N, Chatfield S, Capron A, et al (1994) Construction expression and immunogenicity of the Schistosoma mansoni P28 glutathione S-transferase as a genetic fusion to tetanus toxin fragment C in a live Aro attenuated vaccine strain of *Salmonella*. Proc Natl Acad Sci U S A 91:11261–11265

Lefrancois L (2006) Development trafficking and function of memory T-cell subsets. Immunol Rev 211:93–103

Lo WF, Ong H, Metcalf ES, Soloski MJ (1999) T cell responses to Gram-negative intracellular bacterial pathogens: a role for CD8+ T cells in immunity to *Salmonella* infection and the involvement of MHC class Ib molecules. J Immunol 162:5398–5406

Lo WF, Woods AS, DeCloux A, Cotter RJ, Metcalf ES, Soloski MJ (2000) Molecular mimicry mediated by MHC class Ib molecules after infection with gram-negative pathogens. Nat Med 6:215–218

Luu RA, Gurnani K, Dudani R, Kammara R, van Faassen H, Sirard JC, Krishnan L, Sad S (2006) Delayed expansion and contraction of CD8+ T cell response during infection with virulent *Salmonella* typhimurium. J Immunol 177:1516–1525

Martinez-Cajas JL, Wainberg MA (2007) Protease inhibitor resistance in HIV-infected patients: molecular and clinical perspectives. Antiviral Res 76:203–221

Marzo AL, Klonowski KD, Le Bon A, Borrow P, Tough DF, Lefrancois L (2005) Initial T cell frequency dictates memory CD8+ T cell lineage commitment. Nat Immunol 6:793–799

Maskell DJ, Sweeney KJ, O'Callaghan D, Hormaeche CE, Liew FY, Dougan G (1987) *Salmonella* typhimurium aroA mutants as carriers of the Escherichia coli heat-labile enterotoxin B subunit to the murine secretory and systemic immune systems. Microb Pathog 2:211–221

Mastroeni P (2002) Immunity to systemic *Salmonella* infections. Curr Mol Med 2:393–406

Mastroeni P, Chabalgoity JA, Dunstan SJ, Maskell DJ, Dougan G (2001) *Salmonella*: immune responses and vaccines. Vet J 161:132–164

Mastroeni P, Harrison JA, Chabalgoity JA, Hormaeche CE (1996) Effect of interleukin 12 neutralization on host resistance and gamma interferon production in mouse typhoid. Infect Immun 64:189–196

Mastroeni P, Sheppard M (2004) *Salmonella* infections in the mouse model: host resistance factors and *in vivo* dynamics of bacterial spread and distribution in the tissues. Microbes Infect 6:398–405

Mastroeni P, Simmons C, Fowler R, Hormaeche CE, Dougan G (2000) Igh-6(-/-) (B-cell-deficient) mice fail to mount solid aquired resistance to oral challenge with virulent *Salmonella* enterica serovar typhimurium and show impaired Th1 T-cell responses to *Salmonella* antigens. Infect Immun 68:46–53

Mastroeni P, Villarreal-Ramos B, Hormaeche CE (1992) Role of T cells TNF alpha and IFN gamma in recall of immunity to oral challenge with virulent *salmonella*e in mice vaccinated with live attenuated aro- *Salmonella* vaccines. Microb Pathog 13:477–491

Matsui H, Eguchi M, Kikuchi Y (2000) Use of confocal microscopy to detect *Salmonella* typhimurium within host cells associated with Spv-mediated intracellular proliferation. Microb Pathog 29:53–59

McClelland M, Sanderson KE, Spieth J, Clifton SW, Latreille P, Courtney L, Porwollik S, Ali J, Dante M, Du F, et al (2001) Complete genome sequence of *Salmonella* enterica serovar Typhimurium LT2. Nature 413:852–856

McSorley SJ, Asch S, Costalonga M, Rieinhardt RL, Jenkins MK (2002) Tracking *Salmonella*-specific CD4 T cells *in vivo* reveals a local mucosal response to a disseminated infection. Immunity 16:365–377

McSorley SJ, Cookson BT, Jenkins MK (2000) Characterization of CD4+ T cell responses during natural infection with *Salmonella* typhimurium. J Immunol 164:986–993

McSorley SJ, Jenkins MK (2000) Antibody is required for protection against virulent but not attenuated *Salmonella* enterica serovar typhimurium. Infect Immun 68:3344–3348

McSorley SJ, Xu D, Liew FY (1997) Vaccine efficacy of *Salmonella* strains expressing glycoprotein 63 with different promoters. Infect Immun 65:171–178

Mead PS, Slutsker L, Dietz V, McCaig LF, Bresee JS, Shapiro C, Griffin PM, Tauxe RV (1999) Food-related illness and death in the United States. Emerg Infect Dis 5:607–625

Misfeldt ML, Johnson W (1979) Identification of protective cell surface proteins in ribosomal fractions from *Salmonella* typhimurium. Infect Immun 24:808–816

Mittrucker H (1999) Critical role of CD28 in protective immunity against *salmonella* typhimurium. J Immunol 163:6769–6776

Mittrucker HW, Kaufmann SH (2000) Immune response to infection with *Salmonella* typhimurium in mice. J Leukoc Biol 67:457–463

Mittrucker HW, Raupach B, Kohler A, Kaufmann SH (2000) Cutting edge: role of B lymphocytes in protective immunity against *Salmonella* typhimurium infection. J Immunol 164:1648–1652

Moon JJ, Chu HH, Pepper M, McSorley SJ, Jameson SC, Kedl RM, Jenkins MK (2007) Naive CD4(+) T cell frequency varies for different epitopes and predicts repertoire diversity and response magnitude. Immunity 27:203–213

Moran GJ, Talan DA, Abrahamian FM (2008) Biological terrorism. Infect Dis Clin North Am 22:145–187: vii

Musson JA, Hayward RD, Delvig AA, Hormaeche CE, Koronakis V, Robinson JH (2002) Processing of viable *Salmonella* typhimurium for presentation of a CD4 T cell epitope from the *Salmonella* invasion protein C (SipC) Eur J Immunol 32:2664–2671

Nauciel C (1990) Role of CD4+ T cells and T-independent mechanisms in aquired resistance to *Salmonella* typhimurium infection. J Immunol 145:1265–1269

Nauciel C, Espinasse-Maes F (1992) Role of gamma interferon and tumor necrosis factor alpha in resistance to *Salmonella* typhimurium infection. Infect Immun 60:450–454

Negrin RS, Contag CH (2006) *In vivo* imaging using bioluminescence: a tool for probing graft-versus-host disease. Nat Rev Immunol 6:484–490

Newton SM, Jacob CO, Stocker BA (1989) Immune response to cholera toxin epitope inserted in *Salmonella* flagellin. Science 244:70–72

Okeke IN, Aboderin OA, Byarugaba DK, Ojo KK, Opintan JA (2007) Growing problem of multidrug-resistant enteric pathogens in Africa. Emerg Infect Dis 13:1640–1646

Pape KA, Kearney ER, Khoruts A, Mondino A, Merica R, Chen ZM, Ingulli E, White J, Johnson JG, Jenkins MK (1997) Use of adoptive transfer of T-cell-antigen-receptor-transgenic T cell for the study of T-cell activation *in vivo*. Immunol Rev 156:67–78

Parkhill J, Dougan G, James KD, Thomson NR, Pickard D, Wain J, Churcher C, Mungall KL, Bentley SD, Holden MT, et al (2001) Complete genome sequence of a multiple drug resistant *Salmonella* enterica serovar Typhi CT18. Nature 413:848–852

Parry CM, Hien TT, Dougan G, White NJ, Farrar JJ (2002) Typhoid fever. N Engl J Med 347:1770–1782

Pasetti MF, Levine MM, Sztein MB (2003) Animal models paving the way for clinical trials of attenuated *Salmonella* enterica serovar Typhi live oral vaccines and live vectors. Vaccine 21:401–418

Pasetti MF, Salerno-Goncalves R, Sztein MB (2002) *Salmonella* enterica serovar Typhi live vector vaccines delivered intranasally elicit regional and systemic specific CD8+ major histocompatibility class I-restricted cytotoxic T lymphocytes. Infect Immun 70:4009–4018

Paul C, Shalala K, Warren R, Smith R (1985) Adoptive transfer of murine host protection to salmonellosis with T-cell growth factor-dependent *Salmonella*-specific T-cell lines. Infect Immun 48:40–43

Paul CC, Norris K, Warren R, Smith RA (1988) Transfer of murine host protection by using interleukin-2-dependent T-lymphocyte lines. Infect Immun 56:2189–2192

Peters B, Sette A (2007) Integrating epitope data into the emerging web of biomedical knowledge resources. Nat Rev Immunol 7:485–490

Pope C, Kim SK, Marzo A, Williams K, Jiang J, Shen H, Lefrancois L (2001) Organ-specific regulation of the CD8 T cell response to Listeria monocytogenes infection. J Immunol 166:3402–3409

Qimron U, Madar N, Mittrucker HW, Zilka A, Yosef I, Bloushtain N, Kaufmann SH, Rosenshine I, Apte RN, Porgador A (2004) Identification of *Salmonella* typhimurium genes responsible for interference with peptide presentation on MHC class I molecules: Deltayej *Salmonella* mutants induce superior CD8 T-cell responses. Cell Microbiol 6:1057–1070

Rabsch W, Tschape H, Baumler AJ (2001) Non-typhoidal salmonellosis: emerging problems. Microbes Infect 3:237–247

Ravindran R, Foley J, Stoklasek T, Glimcher LH, McSorley SJ (2005) Expression of T-bet by CD4 T cells is essential for resistance to *Salmonella* infection. J Immunol 175:4603–4610

Ravindran R, McSorley SJ (2005) Tracking the dynamics of T-cell activation in response to *Salmonella* infection. Immunology 114:450–458

Reiley WW, Calayag MD, Wittmer ST, Huntington JL, Pearl JE, Fountain JJ, Martino CA, Roberts AD, Cooper AM, Winslow GM, Woodland DL (2008) ESAT-6-specific CD4 T cell responses to aerosol Mycobacterium tuberculosis infection are initiated in the mediastinal lymph nodes. Proc Natl Acad Sci U S A

Richter-Dahlfors A, Buchan AM, Finlay BB (1997) Murine salmonellosis studied by confocal microscopy: *Salmonella* typhimurium resides intracellularly inside macrophages and exerts a cytotoxic effect on phagocytes *in vivo*. J Exp Med 186:569–580

Rollenhagen C, Bumann D (2006) *Salmonella* enterica highly expressed genes are disease specific. Infect Immun 74:1649–1660

Rollenhagen C, Sorensen M, Rizos K, Hurvitz R, Bumann D (2004) Antigen selection based on expression levels during infection facilitates vaccine development for an intracellular pathogen. Proc Natl Acad Sci U S A 101:8739–8744

Salazar-Gonzalez RM, McSorley SJ (2005) *Salmonella* flagellin a microbial target of the innate and adaptive immune system. Immunol Lett 101:117–122

Salazar-Gonzalez RM, Niess JH, Zammit DJ, Ravindran R, Srinivasan A, Maxwell JR, Stoklasek T, Yadav R, Williams IR, Gu X, et al (2006) CCR6-Mediated Dendritic Cell Activation of Pathogen-Specific T Cells in Peyer's Patches. Immunity 24:623–632

Salcedo SP, Noursadeghi M, Cohen J, Holden DW (2001) Intracellular replication of *Salmonella* typhimurium strains in specific subsets of splenic macrophages *in vivo*. Cell Microbiol 3:587–597

Santos RL, Zhang S, Tsolis RM, Kingsley RA, Adams LG, Baumler AJ (2001) Animal models of *salmonella* infections: enteritis versus typhoid fever. Micrtobes Infect 3:1335–1344

Sheppard M, Webb C, Heath F, Mallows V, Emilianus R, Maskell D, Mastroeni P (2003) Dynamics of bacterial growth and distribution within the liver during *Salmonella* infection. Cell Microbiol 5:593–600

Sinha K, Mastroeni P, Harrison J, de Hormaeche RD, Hormaeche CE (1997) *Salmonella* typhimurium aroA, htrA, and AroD htrA mutants cause progressive infections in athymic (nu/nu) BALB/c mice. Infect Immun 65:1566–1569

Sobel J, Khan AS, Swerdlow DL (2002) Threat of a biological terrorist attack on the US food supply: the CDC perspective. Lancet 359:874–880

Srinivasan A, Foley J, McSorley SJ (2004a) Massive Number of Antigen-Specific CD4 T Cells during Vaccination with Live Attenuated *Salmonella* Causes Interclonal Competition. J Immunol 172:6884–6893

Srinivasan A, Foley J, Ravindran R, McSorley SJ (2004b) Low-dose *salmonella* infection evades activation of flagellin-specific CD4 T cells. J Immunol 173:4091–4099

Srinivasan A, McSorley SJ (2006) Activation of *Salmonella*-specific immune responses in the intestinal mucosa. Arch Immunol Ther Exp (Warsz) 54:25–31

Srinivasan A, McSorley SJ (2007) Pivotal advance: exposure to LPS suppresses CD4+ T cell cytokine production in *Salmonella*-infected mice and exacerbates murine typhoid. J Leukoc Biol 81:403–411

Srinivasan A, Salazar-Gonzalez RM, Jarcho M, Sandau MM, Lefrancois L, McSorley SJ (2007) Innate immune activation of CD4 T cells in *salmonella*-infected mice is dependent on IL-18. J Immunol 178:6342–6349

Strindelius L, Degling Wikingsson L, Sjoholm I (2002) Extracellular antigens from *Salmonella* enteritidis induce effective immune response in mice after oral vaccination. Infect Immun 70:1434–1442

Su LH, Chiu CH (2007) *Salmonella*: clinical importance and evolution of nomenclature. Chang Gung Med J 30:210–219

Subramanian, N, Qadri A (2006) Lysophospholipid sensing triggers secretion of flagellin from pathogenic *salmonella*. Nat Immunol 7:583–589

Svensson M, Johansson C, Wick MJ (2000) *Salmonella* enterica serovar typhimurium-induced maturation of bone marrow-derived dendritic cells. Infect Immun 68:6311–6320

Swain SL, Agrewala JN, Brown DM, Jelley-Gibbs DM, Golech S, Huston G, Jones SC, Kamperschroer C, Lee WH, McKinstry KK, et al (2006) CD4+ T-cell memory: generation and multi-faceted roles for CD4+ T cells in protective immunity to influenza. Immunol Rev 211:8–22

Tabaraie B, Sharma BK, Sharma PR, Sehgal R, Ganguly NK (1994) Evaluation of *Salmonella* porins as a broad spectrum vaccine candidate. Microbiol Immunol 38:553–559

Tarr PE, Kuppens L, Jones TC, Ivanoff B, Aparin PG, Heymann DL (1999) Considerations regarding mass vaccination against typhoid fever as an adjunct to sanitation and public health measures: potential use in an epidemic in Tajikistan. Am J Trop Med Hyg 61:163–170

Tobar JA, Carreno LJ, Bueno SM, Gonzalez PA, Mora JE, Quezada SA, Kalergis AM (2006) Virulent *Salmonella* enterica serovar typhimurium evades adaptive immunity by preventing dendritic cells from activating T cells. Infect Immun 74:6438–6448

Tobar JA, Gonzalez PA, Kalergis AM (2004) *Salmonella* escape from antigen presentation can be overcome by targeting bacteria to Fcgamma receptors on dendritic cells. J Immunol 173:4058–4065

Ugrinovic S, Menager N, Goh N, Mastroeni P (2003) Characterization and development of T-Cell immune responses in B-cell-deficient (Igh-6(-/-)) mice with *Salmonella* enterica serovar Typhimurium infection. Infect Immun 71:6808–6819

van der Velden AW, Velasquez M, Starnbach MN (2003) *Salmonella* rapidly kill dendritic cells via a caspase-1-dependent mechanism. J Immunol 171:6742–6749

VanCott JL, Chatfield SN, Roberts M, Hone DM, Hohmann EL, Pascual DW, Yamamoto M, Kiyono H, McGhee JR (1998) Regulation of host immune responses by modification of *Salmonella* virulence genes. Nat Med 4:1247–1252

Verma NK, Ziegler HK, Stocker BA, Schoolnik GK (1995) Induction of a cellular immune response to a defined T-cell epitope as an insert in the flagellin of a live vaccine strain of *Salmonella*. Vaccine 13:235–244

Walker RI (2005) New vaccines against enteric bacteria for children in less developed countries. Expert Rev Vaccines 4:807–812

Weintraub BC, Eckmann L, Okamoto S, Hense M, Hedrick SM, Fierer J (1997) Role of alphabeta and gammadelta T cells in the host response to *Salmonella* infection as demonstrated in T-cell-receptor-deficient mice of defined Ity genotypes. Infect Immun 65:2306–2312

Wick MJ (2003) The role of dendritic cells in the immune response to *Salmonella*. Immunol Lett 85:99–102

Wick MJ, Pfeifer JD (1996) Major histocompatibility complex class I presentation of ovalbumin peptide 257–264 from exogenous sources: protein context influences the degree of TAP-independent presentation. Eur J Immunol 26:2790–2799

Wijburg OL, Uren TK, Simpfendorfer K, Johansen FE, Brandtzaeg P, Strugnell RA (2006) Innate secretory antibodies protect against natural *Salmonella* typhimurium infection. J Exp Med 203:21–26

Williams MA, Ravkov EV, Bevan MJ (2008) Rapid culling of the CD4+ T cell repertoire in the transition from effector to memory. Immunity 28:533–545

Wolf AJ, Desvignes L, Linas B, Banaiee N, Tamura T, Takatsu K, Ernst JD (2008) Initiation of the adaptive immune response to Mycobacterium tuberculosis depends on antigen production in the local lymph node not the lungs. J Exp Med 205:105–115

Woolhouse ME, Haydon DT, Antia R (2005) Emerging pathogens: the epidemiology and evolution of species jumps. Trends Ecol Evol 20:238–244

Yrlid U, Svensson M, Johansson C, Wick MJ (2000) *Salmonella* infection of bone marrow-derived macrophages and dendritic cells: influence on antigen presentation and initiating an immune response. FEMS Immunol Med Microbiol 27:313–320

Zinkernagel RM (1976) Cell-mediated immune response to *Salmonella* typhimurium infection in mice: development of nonspecific bactericidal activity against Listeria monocytogenes. Infect Immun 13:1069–1073

Imaging *Listeria monocytogenes* Infection *In Vivo*

Vjollca Konjufca and Mark J. Miller

Contents

Abstract *Listeria monocytogenes* infection in mice is a highly prolific model of bacterial infection. Several *in vivo* imaging approaches have been used to study host cell dynamics in response to infection including bioluminescence imaging, confocal microscopy and two-photon microscopy. The application of *in vivo* imaging to study transgenic mouse models is providing unprecedented opportunities to test specific molecular mechanistic theories about how the host immune response unfolds. In complementary studies, *in vivo* imaging can be performed using genetically engineered bacterial mutants to assess the impact of specific virulence factors in host

---

V. Konjufca and M.J. Miller (✉)
Department of Pathology and Immunology, Washington University School of Medicine, 660 South Euclid Avenue, St. Louis, MO, 63110-1093, USA
e-mail: miller@pathology.wustl.edu

D. McGavern and M. Dustin (eds.), *Visualizing Immunity,* Current Topics in Microbiology and Immunology 334,
DOI: 10.1007/978-3-540-93864-4_9,

cell invasion and pathogenesis. The purpose of this chapter is to provide a general rationale for why *in vivo* imaging is important, provide an overview of various techniques highlighting the strengths and weaknesses of each, and provide examples of how various imaging techniques have been used to study *Listeria* infection. Lastly, our goal is to make the reader aware of the tremendous potential these approaches hold for studying host–pathogen interactions.

1 Introduction

Major advances in the fields of bacterial pathogenesis and immunology have been driven by the adoption of new technologies such as molecular biology, biochemistry, and X-ray crystallography. Microscopy of *in vitro* cultured cells and fixed tissue specimens has led to many fundamental discoveries in the fields of immunology and microbiology. *In vitro* approaches permit the study of individual cells and bacteria in a highly controlled environment, but the artificial conditions under which these cells and tissues are imaged may raise concerns about the physiological relevance of the observations. On the other hand, histological examination allows one to observe cells in their native tissue environment, but tissue fixation and staining also have associated caveats and, more importantly, this approach only allows one to glimpse a snapshot in time (i.e., information regarding cell dynamics is lost). The convergence of technical advances in manipulating mammalian and bacterial genetics and advances in imaging technology have lead to a watershed moment in the study of host–pathogen interactions. For the first time, the molecular mechanisms and cellular dynamics that determine the outcome of an infection can be studied as they unfold in real-time in a living animal.

The host anatomy plays an important role in immune responses to pathogens. Understanding how infections unfold in tissues such as the spleen, liver, lung, and gut, and how the diverse mixture of resident cells in theses tissues work together to combat infection, has important implications for improving our understanding of microbial pathogenesis and guiding the development of more effective vaccines and therapeutics to target bacterial infections. For example, clinical observations and experimental animal models implicate the spleen as an important site for host immune responses to bacteria, yet our understanding of how the spleen's unique structure and various resident myeloid and lymphoid cells work together to combat infection remains poorly described in terms of cell dynamics. While biochemical and genetic studies provide insight into molecular mechanisms of microbial pathogenesis, *in vivo* imaging approaches are crucial for understanding the cell dynamics that underlie host immune responses and the impact of tissue anatomy on host–pathogen interactions. Understanding host cell dynamics during an infection with a model pathogen such as *Listeria monocytogenes* will enhance our fundamental knowledge of how the host responds to bacterial infection and provide a framework for future studies of other more virulent human pathogens.

2 *Listeria* Infection in Mice as a Model of Host Pathogen Interactions

Listeria monocytogenes (*Listeria*) is a Gram-positive facultative intracellular bacterium and a common cause of food poisoning (Cossart and Toledo-Arana 2008). In healthy individuals, *Listeria* infection typically causes acute vomiting and diarrhea with most patients making a full recovery in 7–10 days. However, *Listeria* infection can present a significant risk to immune-compromised individuals and can trigger miscarriage in pregnant women. *Listeria* is commonly used to study host immune responses to bacterial infection in mice (Pamer 2004, p. 199). Infecting mice with *Listeria* orally is inefficient since the virulence factors InlA and InlB, which normally bind to E-cadherin and facilitate invasion of the human intestinal epithelium, do not bind the murine E-cadherin (Lecuit 2005). Therefore, in the laboratory setting, mice are most often infected by injecting bacteria i.v. or intraperitoneally (i.p.), routes by which bacteria rapidly enter the circulation and infect the spleen and liver.

The popularity of *Listeria* infection as an experimental model system is in large part due to its many practical advantages: the bacteria are easy to grow in culture, they are amenable to genetic manipulation, and *Listeria* infections in mice are highly reproducible. The ability to engineer mutations in *Listeria* and express model antigens such as Ova (Foulds et al., 2002) and fluorescent labels such as GFP (green fluorescent protein) for visualizing bacteria have greatly facilitated the study of innate (Rogers and Unanue 1993; Unanue 1997) and adaptive immunity (Busch and Pamer 1999; Busch et al., 1998; Harty et al., 2000; Mackaness 1962; Pamer 2004).

3 Innate Immune Responses to *Listeria*

After intraveneous (i.v.) inoculation, *Listeria* is initially trapped by cells in the marginal zone (MZ) of the spleen (Aichele et al. 2003; Conlan 1996; Muraille et al. 2005), but which host cells play a dominant role in trapping *Listeria* is controversial (Jablonska et al. 2007; Neuenhahn et al. 2006). Once inside host cells, *Listeria* can escape from the phagosome by secreting the virulence factor listeriolysin O (LLO) (Dramsi and Cossart 2002; Glomski et al. 2003; Michel et al. 1990). The entry of *Listeria* into the cytosol triggers inflammatory responses through an NF-κB-dependent mechanism (McCaffrey et al. 2004; O'Connell et al. 2005; O'Riordan et al. 2002). MyD88 knockout mice are highly susceptible to *Listeria* infection, implicating toll-like receptor (TLR) signaling in mediating early host responses (Edelson and Unanue 2002; Seki et al. 2002). By 24 h, *Listeria* enters the periarteriolar lymphiod sheath (PALS) (Aichele et al. 2003; Conlan 1996; Muraille et al. 2005) where it proliferates for several days, before its numbers begin to decline and bacteria are eventually cleared.

In the spleen, bacteria encounter various tissue resident phagocytes including macrophages, dendritic cells (DCs), and neutrophils (Kraal 1992; Mebius and

Kraal 2005). The contribution of each phagocyte type to the capture and control of bacteria is likely dependent on the pathogen itself and the route of infection; however, a comprehensive picture of these processes is lacking.

4 Adaptive Immunity to *Listeria*

Immune responses to *Listeria* have been extensively characterized in mice. *Listeria* reproducibly infects the liver and spleen where it persists for several days before being cleared by *Listeria*-specific T cell responses. Mice that survive *Listeria* challenge develop long-lasting CD8 T cell immunity (Busch and Pamer 1999; Busch et al. 1998; Lara-Tejero and Pamer 2004; Lauvau et al. 2001; Shedlock and Shen 2003; Sun and Bevan 2003). While CD8 T cells provide robust protection to lethal challenge (Pamer 2004), CD4 T cell responses appear to be required for generating long-lived CD8 memory (Shedlock and Shen 2003; Sun and Bevan 2003). Although, splenectomy increases initial resistance to *Listeria* infection, asplenic mice are susceptible to secondary challenge, suggesting that the spleen is a critical site for the development of adaptive immunity (Skamene and Chayasirisobhon 1977). CD8 T cell activation kinetics are surprisingly transient in the spleen (Mercado et al. 2000); priming peaks within the first day after *Listeria* infection and then sharply decreases by 72 h.

In contrast to immunization with virulent *Listeria*, heat-killed and LLO-deficient *Listeria* provide only partial protection to challenge (Glomski et al. 2003; Lauvau et al. 2001). *Listeria* mutants that are unable to replicate *in vivo* have shown promise as vaccine vectors (Brockstedt et al. 2005, 2004) as these organisms, while attenuated, still retain their antigenicity and elicit superior cell-mediated immune responses. More recently, Datta et al. reported that irradiated *Listeria* efficiently stimulated antigen-specific CD4 T cells and primed CD8 T cells by cross-presentation (Datta et al. 2006).

The importance of DCs in promoting adaptive immunity to *Listeria* was shown by experiments using low-dose clodronate-liposome treatment to selectively deplete MZM (but not DCs) (Aichele et al. 2003) and more formally demonstrated using the CD11c-DTR (CD11c-driven expression of diptheria toxin receptor) system to deplete DCs (Jung et al. 2002; Muraille et al. 2005; Neuenhahn et al. 2006). However, this approach has been shown by others to deplete MZ macrophages as well as DCs (Probst et al. 2005), making it difficult to exclude a role for macrophages in antigen presentation. Moreover, clodronate-liposome treatment can deplete DCs in the MZ and PALS at 48 h after treatment (Aoshi et al. 2008), making this approach problematic as well for elucidating the individual roles of macrophages and DCs in antigen presentation.

The impact of the tissue architecture on T cell priming is an area of great interest (Khanna et al. 2007), but further study is needed to determine how the local tissue microenvironment and tissue-specific antigen presenting cells (APCs) influence T cell priming. Kursar and coworkers suggested that the increased susceptibility to

Listeria infection of CCR7-deficient mice might be due to the improper colocalization of T cells and DCs (Kursar et al. 2005). Furthermore, Muraille et al. suggested that the entry of *Listeria* to the PALS could be required for efficient antigen presentation to T cells (Muraille et al. 2005). Recently, it was reported that antigen presentation to CD8 T cells was initiated by the entry of *Listeria* into the PALS in the spleen, suggesting that T cell priming occurs in specific microcompartments of the spleen (Aoshi et al. 2008).

5 *In Vivo* Bioluminescence Imaging Studies of *Listeria* Infection

Bioluminescence imaging (BLI) is a noninvasive technique that allows real-time *in vivo* imaging of the whole living animal (Contag and Bachmann 2002; Contag and Ross 2002; Doyle et al. 2004). The greatest advantage of BLI is that it allows one to study the location, dissemination, and distribution of an infectious agent throughout the body over the course of an infection (Doyle et al. 2004; Hardy et al. 2004, 2006). In recent years, this method has been used to study a variety of pathogens, such as *Salmonella enterica* serovar Typhimurium (Burns-Guydish et al. 2005; Beeston and Surette 2002; Contag et al. 1995), *Staphylococcus aureus* (Francis et al. 2000), *Streptococcus pneumoniae* (Francis et al. 2001), and *Listeria* (Hardy et al. 2004, 2006; Riedel et al. 2007). BLI consists of a cooled charge-coupled device (CCD) camera, capable of imaging very low levels of visible light emitted from internal body organs of small animals, a dark box in which the animal is placed during imaging, and the software that serves to operate the system and analyze the collected data (Contag and Bachmann 2002; Contag and Ross 2002).

Bioluminescence in itself is a light-emitting chemical process that naturally occurs in many invertebrate organisms and serves a variety of purposes, such as locating food sources or potential mates (Hastings 1996; Hastings and Johnson 2003; Wilson and Hastings 1998). During a bioluminescent reaction, a luciferase enzyme in the presence of oxygen catalyzes the conversion of a substrate (commonly termed luciferin) into a luciferase-bound peroxy-luciferin intermediate that releases photons of visible light with emission spectra between 400 and 620 nm (Doyle et al. 2004; Hastings 1996). Therefore, for BLI *in vivo*, the presence of oxygen in the tissues is critical. Several bioluminescent reporter proteins have been chemically characterized and used for BLI. These include firefly luciferase, click beetle luciferase, renilla luciferase, and the photorhabdus luminescens luciferase (de Wet et al. 1985; Ruby and Nealson 1976; Wilson and Hastings 1998). Bioluminescent reporters are commonly incorporated into genomes of animals, bacteria, or cell lines, and can be used to simply mark cells or tissues, and to study gene expression, protein–protein or host–pathogen interactions (Contag and Bachmann 2002; Contag et al. 1995; Contag and Ross 2002). In applications where luciferases are incorporated into the genome of an animal, a drawback is that a substrate needs to be injected into the animal for the bioluminescent reaction to occur.

However, bacterial luciferase reporters in which the substrate and the luciferase are expressed simultaneously are very useful, although only about 10% of the emitted light is above 600 nm. An advantage of these reporters is that there is no need to inject a substrate into the animal for imaging.

In vivo BLI, sensitivity depends on the strength of the generated signal (type of reporter used), type of tissue being imaged, and the depth of the light source within an animal (Doyle et al. 2004). All mammalian tissues absorb or scatter light as it passes through them. Short wavelength light (blue or green) is scattered more, while red light is scattered less (e.g., hemoglobin absorbs most of the light emitted within the blue–green, but not the red, spectrum) (Doyle et al. 2004). This property makes the red light very desirable for BLI applications, and discovery or development of reporters with this emission spectrum would greatly expand the applications of the BLI technique. For example, renilla luciferase emits light at wavelengths less than 600 nm, while only 10% of the light from bacterial luciferase is above 600 nm; consequently neither luciferase efficiently transmit light from deep tissues (Rice et al. 2001). By comparison, about 30% of the emission spectrum of firefly luciferase falls above 600 nm, making this reporter a better choice for deep tissue imaging (Doyle et al. 2004; Rice et al. 2001).

Although *Listeria* has been one of the most studied bacterial pathogens and its interactions with the host rather well defined, introduction of BLI led to novel discoveries regarding *Listeria* pathogenesis (Hardy et al. 2004, 2006). Using sequential analysis of the whole mouse body over time, Hardy and coworkers found that bioluminescent *Listeria* can colonize and replicate in the gall bladder of infected mice, whether mice were inoculated with a very high dose or a low dose of *Listeria* (Hardy et al. 2004). This finding raised new questions regarding previously uncharacterized interactions between *Listeria* and its host such as the possibility that asymptomatic hosts may harbor *Listeria* and may contribute to the spread of *Listeria* to other hosts. By analyzing dissected tissues, the authors were able to identify the gall bladder lumen as the location where *Listeria* replicates extracellularly (Hardy et al. 2006). From this location, *Listeria* would be secreted into the intestinal tract to either reinfect the same host, or shed into the outside environment to infect other hosts, much like *Salmonella enterica* serovar Typhi. Indeed, using BLI and *Listeria* harboring luciferase, the authors showed that bacteria replicating in the gall bladder are efficiently expelled into the intestines upon stimulation of gall bladder contractions by either injecting cholecystokinin i.p. or by feeding fasted mice a milk meal (Hardy et al. 2006). BLI, in combination with bacterial colony counts from tissues, was critical for detecting these previously unrecognized "hot spots" of bacterial replication (Hardy et al. 2006).

However, for a variety of applications BLI has disadvantages. First, while BLI is useful for showing the spread of an infection throughout the whole animal, it does not provide information about the host–pathogen interactions at a cellular level because the resolution is typically too low to image individual host cells and bacteria. Secondly, although time-lapse BLI allows one to observe the kinetics of infection, it often requires exposure times greater than several minutes making it impractical for investigating dynamic phenomena that occur in the order of seconds to minutes, such as leukocyte recruitment. Moreover, the luciferase gene and its substrate need to be expressed in bacteria, and this entails substantial effort in cloning these genes and optimizing their expression in bacterial vectors. In addition, the expression of reporters can often change the physiology of the bacteria which in some cases can lead to either attenuation or increased virulence of the bacterial strains (Riedel et al. 2007). For example, to introduce the luciferase into *Listeria* via a transposon (Hardy et al. 2004) requires screening of a bank of transposon integrants to identify clones that exhibit the highest luminescence (Riedel et al. 2007). In addition, the high-fluorescing clones that would be selected for BLI applications might exhibit features different from parental strains of interest (Riedel et al. 2007); such was the case in the gall bladder studies where the bioluminescent *Listeria* strain exhibited a four-fold-higher LD_{50} compared to the parental strain 10403S (Hardy et al. 2006). This observation raises questions about whether the insertion of luciferase gene has inactivated other virulence genes of *Listeria* and whether the parental strain would behave similarly in the gall bladder environment.

To enable reproducible labeling of various *Listeria* strains with bioluminescent markers, a method was developed which allows a stable constitutive expression of high levels of luminescence in *Listeria* (Riedel et al. 2007). To monitor *Listeria* gene expression, synthetic lux operon derived from *Photorhabdus luminescens* was cloned into a chromosomal integration vector pPL2lux, that was optimized for use with Gram positive bacteria (Riedel et al. 2007). The advantage of this labeling system is that it allows site-specific integration of the pPL2, constitutive expression of high levels of bioluminescence, it can be used in other *Listeria* strains, and it does not alter the growth rate or the virulence of the bioluminescent strain compared to the parental strain.

BLI has contributed to our understanding of the disease processes in real-time in a living animal. However, the weakness of the generated signals and, most importantly, the weakness of resolution is still one of its main drawbacks. BLI can detect a general region in the body with a high burden of bacteria and, without using other supplemental techniques, it is not powerful enough to conclusively identify the individual organs and much less smaller regions within organs that harbor replicating bacteria. Many new biotechnological advances have and will improve BLI over time, making it an even more powerful approach for *in vivo* imaging.

6 *In Vivo* Fluorescence Imaging Techniques

Fluorescence imaging techniques have gained widespread use due to their high degree of spatial resolution and the ability to simultaneously identify multiple tissue structures, host cells, and pathogens (Day and Schaufele 2008; Schulz and Semmler 2008; Terai and Nagano 2008). Fluorescence occurs when the energy of a photon is absorbed by a fluorophore, thus promoting an electron to an excited state. The electron undergoes nonradiative energy loss before returning to the ground state and releasing the remaining energy as a longer wavelength photon. Fluorescence emission takes place rapidly (femto–nanosecond time scales) and can occur tens of thousands of times before the fluorophore becomes photobleached. Fluorescence can be highly efficient with many common dyes, such as FITC, having quantum yields of better than 90%. Fluorescence microscopy requires that cells and structures of interest be labeled with fluorescent molecules (See Table 1). These labels can take many forms including aromatic dyes, fluorescent proteins, or fluorescent nanocrystals, often called quantum dots. A particularly useful approach is to load the cells of interest with AM-ester fluorescent dyes such as CFSE (Invitrogen). These esterified dyes diffuse rapidly across the cell membrane, but once inside, they are cleaved by intracellular esterases and react with amine or sulfhydryl groups on proteins to form long-lived stable adducts. This approach works well for labeling both host cells and bacteria, but has the disadvantage that, as cells proliferate, the label is diluted making it difficult to tack the cells after several rounds of division.

For long-term cell tracking, host cells can be engineered to express fluorescent proteins. Several mouse models exist in which fluorescent proteins are expressed in myeloid and lymphoid cell lineages. We have used extensively the CD11c-YFP mouse (Lindquist et al. 2004) and the eGFP-MHCII knockin mouse (Boes et al. 2002) to visualize antigen presenting cells during *Listeria* infection. A third model that has been used to study host responses to injury and infection is the fractalkine receptor knock-in mouse (CX3CR1-eGFP) (Geissmann et al. 2003), which labels resident macrophages in peripheral tissues as well as DCs in the gut (Chieppa et al. 2006) and microglia in the brain (Kim and Dustin 2006). Finally, in the LysM-eGFP

Table 1 Excitation and emission wavelengths for fluorescent proteins and dyes commonly used in confocal and two-photon microscopy. The excitation wavelength optimum is given for single photon excitation and a range of useful excitation wavelengths is given for 2P, since 2P excitation is comparatively broad. Q-dots are not listed in the table but they excite well with 2P at a broad range of wavelengths and have narrow emission spectra. Information in this table has been complied from (Shaner et al. 2005) the Invitrogen web site and our own unpublished observations

	1 P excitation optima (nm)	2P excitation range (nm)	Emission (nm)
Fluorescent proteins			
CFP	434	860–890	477
Cerulean	425	860–890	505
C-Pet	425	860–890	505
eGFP	477	850–910	508
eYFP (Venus)	513	850–910	527
Y-Pet	490	890–910	550
DsRed T3	558	890–910	583
tdTomato	535	890–910	615
mStrawberry	550	890–910	630
mCherry	560	890–910	640
Fluorescent dyes			
DAPI	358	750–800	462
CMAC	353	750–800	466
CFSE	492	780–900	517
CMTMR	541	800–820	565
CMTPX	577	890–910	602

knockin mouse, GFP is highly expressed in neutrophils and to a lesser degree in monocytes (Chtanova et al. 2008; Peters et al. 2008; Egen et al. 2008), and therefore this model has great potential to provide insight into numerous diseases since these cell types play central roles in chronic inflammation and host responses to infection.

Fluorescent proteins can also be expressed in bacteria in order to visualize them *in vivo*. Genes encoding fluorescent proteins can be integrated into the bacterial genome via homologous recombination. Such an approach is tedious, time-consuming, and lacks flexibility. Alternatively, expression of foreign antigens by bacteria can be accomplished by introducing into bacteria recombinant plasmids that harbor foreign genes. This approach is very practical and flexible: plasmids can be easily introduced into different strains of bacteria and the amount of heterologous antigen being expressed can be controlled by using plasmids with different origins of replication. For stable plasmid maintenance, antibiotic resistance markers have been used extensively. However, the emergence of antibiotic-resistant bacterial strains has made this option less desirable, especially in recombinant bacteria that are used as vaccine vectors. In addition to being impractical, the use of antibiotic resistance as selective determinants in recombinant bacterial vaccines is prohibited by FDA regulations (Paterson and Johnson 2004).

To circumvent these issues, balanced lethal host–vector systems have been developed for both Gram negative and Gram positive bacteria for stable plasmid maintainance (Curtiss et al. 1989; Paterson and Johnson 2004). In balanced lethal host–vector approach an essential gene of bacteria that confers auxotrophy is deleted from the chromosome. This chromosomal deletion is then complemented by a plasmid that harbors the deleted gene (Curtiss et al. 1989; Paterson and Johnson 2004). Verch and coworkers (Verch et al. 2004) developed a balanced lethal host–vector system in which two genes of *Listeria* that control D-alanine metabolism, namely *dal* and *dat*, were deleted from the chromosome of *Listeria*, imparting a requirement for D-alanine on *Listeria*. *Listeria dal* gene was then introduced into a plasmid that complements D-alanine racemase-deficient *Listeria*, which would die without the plasmid.

The creation of GFP-expressing *Listeria* strains have been reported by several authors (Bubert et al. 1999; Fortinea et al. 2000; Freitag and Jacobs 1999). In many of the early reports, the construction of vectors and visualization of *Listeria* in cultured cells by fluorescence microscopy was described (Bubert et al. 1999; Freitag and Jacobs 1999).

Fortineau and coworkers reported construction of several vectors containing a gene encoding a mutant GFP protein designated GFP-mut1 with red-shifted excitation maxima (Fortinea et al. 2000). Using these vectors, it was possible to detect *Listeria* in cultured cells and in tissue sections of infected mice. In addition, the vector was very stable *in vivo* and did not affect the virulence of *Listeria* in a mouse model of infection (Fortinea et al. 2000). More recently, a fluorescent-labeling system for *Listeria* was reported that allows visualizing of six different *Listeria* strains simultaneously. This system can be particularly useful for comparing the invasion potential of various *Listeria* strains in coinfection studies, as well as discriminating *Listeria* from other bacteria in an unsterile environment (Andersen et al. 2006).

7 Confocal Laser Scanning Microscopy

Confocal laser scanning microscopy (CLSM) has advanced tremendously our understanding of not only cell biology, but also bacterial gene functions and host–pathogen interactions in general. Numerous excellent review articles and books describe the principles and applications of CLSM (Amos and White 2003; Olympus 2008; Pawley 2006; Schulz and Semmler 2008). In CLSM, an objective lens focuses the excitation light to a point in the specimen. The laser is scanned through the specimen using a set of mirrors to deflect the beam and, at each point in the scan, the emitted photons are detected by photomultiplier tubes (PMTs). Localizing the fluorescence emission in 3D space is achieved by placing a pinhole aperture in front of the detector confocal to the sample plane to reject out-of-focus light coming from planes above and below the focal plane. The output from the PMTs is collected and digitized to build an image that is displayed by the computer. Applications of CLSM are far superior to conventional optical microscopy, primarily because they make it possible to obtain high contrast images and provide 3D information within living or fixed tissues. By sequentially advancing the focus of the objective with a z-stepper motor or piezoelectric actuator, a series of optical sections through the sample can be collected. A z-series of optical sections can then be rendered to generate three-dimensional, sharp, subcellular resolution images.

A wide range of fluorescent probes have been used in CLSM applications. DNA intercalating dyes such as DAPI or ethidium bromide can be used to visualize nuclei. Dye-conjugated nucleic acid probes are often used in fluorescence *in situ* hybridization (FISH) approaches for visualizing chromosomes and genetic mapping studies. Fluorochrome-conjugated antibodies are commonly used to label cell surface markers, as well as intracellular structures, organelles, and even proteins. Some fluorescent dyes bind directly to cellular proteins, such as phalloidin which binds F-actin with high specificity, while other probes have been designed to selectively accumulate in various organelles such as lysozomes and mitochondria (LysoTracker and MitoTracker dyes from Invitrogen). In addition to labeling various cell and tissue components, fluorescent probes can be used to monitor dynamic processes in cells or tissues such as changes in pH, intracellular calcium ions reactive oxygen species (ROS), cell membrane potential, cellular integrity, endocytosis, exocytosis, membrane fluidity, protein trafficking, signal transduction, and enzymatic activity.

The fluorophores typically used in CLSM applications have been selected based on their superior quantum yield, Stoke's shift characteristics, inherent photostability and whether or not they can be excited by a common illumination source, for example a Hg arc lamp or Argon ion laser. In CLSM, multiple fluorophores can be used simultaneously, assuming that they have sufficiently different emission spectra to be separated or sufficiently different absorption spectra to be selectively excited. The use of multiple fluorescent probes enables researchers to examine simultaneously several structures or proteins within a single cell or, alternatively, the interactions of different cell types in a tissue.

Although imaging specimens by CLSM does produce images of remarkable clarity and detail, the obtained images are susceptible to artifacts associated with staining and fixing of the specimen or in live cell imaging, because the high power lasers used for excitation can rapidly induce photodamage, leading to altered cell behavior and morphology (Flaberg et al. 2008). To address some of these limitations, extended field laser confocal microscopy (EFLCM) was developed (Flaberg et al. 2008). EFLCM uses a microlens-enhanced Nipkow spinning disc to achieve parallel beam illumination which, when combined with high quantum efficiency CCD camera-based image capture, results in reduced specimen photodamage and increased speed of acquisition (Flaberg et al. 2008).

8 Application of CLSM in Studying *Listeria* Pathogenesis

CLSM has been used widely to study *Listeria* infection in cultured cell or fixed tissue sections. This imaging approach has produced novel findings regarding *Listeria* distribution in host cells and tissues, *Listeria* gene functions, and host–pathogen interactions in general (Braun et al. 1998; Pron et al. 2001; Walch et al. 2007). In contrast, there have been relatively few studies in which CLSM has been used to image *Listeria*–host interactions *in vivo*.

In a series of elegant *in vivo* studies, a combination of intravital confocal microscopy and a novel transgenic mouse model were used to examine the trafficking behavior of monocytes (Auffray et al. 2007). In uninfected mice, CX3CR1^{high} Gr1^{low} monocytes migrated along the lumen of blood vessels, in what was termed a steady-state patrolling behavior, but did not extravasate into the tissue. After peritoneal infection with *Listeria*, Gr1^{low} monocytes rapidly underwent extravasation, peaking at 1–2 h, a time when neutrophil recruitment was just starting and inflammatory monocyte homing was undetectable (CX3CR1^{low} Gr1^{high}). Patrolling behavior appears to be necessary for the efficient extravasation of Gr1^{low} monocytes into tissues, since *Cx3cr1*$^{-/-}$ Gr1^{high} monocytes exhibited decreased crawling and patrolling behavior and had delayed recruitment kinetics. Moreover, Gr1^{low} monocytes were the primary producers of tumor necrosis factor-α (TNFα) early after infection, suggesting that this monocyte subset plays an important role in the initiation of the inflammatory response (Auffray et al. 2007).

Despite the significant limitations of confocal microscopy in terms of tissue viability and imaging depth, it still remains an extremely valuable tool for *in vivo* studies and provides far superior image resolution to BLI.

9 Introduction to Two-Photon Microscopy

Two-photon (2P) microscopy is a submicron resolution imaging technique (Denk et al. 1990) that permits the optical sectioning of diverse tissues including brain, skin, gut, bone, and lymphoid organs. Imaging preparations often involve explanted tissues/organs placed under a flow of warm oxygenated medium to preserve tissue

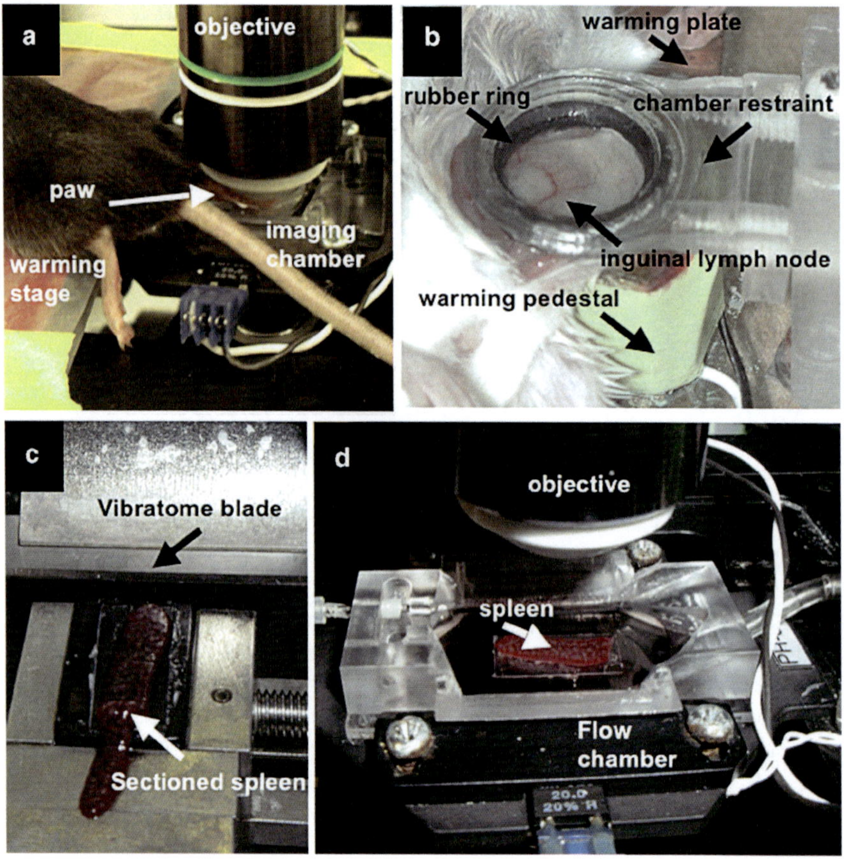

Fig. 1 2P imaging preparations for studying *Listeria* infection. (**a**) A noninvasive imaging of infected footpad. (**b**) An intravital preparation for studying antigen presentation in the inguinal lymph node. (**c**) The explanted spleen is sectioned with a Vibratome (Pelco) to expose the marginal zone and white pulp regions. (**d**) The spleen is then placed in an imaging chamber and submerged under the flow of warm oxygenated medium for imaging infection *in situ*

viability during imaging, or alternatively, imaging can be performed *in vivo* in anesthetized mice (Fig. 1). 2P imaging requires that specific cell populations are fluorescently labeled. This can be accomplished with vital dyes such as the CellTracker probes (Invitrogen) or by using transgenic mice that express fluorescent protein (GFP, YFP, or CFP). In some cases, tissues and cells give off intrinsic signals, such as the second harmonic generation signal produced by collagen (Gauderon et al. 2001) or the auto-fluorescence given off by macrophages. In combination with an appropriate fluorescence reporter, 2P microscopy can serve as an *in situ* assay for many immune cell functions including migration and chemotaxis (Castellino et al. 2006; Okada et al. 2005; Witt et al. 2005; Zinselmeyer et al. 2008), antigen presentation (Bousso et al. 2002; Bousso and Robey 2003; Lindquist et al. 2004; Mempel et al. 2004; Miller et al. 2004a, b, 2002; Shakhar et al. 2005; Witt et al. 2005;

Zinselmeyer et al. 2005), cell signaling (Bhakta et al. 2005), and T cell effector functions (Boissonnas et al. 2007; Kawakami et al. 2005; Mrass et al. 2006).

2P microscopy works by exciting a fluorophore with two longer-wavelength lower-energy photons, instead of a single high-energy photon as is the case in standard fluorescent approaches. The two photons are delivered by a femtosecond-pulsed Ti:Sapphire laser and absorbed nearly simultaneously to stimulate fluorescence in a diffraction-limited focal volume (<1 μm). This laser "spot" is then scanned rapidly through the sample and the fluorescence emission is collected by PMTs to build the digital image pixel by pixel. In a typical experiment, an image z-stack is acquired repeatedly from the sample to generate a time-lapse data set. These multidimensional data sets are then rendered and quantitatively analyzed for cell velocity, colocalization, cell shape, volume, number, intensity, and color. Although this approach is computationally intensive and time consuming, it provides high-resolution single-cell spatiotemporal information that other imaging techniques cannot. The main technical advantages of 2P microscopy are: (1) significantly less sample photodamage than confocal microscopy; (2) increased detection sensitivity because even scattered photons emitted by the sample contribute to the overall signal; and (3) images can be acquired several hundred microns deep in scattering tissues (compared to <80μ for standard confocal microscopes) (Table 2).

10 2P Microscope Design

In the last several years, 2P microscopes have become commercially available. The advantages of these turn-key systems are that they are installed for the user and service and maintenance are provided by the vendor. Moreover, these systems are designed to support a number of different imaging applications and often come with extensive and polished acquisition software and in some cases basic analysis software. Some researchers have chosen to build their own systems from commercially available components. The advantage of this approach is that the system can be tailored to specific applications, such as intravital imaging, and optimized for sensitivity. Moreover, once the system is built, the investigator is in the position of being able to further modify the equipment for new applications or quickly repair the system as needed, without having to rely on the microscope vendor for service, which can at times be frustratingly slow to arrange. The 2P microscope built in our laboratory is based on Ian Parker's prototype instrument at UC Irvine (Nguyen et al. 2001); however, our system incorporates several significant design improvements. First, the system is capable of capturing up to four fluorescent channels simultaneously at video-rate (30 frames s^{-1}) acquisition speeds (commercial systems typically have two acquisition channels and operate at ~2 frames s^{-1}). A fast scanning speed is especially useful when analyzing phenomena such as lymphocyte homing from the circulation or DCs dendrite dynamics that are difficult to detect with slower systems. A second novel feature of our system is that it is equipped with two

Table 2 A comparison of *in vivo* imaging techniques used to study Listeria infection

In vivo imaging technique	Spatial resolution	Time resolution	Pros	Cons
Biolminescence imaging	Typically >mm in x,y No z information	minutes	No photodamage Noninvasive Whole animal imaging Longitudinal studies	Luminescence is dependent on O_2 Typically one probe is used at a time Low spatial temporal resolution No optical sectioning possible
Confocal microscopy	<1 μm in x,y,z	500 ms with typical scan head <33 ms with resonant mirror or spinning disk scanning	Superior 3D resolution Multiple probes can be excited independently or simultaneously	High photodamage Rapid photobleaching Limited depth of imaging <80 μ m Surgery required for reaching deep tissues
Two-photon microscopy	<1 μm in x,y ~1.5 μm in z	500 ms with typical scan head <33 ms with resonant mirror scanning	Lower photodamage than confocal Deep tissue imaging up to 500 μ m depending on the tissue Scattered photons are detected Improving sensitivity Noninvasive longitudinal studies are possible depending on the prep Broad excitation allows multiple fluorophores to be excited with a single laser UV dyes can be excited without special UV- transparent optics	Systems are expensive to set up and maintain Z-resolution is less than confocal Broad excitation characteristics make it difficult to excite fluorescent proteins independently

tunable Ti:Sapphire lasers, which permits fluorescent probes with different optimal excitation wavelengths to be used together; for example, YFP and CMAC, which excite optimally at 900 nm and 770 nm, respectively. This makes it possible to use complex mixtures of cells that can not be imaged with other 2P microscopes and run side-by-side internal controls (i.e., cotransfer nonantigen specific T cells when studying antigen-specific T cell responses). Recently, we have upgraded our system with more efficient optical filters (SemRock) and have added a pulse compensation unit (FemtoControl, Coherent). These modifications have substantially improved detection sensitivity and the depth of tissue imaging. Our laboratory has worked with A&B software to develop a custom acquisition platform (ImageWarp, in collaboration with Boris Nalibotski at A&B software) to perform on-the-fly image correction and control hardware components, such as the z-focus motor and laser shutters. ImageWarp allows multidimensional data files (time and three spatial dimensions) to be streamed to our laboratory server (X-Serve with 3.75 Tb capacity RAID5 array; Apple), where the data can be immediately accessed by computer workstations running Volocity (Improvision) and Imaris (Bitplane) software for rendering and analysis.

11 Live Tissue Imaging and *In Vivo* Cell Dynamics

2P microscopy has been broadly adopted by immunologists (Germain et al. 2006) to study a wide range of immunological phenomena including positive selection (Bhakta et al. 2005; Bousso et al. 2002; Bousso and Robey 2004; Witt et al. 2005), T cell priming (Bousso and Robey 2003; Lindquist et al. 2004; Mempel et al. 2004; Miller et al. 2004b; Shakhar et al. 2005), immunoregulation and tolerance (Schneider et al. 2006; Tadokoro et al. 2006; Tang et al. 2006; Zinselmeyer et al. 2005), autoimmunity (Kawakami et al. 2005; Nitsch et al. 2004), infection dynamics (Chieppa et al. 2006), and tumor rejection (Boissonnas et al. 2007; Mrass et al. 2006). The value of this approach is that it provides single-cell spatiotemporal information in intact living tissues. However, as the technology develops and new transgenic mice and fluorescent probes become available, 2P microscopy may commonly be used to obtain functional readouts of immune cell function *in vivo* including calcium signaling (Bhakta and Lewis 2005; Bhakta et al. 2005), gene expression, cell proliferation (Miller et al. 2002), chemotaxis (Beuneu et al. 2006; Castellino et al. 2006; Hugues et al. 2006; Okada et al. 2005), cytokine secretion, apoptosis, and CTL killing (Boissonnas et al. 2007; Mempel et al. 2006).

Several 2P studies have addressed the dynamics of T cell priming (Bousso and Robey 2003; Miller et al. 2004b, 2002; Shakhar et al. 2005; Zinselmeyer et al. 2005). In the absence of cognate antigen, naive CD4 T cells approach DCs along random trajectories, and CD4 T cell – DC interactions (several thousand per hour) occur as a natural consequence of T cell motility (Miller et al. 2003) and vigorous dendrite movement (Miller et al. 2004a, 2003). In the case of CD8 T cells, two published studies observed random T cell migration (Bousso and Robey 2003;

Mempel et al. 2004) while a more recent report showed that licensed DCs enhance the recruitment of CD8 T cells by chemotaxis (Castellino et al. 2006).

When T cells recognize cognate antigen, they display stereotypical changes in their behavior and their interactions with DCs. Early antigen recognition events occur typically 'at arms length' on DC dendrites, they are short-lived, and frequently involve serial contacts with multiple DCs (Miller et al. 2004b). At later times, T cell–DC contact durations increase and T cells form stable clusters. These clusters are observed 8–12 h before cells enter a second dynamic interaction stage (Miller et al. 2004b). This late stage of activation is characterized by an increase in T cell motility and a decrease in the duration of T cell–DC contacts. The activation behavior of antigen-specific CD8 and CD4 T cells appears similar (Bousso and Robey 2003; Mempel et al. 2004), suggesting that 2P can be used as a robust *in situ* readout of active antigen presentation. Contact between lymphocytes and APCs is observed in various forms depending on the APC type, activation state, and the activation state of the T cell (Lin et al. 2005). While long-lived T cell contacts are associated with antigen recognition, dynamic serial interactions that last on the order of minutes and interactions that were detected only as a decrease in cell velocity have been reported (Shakhar et al. 2005; Zinselmeyer et al. 2005). Sustained physical contact between cells implies molecular recognition at some level. However, cell clustering has been associated with both priming and tolerization, and it is therefore important to interpret T cell behaviors along with other measures of function, such as cell proliferation, CTL-killing, or delayed type hypersensitivity.

2P microscopy has also been used to examine cognate T cell–B cell interactions following antigen challenge (Okada et al. 2005). It was shown by 2P imaging that antigen-engaged CCR7-sufficient B cells (but not cotransferred CCR7-deficient cells) moved with a strong directional bias toward the edge of the B cell follicle, where they paired with cognate CD4 helper T cells. More recently, 2P microscopy was used to reveal the dynamics of T cell help and B cell trafficking during B cell selection and maturation in the germinal centers (Allen et al. 2007; Hauser et al. 2007; Schwickert et al. 2007).

The next wave of 2P imaging applications will be to study host immune responses to a wide array of pathogens. Recently, several groups have used the technique to investigate the host response to infection with parasites (Chtanova et al. 2008; Peters et al. 2008), virus like particles (Junt et al. 2007), and bacteria (Chieppa et al. 2006; Egen et al. 2008; Zinselmeyer et al. 2008). *Listeria* is a highly prolific model of bacterial infection and in the future 2P imaging may uncover many fundamental details regarding both the early host–pathogen interactions during the innate immune response and the interactions that trigger and regulate the adaptive immune response.

12 2P Imaging Data Analysis

The application of 2P microscopy to the study cell dynamics in lymphoid (and peripheral tissues) has made it possible to quantitatively describe basic parameters of leukocyte behavior *in vivo*. Standard measures often include cell displacement plots to determine whether cell migration is random or has a directional bias. The displacement plot is graphed as a mean cell displacement versus time^2 or more formally as displacement^2 versus time. The slope of a regression line fitted to the data in this plot can then be used to calculate a motility coefficient previously described as $M = X^2/4t$ (Miller et al. 2002) for 2D migration and $M = X^2/6t$ for 3D migration, where X equals displacement in μm and t equals time in minutes. Other basic measures include median instantaneous velocity, shape index (the ratio of cell length to width), meandering or confinement ratios (describes the straightness of a cell's path), and colocalization, which allows contact durations between cells to be measured.

13 2P Microscopy of *Listeria* Infection

Recruitment of leukocytes to the sites of inflammation is an essential host defense mechanism and a prominent feature of autoimmune diseases such as arthritis. Although a large body of work describes this process, there is little quantitative data describing the dynamics of leukocyte recruitment and homing *in vivo*. Zinselmeyer et al. developed a noninvasive 2P microscopy approach to study leukocyte homing in the mouse footpad under steady state conditions or after subcutaneous (s.c.) injection of *Listeria* (Zinselmeyer et al. 2008). Their results reveal that in the absence of inflammation, neutrophils move within blood vessels with a speed of several hundred μm s^{-1} and only rarely adhere to the endothelium or entered the tissue parenchyma. In contrast, when *Listeria* was injected into the tissue, neutrophils exhibited slow movement (5–60 μm min^{-1}), similar to the rolling and tethering behavior described for monocytes (Auffray et al. 2007). Within 15 min of neutrophil transfer, many neutrophils had extravasated and were crawling within tissues with velocities of ~7–8 μm min^{-1}. Moreover, the neutrophil trajectories showed a striking directional bias towards foci of infection, which is indicative of chemotaxis.

In another study, Aoshi et al. used 2P microscopy and the adoptive transfer of *Listeria*-specific T cells to study antigen presentation in the spleen in during *Listeria* infection (Aoshi et al. 2008). Previous reports (Conlan 1996; Muraille et al. 2005) noted a striking change in *Listeria* localization from MZ to PALS over the first 24 h of infection. Muraille et al. showed that ActA-deficient *Listeria*, in the presence of gentamicin, were capable of infecting the PALS and argued that *Listeria* enters the PALS by means of host cells (Muraille et al. 2005). To test this hypothesis, mice were treated with 500 ng pertussis toxin (PTx, Sigma) before *Listeria* challenge (Aoshi et al. 2008). PTx blocks Gαi protein-mediated chemokine receptor signaling and inhibits leukocyte trafficking (Ato et al. 2004; Huang et al. 2007; Itano et al. 2003; Okada and Cyster 2007). PTx treatment did not change the initial tissue distribution or the amount of *Listeria* captured in the MZ, yet it profoundly inhibited *Listeria* entry into the PALS 24 h after challenge.

In order to determine which cells are crucial for *Listeria* invasion of the PALS, the different recovery kinetics of splenic macrophages and DCs following clodronate-liposome depletion (van Rooijen et al. 1989) were exploited to look for an association between host cell recovery and *Listeria* infection of the PALS. When both macrophages and CD11c cells were depleted, the vast majority of bacteria were found in the MZ and few bacteria were present in the PALS. In mice infected after CD11c$^+$ cells had recovered but macrophages had not, widespread *Listeria* growth in the PALS was observed, suggesting that DCs are important for the progression of *Listeria* infection to the PALS.

14 Using 2P Microscopy to Assess the Presentation of Bacterial Antigen *In Situ*

The precise location of T cell priming in the spleen is unknown. T cells are most abundant in the PALS; however, they are also found in the red pulp and MZ and antigen presentation could occur in these compartments as well. Based on the results of clodronate and CD11c–DTR studies, we speculated that the migration of infected DCs to the PALS could trigger antigen presentation to T cells.

At various times after infection, time-lapse 2P microscopy (Germain et al. 2006) was used to assess antigen presentation. This approach relied on dye-labeled *Listeria*-specific CD8 T cells from WP11.12 TCR transgenic mice, (which express a transgenic TCR that recognizes p60$_{449–457}$ peptide in H-2K^d) (Mercado et al. 2000; Wong and Pamer 2003) to detect antigen presentation *in situ*. Antigen-specific and polyclonal control T cells were adoptively transferred into BALB/c mice and allowed to home before mice were infected with *Listeria*. These experiments employed a simple spleen explant preparation. Due to the optical properties of the spleen, whole organ imaging is limited to <150 μm below the surface (unpublished observations; Odoardi et al. 2007). With this limited imaging depth, it is difficult to image the MZ and PALS. One solution is to remove the spleen, secure it to a plastic cover slip with VetBond (3M) and cut away ~400–800 μm of the overlying red pulp using a Vibratome (Pelco) to expose the deeper MZ and white pulp areas for imaging. Explanted spleen sections can then be placed in a flow chamber maintained at 37°C by perfusion with warm, high glucose DMEM bubbled with a mixture of 95% O_2 and 5% CO_2. This preparation is stable for several hours and permits time-lapse imaging of the cell dynamics in the MZ and white pulp. While there are significant caveats to the explant approach since the tissue has been physically damaged by sectioning and because the blood flow and recruitment of circulating cells has been disrupted, this preparation provides a unique opportunity to study cell dynamics in the spleen.

It was observed by 2P imaging that, in uninfected BALB/c mice, WP11.12 T cells moved randomly in the PALS with a median track velocity of ~5 μm min^{-1} and a motility coefficient of ~60 μm^2 min^{-1}, similar to polyclonal CD8 T cells (Aoshi et al. 2008). In the absence of infection, T cells migrated randomly and did not form clusters, in good agreement with behavior observed in lymph nodes (Miller et al. 2002). However, median T cell velocities measured in spleen explants were lower than mean values published for T cell migration in lymph nodes of ~12 mm min^{-1} (Miller et al. 2002). The most likely explanation for this is that the decreased frame rate used in the study (~2 images min^{-1} vs 4–6 images min^{-1} in previous studies) smoothes the characteristically large fluctuations in T cell velocity thus yielding a lower median velocity, since the median is sensitive to the velocity range.

In mice infected with *Listeria*, adoptively transferred *Listeria*-specific WP11.12 T cells showed significantly reduced velocity and motility (Aoshi et al. 2008). WP11.12 T cells formed large aggregates in the PALS, containing approximately 15 cells per cluster. The velocity and motility of polyclonal CD8 T cells remained unchanged with *Listeria* infection, indicating that cluster formation was antigen-specific.

When mice were injected i.p. with 500 ng of PTx (Sigma) 1 day before infection, transport of *Listeria* to the PALS was inhibited and bacteria-specific T cell clustering was completely absent (Aoshi et al. 2008). The lack of clustering is not due to the inhibition of T cell motility, since adoptively transferred polyclonal T cells moved with comparable velocity and motility in PTx pretreated and untreated mice. As a control for PTx effects on T cells, we administered PTx after *Listeria* entered the PALS (10 h after infection). Once the transport of *Listeria* had occurred, PTx treatment failed to inhibit WP11.12 cluster formation. As a second indicator of antigen presentation, we measured WP11.12 T cell proliferation by CFSE dilution assay. PTx pretreatment reproducibly decreased the number of proliferating WP11.12 T cells compared to untreated controls by ~40% but did not completely block it.

15 Single-Cell Tracking of MZ Macrophages and PALS DCs

Host phagocytes in the spleen have also been imaged (Aoshi et al. 2008) using the CD11c-YFP transgenic mouse (Lindquist et al. 2004) (gift of M. Nussenzweig). As previously reported, this mouse expresses high levels of eYFP (Venus) in CD11c^{+} DCs. By immunofluorescence, YFP expression was also found in MARCO^{+} and ER-TR9^{+} MZ macrophages in the spleen. These cells are not recovered efficiently by collagenase digestion and therefore are not detected by flowcytometry. The CD11c-YFP transgenic mouse offers a outstanding opportunity to analyze both DC and macrophage behaviors *in situ* with 2P imaging.

Splenic macrophages and DCs both express YFP, but MZ macrophages can be distinguished from DCs by injecting fluorescent beads before imaging. In proof of principle experiments, we stained cryosections with antibodies to MARCO and CD11c and showed that macrophages ingest dozens of fluorescent beads whereas DCs take up only a few (Aoshi et al. 2008). Also, MZ macrophages have a more compact morphology and, therefore, based on size, shape, and differences in bead uptake, macrophages can be readily distinguished from DCs in 2P images. Injected beads also provide a convenient MZ landmark and adoptively transferred dye-labeled T cells and B cells can be used to demarcate their respective regions in the white pulp. Moreover, the 2P laser produces a second harmonic generation signal in the MZ and bridging channel (presumably from collagen fibers), providing an additional point of orientation.

The motility of YFP^{+} cells in the MZ and PALS was analyzed by tracking individual cells in the absence of infection. CD11c-YFP mice were injected with red beads and, 30 min later, spleens were harvested and prepared for 2P microscopy.

Spleen explants were imaged every 30 s for ~30 min and individual cells were tracked in time-lapse movies. The analysis focused on YFP^+ bead-containing macrophages in the MZ and YFP^+ DCs in the PALS. Under steady state conditions, MZ macrophages rapidly extended multiple short pseudopodia, but were essentially nonmotile at the time-scale of the imaging experiments (median velocity <2 μm min^{-1}). In contrast, YFP^+ cells in the PALS displayed greater changes in membrane shape and had significantly greater cell displacement and median instantaneous velocities (ranging from 3 to 6 μm min^{-1} in different experiments).

Although the explanted spleen preparation (Aoshi et al. 2008) is a significant improvement over existing *in vitro* models, it is possible that the loss of blood flow and innervation alter host cell behavior such as phagocytosis or chemotaxis. This was also an initial criticism of explanted lymph node imaging studies (von Andrian 2002). However, T cell migration and priming in lymph node explants was shown to be remarkably similar to cell behaviors observed in intravital preparations (Mempel et al. 2004; Miller et al. 2003; Shakhar et al. 2005). The biggest concern is that sectioning the spleen damages the tissue and alters cell behavior. One precaution that minimizes this problem is to image deeper white pulp regions that appear to be intact based on the presence of a "cap" of undamaged MZ macrophages.

Because of the various caveats associated with sectioning the spleen, developing an intravital approach to image the MZ and T cell areas in living mice is a high priority. Recently, intravital 2P microscopy was used to study T cell activation in the spleen, but this was done in superficial regions 80–100 μm below the surface and presumably in the red pulp (Odoardi et al. 2007). Achieving sufficient imaging depth to acquire pictures of the MZ and PALS regions has been a significant technical challenge. In preliminary experiments, we found that using a pulse compressor (FemtoControl, Coherent) to correct for 2P pulse broadening and increase peak laser power was highly beneficial for increasing the depth of imaging in the spleen. We expect it will soon be possible to image spleen at depths of ~200–250 μm using a combination of pulse compensation and more efficient optics (SemRock) to enhance detection sensitivity. However, explanted preparations will still be useful since intravital 2P imaging of the spleen is likely to be more limited in terms of the duration of imaging and the amount of tissue available for examination. Moreover, intravital 2P imaging has a higher animal pain category designation, compared to the explant preparation, and therefore explant imaging should be used as a primary approach and a limited number of intravital 2P studies should be performed to confirm the results obtained from tissue explants.

16 2P Microscopy can Detect Individual Bacteria in Infected Host Cells

In unpublished work, we tested the feasibility of detecting individual bacteria in the spleens of CD11c-YFP mice. *Listeria* was visualized using the GFP-labeled *Listeria* strain (Fortinea et al. 2000). GFP (em = 508 nm) is distinguished from YFP (em = 527 nm) by using a dichroic filter at 505 nm; *Listeria*-GFP signal falls between the blue and green channels, hence its cyan color, and YFP signal falls primarily in the green channel (Fig. 2). At 15 h postchallenge, GFP-*Listeria* were clearly visible as bright cyan spots within YFP^+ cells of varying brightness (Fig. 2, see arrows). This strain expresses constitutively high levels of GFP and when these bacteria are reisolated at day 3 from infected spleens, nearly 80% express GFP (data not shown), demonstrating that the plasmid is retained remarkably well *in vivo*.

In our hands, the pNF8-containing strain is attenuated compared to the parental EGD strain, but shows similar growth kinetics early during infection. Because the LD_{50} for GFP-expressing strains are 10–50-fold higher than the LD_{50} for EGD, the typical doses of fluorescent bacteria used in our imaging experiments (10^5–10^6) are not lethal. This is an important point since one of the main limitations of imaging approaches for studying infection is detecting a relatively rare population of bacteria in a large volume of tissue. Mice are typically inoculated with doses well above the LD_{50} to facilitate the detection of bacteria. This raises the concern that high doses of bacteria might overwhelm the capture capacity of the spleen or liver and alter the initial host cell distribution. In this regard, 2P imaging may have significant advantages for detecting low doses of bacteria since substantially more tissue can be captured in the images. For example, a typical 2P image might consist of 33

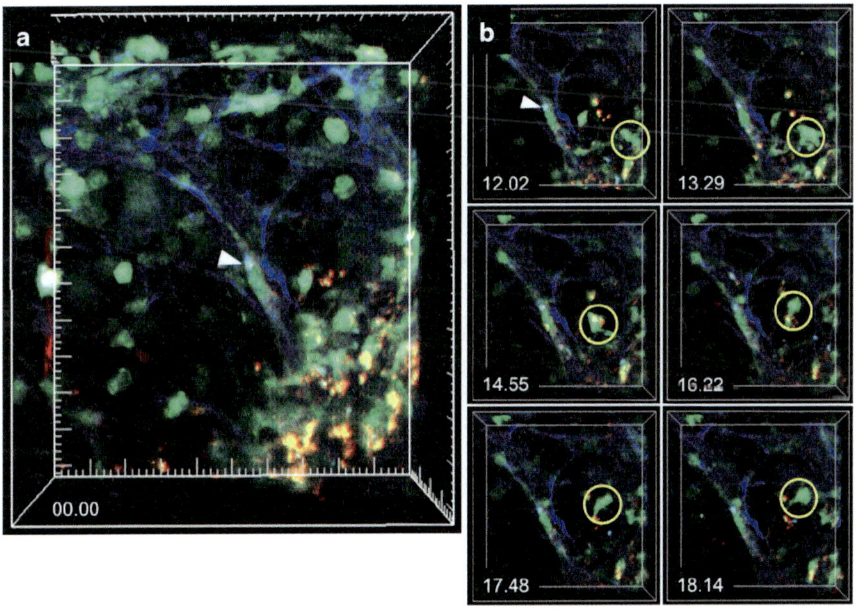

Fig. 2 2P imaging of *Listeria* infection in explanted spleen. (**a**) CD11c-YEP^+ marginal zone phagocytes (green) infected with *Listeria* (cyan, white arrow). Fluorescent 1 μm beads (red) were injected i.v. before imaging to identify the MZ. MZ macrophages ingest large numbers of beads while PALS DCs contain only a few beads. Collagen fibers appear blue due to the 2P second harmonic generation signal. The large ticks on the grid are equal to 20 μm. (**b**) Time-lapse images from a sub-region of (**a**) shows a migrating YEP^+ cell (yellow circle). Times is displayed as min:sec

individual z-steps each ~3 μm thick. This represents roughly the same volume as 20 × 5 μm histological section of the same x, y dimensions. Theoretically, this would increase the detection by 20-fold. At 5×10^4 CFU, the amount of bacteria in individual cryosections (cut from the same spleen) varied from no detectable bacteria to tens of bacteria. This variability is likely due to peculiarities of the splenic circulation. In our experience, bacteria do not distribute uniformly through the spleen, but often aggregate in local regions of the tissue. With video-rate 2P microscopes, the entire tissue can be rapidly scanned to identify regions that contain bacteria.

17 Future Directions

Listeria has been a highly prolific model for studying bacterial infection and, in the future, 2P imaging may uncover many fundamental details regarding both the early host–pathogen interactions during the innate immune response and the interactions that trigger and regulate the adaptive immune response. Recently, several groups have used the technique to investigate the host response to infection with parasites (Chtanova et al. 2008; Peters et al. 2008), virus-like particles (Junt et al. 2007), and bacteria (Chieppa et al. 2006; Egen et al. 2008; Zinselmeyer et al. 2008). Continued growth in the use of 2P imaging techniques will hinge on whether fluorescent probes for molecular interactions (Seveau et al. 2007; Shaughnessy et al. 2007; Theriot et al. 1992) can be implemented in the *in vivo* environment and whether functional readouts such as phagocytosis and killing can be linked to live cell dynamics. Moreover, combining 2P imaging with gene knockout and transgenic mouse models will provide unprecedented opportunities to test specific molecular mechanistic theories about how the host immune response unfolds during infection. From the side of the pathogen, complementary approaches can be taken by creating mutant bacteria to assess the role of various virulence factors in the pathogenesis of infection. 2P microscopy holds great promise for enhancing our understanding of a wide range of host-pathogen interactions by providing a detailed spatiotemporal picture of how individual cells function during infection and immunity.

References

Aichele P, Zinke J, Grode L, Schwendener RA, Kaufmann SH, Seiler P (2003) Macrophages of the splenic marginal zone are essential for trapping of blood-borne particulate antigen but dispensable for induction of specific T cell responses. J Immunol 171:1148–1155

Allen CD, Okada T, Tang HL, Cyster JG (2007) Imaging of germinal center selection events during affinity maturation. Science 315:528–531

Amos WB, White JG (2003) How the confocal laser scanning microscope entered biological research. Biol cell (Eur Cell Biol Org) 95:335–342

Andersen JB, Roldgaard BB, Lindner AB, Christensen BB, Licht TR (2006) Construction of a multiple fluorescence labelling system for use in co-invasion studies of Listeria monocytogenes. BMC Microbiol 6:86

Aoshi T, Zinselmeyer BH, Konjufca V, Lynch JN, Zhang X, Koide Y, Miller MJ (2008) Bacterial entry to the splenic white pulp initiates antigen presentation to CD8 + T cells. Immunity 29:476–486

Ato M, Nakano H, Kakiuchi T, Kaye PM (2004) Localization of marginal zone macrophages is regulated by C-C chemokine ligands 21/19. J Immunol 173:4815–4820

Auffray C, Fogg D, Garfa M, Elain G, Join-Lambert O, Kayal S, Sarnacki S, Cumano A, Lauvau G, Geissmann F (2007) Monitoring of blood vessels and tissues by a population of monocytes with patrolling behavior. Science 317:666–670

Beeston AL, Surette MG (2002) pfs-dependent regulation of autoinducer 2 production in Salmonella enterica serovar Typhimurium. Journal of bacteriology 184:3450–3456

Beuneu H, Garcia Z, Bousso P (2006) Cutting edge: cognate CD4 help promotes recruitment of antigen-specific CD8 T cells around dendritic cells. J Immunol 177:1406–1410

Bhakta NR, Lewis RS (2005) Real-time measurement of signaling and motility during T cell development in the thymus. Semin Immunol 17:411–420

Bhakta NR, Oh DY, Lewis RS (2005) Calcium oscillations regulate thymocyte motility during positive selection in the three-dimensional thymic environment. Nat Immunol 6:143–151

Boes M, Cerny J, Massol R, Op den Brouw M, Kirchhausen T, Chen J, Ploegh HL (2002) T-cell engagement of dendritic cells rapidly rearranges MHC class II transport. Nature 418:983–988

Boissonnas A, Fetler L, Zeelenberg IS, Hugues S, Amigorena S (2007) *In vivo* imaging of cytotoxic T cell infiltration and elimination of a solid tumor. J Exp Med 204:345–356

Bousso P, Bhakta NR, Lewis RS, Robey E (2002) Dynamics of thymocyte-stromal cell interactions visualized by two-photon microscopy. Science 296:1876–1880

Bousso P, Robey E (2003) Dynamics of CD8(+) T cell priming by dendritic cells in intact lymph nodes. Nat Immunol 4:579–585

Bousso P, Robey EA (2004) Dynamic behavior of T cells and thymocytes in lymphoid organs as revealed by two-photon microscopy. Immunity 21:349–355

Braun L, Ohayon H, Cossart P (1998) The InlB protein of Listeria monocytogenes is sufficient to promote entry into mammalian cells. Mol Microbiol 27:1077–1087

Brockstedt DG, Bahjat KS, Giedlin MA, Liu W, Leong M, Luckett W, Gao Y, Schnupf P, Kapadia D, Castro G, et al. (2005) Killed but metabolically active microbes: a new vaccine paradigm for eliciting effector T-cell responses and protective immunity. Nat Med 11:853–860

Brockstedt DG, Giedlin MA, Leong ML, Bahjat KS, Gao Y, Luckett W, Liu W, Cook DN, Portnoy DA, Dubensky TW Jr (2004) Listeria-based cancer vaccines that segregate immunogenicity from toxicity. Proc Natl Acad Sci USA 101:13832–13837

Bubert A, Sokolovic Z, Chun SK, Papatheodorou L, Simm A, Goebel W (1999) Differential expression of Listeria monocytogenes virulence genes in mammalian host cells. Mol Gen Genet 261:323–336

Burns-Guydish SM, Olomu IN, Zhao H, Wong RJ, Stevenson DK, Contag CH (2005) Monitoring age-related susceptibility of young mice to oral Salmonella enterica serovar Typhimurium infection using an *in vivo* murine model. Pediatr Res 58:153–158

Busch DH, Pamer EG (1999) T lymphocyte dynamics during Listeria monocytogenes infection. Immunol Lett 65:93–98

Busch DH, Pilip IM, Vijh S, Pamer EG (1998) Coordinate regulation of complex T cell populations responding to bacterial infection. Immunity 8:353–362

Castellino F, Huang AY, Altan-Bonnet G, Stoll S, Scheinecker C, Germain RN (2006) Chemokines enhance immunity by guiding naive CD8(+) T cells to sites of CD4 T cell-dendritic cell interaction. Nature 440:890–895

Chieppa M, Rescigno M, Huang AY, Germain RN (2006) Dynamic imaging of dendritic cell extension into the small bowel lumen in response to epithelial cell TLR engagement. J Exp Med 203:2841–2852

Chtanova T, Schaeffer M, Han SJ, van Dooren GG, Nollmann M, Herzmark P, Chan SW, Satija H, Camfield K, Aaron H, et al (2008) Dynamics of neutrophil migration in lymph nodes during infection. Immunity 29:487–496

Conlan JW (1996) Early pathogenesis of Listeria monocytogenes infection in the mouse spleen. J Med Microbiol 44:295–302

Contag CH, Bachmann MH (2002) Advances in *in vivo* bioluminescence imaging of gene expression. Annu Rev Biomed Eng 4:235–260

Contag CH, Contag PR, Mullins JI, Spilman SD, Stevenson DK, Benaron DA (1995) Photonic detection of bacterial pathogens in living hosts. Mol Microbiol 18:593–603

Contag CH, Ross BD (2002) It's not just about anatomy: *in vivo* bioluminescence imaging as an eyepiece into biology. J Magn Reson Imag 16:378–387

Cossart P, Toledo-Arana A (2008) Listeria monocytogenes, a unique model in infection biology: An overview. Microbes and infection. Institut Pasteur, Paris, France

Curtiss R, 3rd, Nakayama K, Kelly SM (1989) Recombinant avirulent Salmonella vaccine strains with stable maintenance and high level expression of cloned genes *in vivo*. Immunol Invest 18:583–596

Datta SK, Okamoto S, Hayashi T, Shin SS, Mihajlov I, Fermin A, Guiney DG, Fierer J, Raz E (2006) Vaccination with irradiated Listeria induces protective T cell immunity. Immunity 25:143–152

Day RN, Schaufele F (2008) Fluorescent protein tools for studying protein dynamics in living cells: a review. J Biomed Opt 13:031202

de Wet JR, Wood KV, Helinski DR, DeLuca M (1985) Cloning of firefly luciferase cDNA and the expression of active luciferase in *Escherichia coli*. Proc Natl Acad Sci USA 82:7870–7873

Denk W, Strickler JH, Webb WW (1990) 2-Photon Laser scanning fluorescence microscopy. Science 248:73–76

Doyle TC, Burns SM, Contag CH (2004) *In vivo* bioluminescence imaging for integrated studies of infection. Cell Microbiol 6:303–317

Dramsi S, Cossart P (2002) Listeriolysin O: a genuine cytolysin optimized for an intracellular parasite. J Cell Biol 156:943–946

Edelson BT, Unanue ER (2002) MyD88-dependent but Toll-like receptor 2-independent innate immunity to Listeria: no role for either in macrophage listericidal activity. J Immunol 169:3869–3875

Egen JG, Rothfuchs AG, Feng CG, Winter N, Sher A, Germain RN (2008) Macrophage and T cell dynamics during the development and disintegration of mycobacterial granulomas. Immunity 28:271–284

Flaberg E, Sabelstrom P, Strandh C, Szekely L (2008) Extended Field Laser Confocal Microscopy (EFLCM): combining automated Gigapixel image capture with in silico virtual microscopy. BMC Med Imag 8:13

Fortinea N, Trieu-Cuot P, Gaillot O, Pellegrini E, Berche P, Gaillard JL (2000) Optimization of green fluorescent protein expression vectors for *in vitro* and *in vivo* detection of *Listeria monocytogenes*. Res Microbiol 151:353–360

Foulds KE, Zenewicz LA, Shedlock DJ, Jiang J, Troy AE, Shen H (2002) Cutting edge: CD4 and CD8 T cells are intrinsically different in their proliferative responses. J Immunol 168:1528–1532

Francis KP, Joh D, Bellinger-Kawahara C, Hawkinson MJ, Purchio TF, Contag PR (2000) Monitoring bioluminescent Staphylococcus aureus infections in living mice using a novel luxABCDE construct. Infect Immun 68:3594–3600

Francis KP, Yu J, Bellinger-Kawahara C, Joh D, Hawkinson MJ, Xiao G, Purchio TF, Caparon MG, Lipsitch M, Contag PR (2001) Visualizing pneumococcal infections in the lungs of live mice using bioluminescent Streptococcus pneumoniae transformed with a novel gram-positive lux transposon. Infect Immun 69:3350–3358

Freitag NE, Jacobs KE (1999) Examination of *Listeria monocytogenes* intracellular gene expression by using the green fluorescent protein of *Aequorea victoria*. Infect Immun 67:1844–1852

Gauderon R, Lukins PB, Sheppard CJ (2001) Simultaneous multichannel nonlinear imaging: combined two-photon excited fluorescence and second-harmonic generation microscopy. Micron 32:685–689

Geissmann F, Jung S, Littman DR (2003) Blood monocytes consist of two principal subsets with distinct migratory properties. Immunity 19:71–82

Germain RN, Miller MJ, Dustin ML, Nussenzweig MC (2006) Dynamic imaging of the immune system: progress, pitfalls and promise. Nat Rev Immunol 6:497–507

Glomski IJ, Decatur AL, Portnoy DA (2003) Listeria monocytogenes mutants that fail to compartmentalize listerolysin O activity are cytotoxic, avirulent, and unable to evade host extracellular defenses. Infect Immun 71:6754–6765

Hardy J, Francis KP, DeBoer M, Chu P, Gibbs K, Contag CH (2004) Extracellular replication of Listeria monocytogenes in the murine gall bladder. Science 303:851–853

Hardy J, Margolis JJ, Contag CH (2006) Induced biliary excretion of Listeria monocytogenes. Infect Immun 74:1819–1827

Harty JT, Tvinnereim AR, White DW (2000) CD8 + T cell effector mechanisms in resistance to infection. Annu Rev Immunol 18:275–308

Hastings JW (1996) Chemistries and colors of bioluminescent reactions: a review. Gene 173:5–11

Hastings JW, Johnson CH (2003) Bioluminescence and chemiluminescence. Methods Enzymol 360:75–104

Hauser AE, Junt T, Mempel TR, Sneddon MW, Kleinstein SH, Henrickson SE, von Andrian UH, Shlomchik MJ, Haberman AM (2007) Definition of germinal-center B cell migration *in vivo* reveals predominant intrazonal circulation patterns. Immunity 26:655–667

Huang JH, Cardenas-Navia LI, Caldwell CC, Plumb TJ, Radu CG, Rocha PN, Wilder T, Bromberg JS, Cronstein BN, Sitkovsky M, et al. (2007) Requirements for T lymphocyte migration in explanted lymph nodes. J Immunol 178:7747–7755

Hugues S, Boissonnas A, Amigorena S, Fetler L (2006) The dynamics of dendritic cell-T cell interactions in priming and tolerance. Curr Opin Immunol 18:491–495

Itano AA, McSorley SJ, Reinhardt RL, Ehst BD, Ingulli E, Rudensky AY, Jenkins MK (2003) Distinct dendritic cell populations sequentially present antigen to CD4 T cells and stimulate different aspects of cell-mediated immunity. Immunity 19:47–57

Jablonska J, Dittmar KE, Kleinke T, Buer J, Weiss S (2007) Essential role of CCL2 in clustering of splenic ERTR-9 + macrophages during infection of BALB/c mice by Listeria monocytogenes. Infect Immun 75:462–470

Jung S, Unutmaz D, Wong P, Sano G, De los Santos K, Sparwasser T, Wu S, Vuthoori S, Ko K, Zavala F, et al. (2002) *In vivo* depletion of CD11c(+) dendritic cells abrogates priming of CD8(+) T cells by exogenous cell-associated antigens. Immunity 17:211–220

Junt T, Moseman EA, Iannacone M, Massberg S, Lang PA, Boes M, Fink K, Henrickson SE, Shayakhmetov DM, Di Paolo NC, et al. (2007) Subcapsular sinus macrophages in lymph nodes clear lymph-borne viruses and present them to antiviral B cells. Nature 450:110–114

Kawakami N, Nagerl UV, Odoardi F, Bonhoeffer T, Wekerle H, Flugel A (2005) Live imaging of effector cell trafficking and autoantigen recognition within the unfolding autoimmune encephalomyelitis lesion. J Exp Med 201:1805–1814

Khanna KM, McNamara JT, Lefrancois L (2007) *In situ* imaging of the endogenous CD8 T cell response to infection. Science 318:116–120

Kim JV, Dustin ML (2006) Innate response to focal necrotic injury inside the blood-brain barrier. J Immunol 177:5269–5277

Kraal G (1992) Cells in the marginal zone of the spleen. Int Rev Cytol 132:31–74

Kursar M, Hopken UE, Koch M, Kohler A, Lipp M, Kaufmann SH, Mittrucker HW (2005) Differential requirements for the chemokine receptor CCR7 in T cell activation during Listeria monocytogenes infection. J Exp Med 201:1447–1457

Lara-Tejero M, Pamer EG (2004) T cell responses to *Listeria monocytogenes*. Curr Opin Microbiol 7:45–50

Lauvau G, Vijh S, Kong P, Horng T, Kerksiek K, Serbina N, Tuma RA, Pamer EG (2001) Priming of memory but not effector CD8 T cells by a killed bacterial vaccine. Science 294:1735–1739

Lecuit M (2005) Understanding how Listeria monocytogenes targets and crosses host barriers. Clin Microbiol Infect 11:430–436

Lin J, Miller MJ, Shaw AS (2005) The c-SMAC: sorting it all out (or in). J Cell Biol 170:177–182

Lindquist RL, Shakhar G, Dudziak D, Wardemann H, Eisenreich T, Dustin ML, Nussenzweig MC (2004) Visualizing dendritic cell networks *in vivo*. Nat Immunol 5:1243–1250

Mackaness GB (1962) Cellular resistance to infection. J Exp Med 116:381–406

McCaffrey RL, Fawcett P, O'Riordan M, Lee KD, Havell EA, Brown PO, Portnoy DA (2004) A specific gene expression program triggered by Gram-positive bacteria in the cytosol. Proc Natl Acad Sci USA 101:11386–11391

Mebius RE, Kraal G (2005) Structure and function of the spleen. Nat Rev Immunol 5:606–616

Mempel TR, Henrickson SE, von Andrian UH (2004) T-cell priming by dendritic cells in lymph nodes occurs in three distinct phases. Nature 427:154–159

Mempel TR, Pittet MJ, Khazaie K, Weninger W, Weissleder R, von Boehmer H, von Andrian UH (2006) Regulatory T cells reversibly suppress cytotoxic T cell function independent of effector differentiation. Immunity 25:129–141

Mercado R, Vijh S, Allen SE, Kerksiek K, Pilip IM, Pamer EG (2000) Early programming of T cell populations responding to bacterial infection. J Immunol 165:6833–6839

Michel E, Reich KA, Favier R, Berche P, Cossart P (1990) Attenuated mutants of the intracellular bacterium Listeria monocytogenes obtained by single amino acid substitutions in listeriolysin O. Mol Microbiol 4:2167–2178

Miller MJ, Hejazi AS, Wei SH, Cahalan MD, Parker I (2004a) T cell repertoire scanning is promoted by dynamic dendritic cell behavior and random T cell motility in the lymph node. Proc Natl Acad Sci USA 101:998–1003

Miller MJ, Safrina O, Parker I, Cahalan MD (2004b) Imaging the single cell dynamics of CD4(+) T cell activation by dendritic cells in lymph nodes. J Exp Med 200:847–856

Miller MJ, Wei SH, Cahalan MD, Parker I (2003) Autonomous T cell trafficking examined *in vivo* with intravital two-photon microscopy. Proc Natl Acad Sci USA 100:2604–2609

Miller MJ, Wei SH, Parker I, Cahalan MD (2002) Two-photon imaging of lymphocyte motility and antigen response in intact lymph node. Science 296:1869–1873

Mrass P, Takano H, Ng LG, Daxini S, Lasaro MO, Iparraguirre A, Cavanagh LL, von Andrian UH, Ertl HCJ, Haydon PG, Weninger W (2006) Random migration precedes stable target cell interactions of tumor-infiltrating T cells. J Exp Med 203:2749–2761

Muraille E, Giannino R, Guirnalda P, Leiner I, Jung S, Pamer EG, Lauvau G (2005) Distinct *in vivo* dendritic cell activation by live versus killed *Listeria monocytogenes*. Eur J Immunol 35:1463–1471

Neuenhahn M, Kerksiek KM, Nauerth M, Suhre MH, Schiemann M, Gebhardt FE, Stemberger C, Panthel K, Schroder S, Chakraborty T, et al (2006) CD8alpha + dendritic cells are required for efficient entry of *Listeria monocytogenes* into the spleen. Immunity 25:619–630

Nguyen QT, Callamaras N, Hsieh C, Parker I (2001) Construction of a two-photon microscope for video-rate Ca2 + imaging. Cell Calcium 30:383–393

Nitsch R, Pohl EE, Smorodchenko A, Infante-Duarte C, Aktas O, Zipp F (2004) Direct impact of T cells on neurons revealed by two-photon microscopy in living brain tissue. J Neurosci 24:2458–2464

O'Connell RM, Vaidya SA, Perry AK, Saha SK, Dempsey PW, Cheng G (2005) Immune activation of type I IFNs by *Listeria monocytogenes* occurs independently of TLR4, TLR2, and receptor interacting protein 2 but involves TNFR-associated NF kappa B kinase-binding kinase 1. J Immunol 174:1602–1607

O'Riordan M, Yi CH, Gonzales R, Lee KD, Portnoy DA (2002) Innate recognition of bacteria by a macrophage cytosolic surveillance pathway. Proc Natl Acad Sci USA 99:13861–13866

Odoardi F, Kawakami N, Li Z, Cordiglieri C, Streyl K, Nosov M, Klinkert WE, Ellwart JW, Bauer J, Lassmann H, et al (2007) Instant effect of soluble antigen on effector T cells in peripheral immune organs during immunotherapy of autoimmune encephalomyelitis. Proc Natl Acad Sci USA 104:920–925

Okada T, Cyster JG (2007) CC chemokine receptor 7 contributes to Gi-dependent T cell motility in the lymph node. J Immunol 178:2973–2978

Okada T, Miller MJ, Parker I, Krummel MF, Neighbors M, Hartley SB, O'Garra A, Cahalan MD, Cyster JG (2005) Antigen-engaged B cells undergo chemotaxis toward the T zone and form motile conjugates with helper T cells. PLoS Biol 3:1047–1061

Olympus A, Inc. (2008) Theory of confocal microscopy. http://www.olympusconfocal.com/theory/confocalintro.html

Pamer EG (2004) Immune responses to Listeria monocytogenes. Nat Rev Immunol 4:812–823
Paterson Y, Johnson RS (2004) Progress towards the use of Listeria monocytogenes as a live bacterial vaccine vector for the delivery of HIV antigens. Expert Rev Vaccines 3:S119–S134
Pawley JB (2006) Handbook of biological confocal microscopy, 3rd edn. Springer, New York
Peters NC, Egen JG, Secundino N, Debrabant A, Kimblin N, Kamhawi S, Lawyer P, Fay MP, Germain RN, Sacks D (2008) *In vivo* imaging reveals an essential role for neutrophils in leishmaniasis transmitted by sand flies. Science 321:970–974
Probst HC, Tschannen K, Odermatt B, Schwendener R, Zinkernagel RM, Van Den Broek M (2005) Histological analysis of CD11c-DTR/GFP mice after *in vivo* depletion of dendritic cells. Clin Exp Immunol 141:398–404
Pron B, Boumaila C, Jaubert F, Berche P, Milon G, Geissmann F, Gaillard JL (2001) Dendritic cells are early cellular targets of Listeria monocytogenes after intestinal delivery and are involved in bacterial spread in the host. Cell Microbiol 3:331–340
Rice BW, Cable MD, Nelson MB (2001) *In vivo* imaging of light-emitting probes. J Biomed Opt 6:432–440
Riedel CU, Monk IR, Casey PG, Morrissey D, O'Sullivan GC, Tangney M, Hill C, Gahan CG (2007) Improved luciferase tagging system for *Listeria monocytogenes* allows real-time monitoring *in vivo* and *in vitro*. Appl Environ Microbiol 73:3091–3094
Rogers HW, Unanue ER (1993) Neutrophils are involved in acute, nonspecific resistance to Listeria monocytogenes in mice. Infect Immun 61:5090–5096
Ruby EG, Nealson KH (1976) Symbiotic association of Photobacterium fischeri with the marine luminous fish Monocentris japonica; a model of symbiosis based on bacterial studies. Biol Bull 151:574–586
Schneider H, Downey J, Smith A, Zinselmeyer BH, Rush C, Brewer JM, Wei B, Hogg N, Garside P, Rudd CE (2006) Reversal of the TCR stop signal by CTLA-4. Science 313:1972–1975
Schulz RB, Semmler W (2008) Fundamentals of optical imaging. Handbook of experimental pharmacology. Springer, Berlin, pp 3–22
Schwickert TA, Lindquist RL, Shakhar G, Livshits G, Skokos D, Kosco-Vilbois MH, Dustin ML, Nussenzweig MC (2007) *In vivo* imaging of germinal centres reveals a dynamic open structure. Nature 446:83–87
Seki E, Tsutsui H, Tsuji NM, Hayashi N, Adachi K, Nakano H, Futatsugi-Yumikura S, Takeuchi O, Hoshino K, Akira S, et al. (2002) Critical roles of myeloid differentiation factor 88-dependent proinflammatory cytokine release in early phase clearance of *Listeria monocytogenes* in mice. J Immunol 169:3863–3868
Seveau S, Tham TN, Payrastre B, Hoppe AD, Swanson JA, Cossart P (2007) A FRET analysis to unravel the role of cholesterol in Rac1 and PI 3-kinase activation in the InlB/Met signalling pathway. Cell Microbiol 9:790–803
Shakhar G, Lindquist RL, Skokos D, Dudziak D, Huang JH, Nussenzweig MC, Dustin ML (2005) Stable T cell-dendritic cell interactions precede the development of both tolerance and immunity *in vivo*. Nat Immunol 6:707–714
Shaner NC, Steinbach PA, Tsien RY (2005) A guide to choosing fluorescent proteins. Nat Methods 2(12):905–909
Shaughnessy LM, Lipp P, Lee KD, Swanson JA (2007) Localization of protein kinase C epsilon to macrophage vacuoles perforated by *Listeria monocytogenes* cytolysin. Cell Microbiol 9:1695–1704
Shedlock DJ, Shen H (2003) Requirement for CD4 T cell help in generating functional CD8 T cell memory. Science 300:337–339
Skamene E, Chayasirisobhon W (1977) Enhanced resistance to Listeria monocytogenes in splenectomized mice. Immunology 33:851–858
Sun JC, Bevan MJ (2003) Defective CD8 T cell memory following acute infection without CD4 T cell help. Science 300:339–342
Tadokoro CE, Shakhar G, Shen SQ, Ding Y, Lino AC, Maraver A, Lafaille JJ, Dustin ML (2006) Regulatory T cells inhibit stable contacts between CD4(+) T cells and dendritic cells *in vivo*. J Exp Med 203:505–511

Tang QZ, Adams JY, Tooley AJ, Bi MY, Fife BT, Serra P, Santamaria P, Locksley RM, Krummel MF, Bluestone JA (2006) Visualizing regulatory T cell control of autoimmune responses in nonobese diabetic mice. Nat Immunol 7:83–92

Terai T, Nagano T (2008) Fluorescent probes for bioimaging applications. Curr Opin Chem Biol 12:515–521

Theriot JA, Mitchison TJ, Tilney LG, Portnoy DA (1992) The rate of actin-based motility of intracellular Listeria monocytogenes equals the rate of actin polymerization. Nature 357:257–260

Unanue ER (1997) Inter-relationship among macrophages, natural killer cells and neutrophils in early stages of Listeria resistance. Curr Opin Immunol 9:35–43

van Rooijen N, Kors N, Kraal G (1989) Macrophage subset repopulation in the spleen: differential kinetics after liposome-mediated elimination. J Leukoc Biol 45:97–104

Verch T, Pan ZK, Paterson Y (2004) Listeria monocytogenes-based antibiotic resistance gene-free antigen delivery system applicable to other bacterial vectors and DNA vaccines. Infect Immun 72:6418–6425

von Andrian UH (2002) Immunology. T cell activation in six dimensions. Science 296:1815–1817

Walch M, Latinovic-Golic S, Velic A, Sundstrom H, Dumrese C, Wagner CA, Groscurth P, Ziegler U (2007) Perforin enhances the granulysin-induced lysis of Listeria innocua in human dendritic cells. BMC Immunol 8:14

Wilson T, Hastings JW (1998) Bioluminescence. Annu Rev Cell Dev Biol 14:197–230

Witt CM, Raychaudhuri S, Schaefer B, Chakraborty AK, Robey EA (2005) Directed migration of positively selected thymocytes visualized in real-time. PLoS Biol 3:1062–1069

Wong P, Pamer EG (2003) Feedback regulation of pathogen-specific T cell priming. Immunity 18:499–511

Zinselmeyer BH, Dempster J, Gurney AM, Wokosin D, Miller M, Ho H, Millington OR, Smith KM, Rush CM, Parker I, et al. (2005) *In situ* characterization of CD4(+) T cell behavior in mucosal and systemic lymphoid tissues during the induction of oral priming and tolerance. J Exp Med 201:1815–1823

Zinselmeyer BH, Lynch JN, Zhang X, Aoshi T, Miller MJ (2008) Video-rate two-photon imaging of mouse footpad – a promising model for studying leukocyte recruitment dynamics during inflammation. Inflamm Res 57:93–96

Inflammation on the Mind: Visualizing Immunity in the Central Nervous System

Silvia S. Kang and Dorian B. McGavern

Contents

Abstract The central nervous system (CNS) is a remarkably complex structure that utilizes electrochemical signaling to coordinate activities throughout the entire body. Because the nervous system contains nonreplicative cells, it is postulated that, through evolutionary pressures, this compartment has acquired specialized mechanisms to limit damage. One potential source of damage comes from our immune system, which has the capacity to survey the CNS and periphery for the presence of foreign material. The immune system is equipped with numerous effector mechanisms and can greatly alter the homeostasis and function of the CNS. Degeneration, autoimmunity, and pathogen infection can all result in acute, and sometimes chronic, inflammation within the CNS. Understanding the specialized

S.S. Kang and D.B. McGavern (✉)
National Institutes of Neurological Disorders and Stroke, National Institutes of Health (NIH), 10 Center Drive, Bldg 10, Rm 7C213, Bethesda, MD 20892, USA
e-mail: mcgavernd@mail.nih.gov

D. McGavern and M. Dustin (eds.), *Visualizing Immunity*, Current Topics in Microbiology and Immunology 334,
DOI: 10.1007/978-3-540-93864-4_10,

functionality of innate and adaptive immune cells within the CNS is critical to the design of more efficacious treatments to mitigate CNS inflammatory conditions. Much of our knowledge of CNS-immune interactions stems from seminal studies that have used static and dynamic imaging approaches to visualize inflammatory cells responding to different CNS conditions. This review will focus on how imaging techniques have elevated our understanding of CNS inflammation as well as the exciting prospects that lie ahead as we begin to pursue investigation of the inflamed CNS in real time.

1 Introduction

The CNS is the critical processing center that controls a broad range of functions from breathing to higher thought, making induction of immune responses within this tissue potentially dangerous. Because neurons are held in a postmitotic state and do not replicate readily, injury to this critical population by immune cells can damage neuronal connections that are essential for life. Generally, limited numbers of leukocytes are allowed to gain access into the CNS due to the blood brain barrier (BBB) and blood cerebral spinal fluid barrier, providing one means of protection. However, when a pathogen invades the CNS or an autoimmune response is mounted within this organ, recruited leukocytes can now encounter a variety of cells that are normally anatomically sequestered from the periphery (e.g., astrocytes, microglia, oligodendrocytes, stromal cells, and neurons). It is of great importance that we obtain a more in depth understanding of how CNS immunity is induced, and about the consequences of leukocyte interactions with CNS residents during infection, autoimmunity, and degenerative disease. Importantly, the actual consequences of CNS immunity can sometimes be obscured in systems reliant on tissue culture, because of the intricate and highly complex associations normally found between glial cells and neurons *in vivo*. Moreover, the location of the immune response within the CNS (e.g., the meninges versus the brain parenchyma) can also dictate the type of immune cells recruited and the resident CNS populations that are perturbed. Visualization of immunity in the CNS provides critical information that simply cannot be obtained from *in vitro* studies alone. In this review, we will focus primarily on studies that have used imaging techniques to advance our understanding of CNS immunity. Novel insights into CNS autoimmunity, degenerative diseases, and responses to protein and infectious agents have all been obtained using imaging approaches, and the recent use of multiphoton microscopy to probe CNS immunity is now providing our first glimpse of immune cell dynamics in the nervous system. On the horizon lies a wealth of visual knowledge to be uncovered as we probe the CNS with greater precision and resolution in real time. This review will cover the ground we have trod thus far using imaging approaches and some exciting new prospects that await our discovery in the future.

2 Static Imagery: Lessons Learned from Nondynamic Imaging Approaches

2.1 Initiation of Immune Responses to CNS-Derived Antigens

The induction of adaptive immune responses is often reliant on antigen delivery into the secondary lymphoid organs for priming. Accordingly, the efflux of CNS interstitial fluid and cerebral spinal fluid (CSF) from the CNS into the periphery provides a means for delivering brain-derived antigens predominantly into the cervical lymph nodes (Bradbury and Cole 1980; Bradbury et al. 1981; Bradbury and Westrop 1983; Szentistvanyi et al. 1984; Weller et al. 1996). CT scans monitoring radiolabeled tracers and gross analysis of India ink distribution showed that drainage from the CNS to the periphery occurs rapidly, within minutes to hours (Hunter et al. 1995; Kida et al. 1993). Visualization of CSF-injected India ink drainage by microscopy revealed the passage of CSF through the cribriform plate and into the nasal lymphatics as one major pathway for communication between the CNS and periphery (Kida et al. 1993, 1995). In fact, intracerebral injection and tracking of green fluorescent protein (GFP)-labeled T cells showed that movement through the cribriform plate into the nasal mucosa is a common pathway by which both antigen and T cells gain access to the cervical lymph nodes from the CNS (Goldmann et al. 2006). It is thought that peripheral sampling of CNS-derived antigens through this and other pathways (Cserr et al. 1992) is critical in the initiation of immune responses to pathogens that gain access to the CNS, and anatomical sequestration from these drainage pathways (Stevenson et al. 1997) may be a way for CNS pathogens to avoid detection by the immune system.

To gain additional insights into the role of antigen-presenting cells (APCs) in the initiation of peripheral immune responses to CNS antigens, confocal microscopy was recently employed in a murine system in which fluorescently-tagged ovalbumin (OVA) was used as a model antigen. Intracerebral injection of BODIPY-conjugated OVA confirmed that the rapid efflux of brain-derived antigens into cervical lymph nodes occurs within hours of injection (Ling et al. 2003). Furthermore, using antigen-loaded GFP-tagged dendritic cells (DCs), it was revealed that DCs migrate from the CNS into the cervical lymph nodes (Karman et al. 2004a, b), indicating that antigen can arrive by cellular transport as well as by free drainage through the cerebral extracellular fluid. Initiation of peripheral T cell responses triggered by brain-derived antigens (Ling et al. 2003; Walter and Albert 2007) was subsequently followed by the recruitment of activated T cells back to the regions of the brain where parenchymal antigen retention and uptake was detected (Ling et al. 2003, 2006). T cell homing to CNS tissue was dependent on CNS-derived DCs (Karman et al. 2004b), supporting the emerging idea that tissue DCs can dictate the trafficking patterns of T cells (Mora et al. 2005; Mora and von Andrian 2006). Consistent with this idea are colocalization studies demonstrating that CNS DCs process brain-derived OVA antigen (Karman et al. 2004b) and data from organotypic brain slices, showing that

CD11c^{+} antigen-loaded cells in the CNS continue to stimulate T cell proliferation (Ling et al. 2008). Thus, CNS dendritic cells play an important role in antigen delivery to the periphery, subsequent T cell homing into the CNS, and continued T cell stimulation at the site of antigen retention and presentation. Future imaging studies focused on the dynamics of T cell–DC engagement within the CNS are required to provide novel insights into the consequences of these interactions within the CNS.

2.2 CTL Engagement of Virus-Infected CNS Targets

Generation of an effective antigen-specific adaptive immune response is essential for the clearance of pathogens. CD8^{+} T cells, in particular, are important for mediating antiviral immunity in the periphery. Similarly, CD8^{+} T cells can also dictate CNS immune responses to pathogens, resulting in viral eradication or, in some cases, immunopathology (Chua et al. 2004; McGavern et al. 2002b; Shrestha and Diamond 2007; Shrestha et al. 2006). Primed CD8^{+} cytotoxic T cells (CTL) can purge pathogens from the CNS using cytolytic (e.g., perforin) and/or noncytolytic (e.g., IFNγ) pathways. Both CTL activation and target recognition are reliant on the intercellular communication between T cells and APCs (or virally infected target cells), respectively. For CD8^{+} T cells, critical interactions between the T cell receptor (TCR) and major histocompatibility complex (MHC) I molecules displaying antigens is thought to occur through the anatomical structure referred to as the immunological synapse (Davis et al. 2007; Grakoui et al. 1999; Huse et al. 2006; Monks et al. 1998; Stinchcombe et al. 2006). The hallmark of the synapse is the formation of supramolecular activation clusters (SMAC) with enrichment of TCR molecules within the central SMAC (c-SMAC) and adhesion molecules, including LFA-1, surrounding the peripheral area to form the peripheral SMAC (p-SMAC) (Monks et al. 1998). The concept of the immunological synapse has emerged from extensive *in vitro* data (Davis et al. 2007; Dustin 2008; Grakoui et al. 1999; Huppa and Davis 2003; Monks et al. 1998). However, more recently, viral infection models were used to identify the *in vivo* correlate of the immune synapse between CTL and infected targets residing in the CNS (Barcia et al. 2006; McGavern et al. 2002a). Static imaging studies of the virally-infected CNS have yielded insights into the molecular anatomy of the *in vivo* immune synapse and demonstrated that this structure resembles the synapse observed *in vitro*.

In one study, lymphocytic choriomeningitis virus (LCMV), a natural pathogen for human and murine hosts (Lledo et al. 2003; Roebroek et al. 1994; Schanen et al. 1998), was used to examine the interactions of CTL responses to viral infection within the CNS. LCMV, a prototypic member of the Arenaviridae family, is a noncytopathic virus and; therefore, any pathology that develops after infection is driven by the subsequent immune response, rather than by the virus itself. Intracerebral (i.c.) infection of adult immunocompetent mice with LCMV leads to acute, fatal choriomeningitis within 6–8 days (Kang and McGavern 2008; McGavern et al. 2002b).

Following i.c. inoculation, LCMV localizes to the meninges, ependyma, and choroid plexus within the CNS (McGavern and Truong 2004; Mims 1960; Schwendemann et al. 1983). $CD8^+$ T cells are critical for LCMV-induced meningitis (Fung-Leung et al. 1991; Kang and McGavern 2008) and have been detected in the cerebral spinal fluid (CSF) of infected mice (Ceredig et al. 1987; Zinkernagel and Doherty 1973). Although bulk $CD8^+$ T cells were observed by electron microscopy in the meninges of LCMV-infected mice (Schwendemann et al. 1983), the nature of CTL interactions during fatal meningitis were not defined. B6 mice infected with LCMV mount a robust, polyclonal T cell response directed against the glycoprotein (GP) and nucleoprotein (NP) of LCMV, and most of the MHC I and II restricted epitopes have been identified (Gairin et al. 1995; Gallimore et al. 1998; van der Most et al. 1998). This information facilitated the generation of MHC tetramers that can be used to identify LCMV-specific $CD8^+$ T cells either flow cytometrically or on tissue sections by in situ tetramer staining (McGavern et al. 2002a; Skinner et al. 2000; Skinner and Haase 2005). $CD8^+$ T cell TCR transgenic (tg) mice (referred to as P14 mice) specific to the $D^bGP_{33\text{–}41}$ epitope have also been generated and provide another powerful tool for tracking LCMV-specific CTL responses (Pircher et al. 1989).

Using in situ tetramer staining and confocal microscopy, virus-specific CTL were observed in close proximity to infected target cells residing in the meninges of symptomatic mice infected i.c. with LCMV (McGavern et al. 2002a). Analysis of MHC I tetramer distribution in situ also revealed polarization of the TCR toward infected target cells suggestive of immunological synapse formation. To generate a more robust system to track LCMV-specific CTL, naive TCR-tg P14 cells genetically tagged with GFP were transferred into mice prior to infection. Coronal brain reconstructions obtained from infected mice at day 6 postinfection demonstrated that GFP^+ LCMV-specific CTL localized precisely to the regions of viral infection (McGavern et al. 2002a). To examine the molecular anatomy of the CTL interactions, high-resolution confocal microscopy studies of P14 cells were performed. These studies revealed the redistribution of adhesion (LFA-1) and signaling (Lck) molecules toward the CTL-target cell interface, once again suggestive of immune synapse formation. Visual evidence of perforin deposition onto LCMV-infected cells was consistent with the idea that CTL were actively using lytic effector mechanisms to purge virus from at least some of the CNS targets (McGavern et al. 2002a). Importantly, P14 CTL were found in close juxtaposition of multiple target cells, enforcing the concept that one mechanism for efficient viral clearance is through simultaneous engagement of infected cells. A subsequent *in vitro* study has elaborated on this concept by demonstrating the ability of CTL to interact with multiple target cells of different antigenic potential (Wiedemann et al. 2006). Time-lapse confocal microscopy revealed double polarization of T cell lytic granules when simultaneously engaged with different target cells. Interestingly, TCR clustering and Ca^{2+} signaling was still observed even after target cell lysis. These data suggest that T cells can benefit from continued engagement with annihilated target cells, while making new contacts with subsequent target cells.

In vivo immune synapse formation was formally defined in 3D using a model reliant upon a nonreplicating adenovirus infection (Barcia et al. 2006, 2008). In this model, infection with thymidine kinase (TK)-tagged adenovirus predominantly targeted astrocytes within the CNS. Although both CD4$^+$ and CD8$^+$ T cells were required for clearance of the virus after immunization, CD4$^+$ cells were sequestered in the perivascular compartment whereas CD8$^+$ T cells localized to the parenchyma. Confocal studies visualizing adenovirus and CD8$^+$ T cells demonstrated close interactions between infected targets and CTL. Within the CD8$^+$ T cells, signaling molecules like ZAP-70 and phospho-lck polarized towards the interface with the adenovirus-infected targets. 3D reconstructions of the interface revealed central clustering of TCR molecules surrounded by a peripheral ring of LFA-1 (Barcia et al. 2006). This was consistent with molecular anatomy of synapses defined *in vitro* (Monks et al. 1998). However, further analysis of astrocyte viral clearance in the CNS revealed the presence of mature CTL synapses (defined by formation of the c-SMAC and p-SMAC) as well as CTL interactions that occurred in the absence of adhesion molecule and TCR polarization (Barcia et al. 2008). Previously, directionally distinct pathways for cellular secretion of cytokines were defined *in vitro* with some cytokines, such as TNFα, being released multidirectionally, whereas others, such as IFNγ and IL-2, were preferentially secreted at the synapse (Huse et al. 2006). Consistent with synaptic IFNγ release, static imagery confirmed that IFNγ secretion occurred at the interface of CTL–astrocyte contacts (Barcia et al. 2008). Interestingly, IFNγ secretion was observed with and without mature immunological synapse formation. IFNγ$^+$ polarization without mature synapse formation was common in T cells associated with astrocytic processes, whereas more mature IFNγ$^+$ synapses were observed when T cells were engaged with astrocytic cell bodies (Barcia et al. 2008). These results suggest that T cell interactions with astrocytes can occur through distinct types of synapses that depend on the location at which contact is made; however, definitive proof of this supposition would require real time imaging, as the static imaging approach used in these studies did not permit examination of synapse dynamics and transformation over time. Nevertheless, these studies did provide definitive data demonstrating the formation of mature immunological synapses *in vivo* during the process of CNS viral clearance. Importantly, the localization of signaling molecules and the presence of polarized IFNγ at the CTL–target cell interface suggests that synapse formation can sometimes have a functional outcome. It remains to be determined whether a mature synapse is required for T cell effector functions *in vivo*.

2.3 CTL-Associated Neuronal Damage

Neurological deficits have been associated with generation of CD8$^+$ T cell responses. Because most neurons are held in a postmitotic state (Herrup and Yang 2007), these cells do not readily renew, allowing for induction of irreparable pathology through CTL targeting. Following infection with Theiler's murine encephalomyelitis virus (TMEV), demyelination and axonal damage is accompanied by neurological deficits, which are associated with the presence of CD8$^+$ T cells (Murray et al. 1998a, b;

Rivera-Quinones et al. 1998). This suggests that recruitment of CTL to the virally-infected CNS can lead to significant neuronal damage. CTL–neuronal interactions have also been implicated in human disease. Rasmussen's encephalitis is a progressive disorder characterized by recurrent seizures (Rasmussen et al. 1958) and diminished neurological function. The etiology of this disease is unknown, although viral infection (Iannetti et al. 1991; Power et al. 1990; Vinters et al. 1993) and a humoral response directed against the glutamate receptor 3 (Levite et al. 1999; Rogers et al. 1994; Twyman et al. 1995; Whitney and McNamara 2000) have been suggested as possible explanations for the disease.

Another possible explanation for pathogenesis in Rasmussen's patients was set forth by a recent microscopy study (Bien et al. 2002). Immunohistochemical examination of tissue specimens from Rasmussen's encephalitis patients revealed the presence of lymphocytic infiltrates that were predominantly $CD8^+$. Around 7% of the $CD8^+$ T cells were in close apposition to neurons, and some of these cells were found to have granzyme B particles polarized to the contact surface. The presence of apoptotic neurons also suggested that CTL were specifically targeting these cells during disease (Bien et al. 2002). These findings support the possibility that CTL in Rasmussen's patients can directly engage neurons and induce lysis via a granzyme-mediated pathway. Further studies are required to determine whether these CTL are targeting a viral or self-antigen.

It is thought that one mechanism by which neurons protect themselves from CTL activity is to reduce surface MHC I expression (Joly et al. 1991; Joly and Oldstone 1992). However, elevated levels of MHC I have been observed during disease (Herrup and Yang 2007; Neumann et al. 2002) and under conditions of inflammation and electrical impairment (Neumann et al. 1995, 1997; Rensing-Ehl et al. 1996). MHC I expression is also thought to participate in neuronal function and repair (Boulanger and Shatz 2004; Huh et al. 2000; Oliveira et al. 2004). To examine the impact of elevated neuronal MHC I expression *in vivo*, transgenic mice were generated to express H-2D^b under the neuron-specific enolase promoter (NSE-D^b mice) (Rall et al. 1995). Neurons extracted from NSE-D^b mice were directly targeted by CTL *in vitro*, and transfer of CTL into NSE-D^b mice persistently infected with LCMV resulted in enhanced infiltration into the CNS, BBB breakdown, and neurological deficits. These findings indicate that low neuronal expression of MHC I is at least one mechanism that protects neurons from CTL-mediated injury.

While it is incredibly difficult to detect neuronal MHC I under steady state or inflammatory conditions, recent studies suggest that MHC I expression on neurons can direct a CTL-mediated disease process (Sanchez-Ruiz et al. 2008). Molecular mimicry between a pathogen and a self-antigen is a powerful means to induce tissue pathology and disease. Neuronal expression of OVA followed by intracerebral infection with *Listeria monocytogenes*-secreting OVA results in an atactic–paretic neurological syndrome during which $CD8^+$ T cells are recruited to MHC I^+ OVA^+ neurons, suggestive of cognate interactions between the two cell populations (Sanchez-Ruiz et al. 2008). *In vitro* studies have also supported the idea that CTL can target and injure MHC I expressing neurons. Quantification of Annexin V^+ apoptotic neurons by microscopy revealed that CTL targeting of hippocampal neurons was dependent on neuronal upregulation of MHC I and the presence of antigen (Medana et al. 2000). Cytoskeletal injury, as evidenced by disruptions in β-tubulin III organization, occurred primarily in the neurites but not in the neuronal cell body. In fact, it was observed that CTL-mediated neurite cleavage could occur within 25 min of contact initiation and was peptide–MHC I-dependent (Medana et al. 2001b). Similar findings were observed in sympathetic neuronal cultures (Manning et al. 1987). Collectively, these *in vivo* and *in vitro* microscopic studies indicate that under certain conditions CTL can target neurons in a MHC I-dependent manner and that neuronal projections appear to be more susceptible to injury than the soma.

CTL can injure neurons using some of the same cytopathic effector mechanisms employed in peripheral tissues. *In vitro* visualization of CTL-induced neuronal intracellular Ca^{2+} flux ($[Ca^{2+}]i$), used to assess neuronal injury, revealed delayed kinetics for $[Ca^{2+}]i$ flux on the order of hours rather than minutes. This timing was consistent with Fas- rather than perforin-mediated neuronal injury (Medana et al. 2000). However, direct application of purified cytotoxic granules to neurons can induce rapid neuronal $[Ca^{2+}]i$ flux characteristic of perforin/granzyme mediated death, and perforin mediated $[Ca^{2+}]i$ flux was observed in FasL deficient neurons because of their inability to inhibit CTL degranulation via engagement of the FasL/Fas pathway (Medana et al. 2001a). Granule mediated lysis of MHC I^+ cerebellar granule neurons was also observed *in vitro* (Rensing-Ehl et al. 1996). Therefore, at least *in vitro* neurons can be damaged by both the perforin/granzyme and FasL/Fas pathways.

Perforin/granzyme-mediated mechanisms of neuronal damage have been implicated in the TMEV model of CNS demyelination and in Rasmussen's encephalitis patients, suggesting that this cytopathic pathway might also be important in injuring neurons *in vivo* (Bien et al. 2002; Murray et al. 1998a). While the majority of published studies suggest that engagement of cognate peptide–MHC I complexes are required for CTL to engage and injure neurons, an *in vitro* study has shown that human fetal neurons can be injured by T cells in a contact-dependent, but peptide–MHC I-independent manner (Giuliani et al. 2003). While this is an interesting possibility, it is not clear whether this mechanism is relevant for immune interactions with mature neurons. Presently, most *in vitro* and *in vivo* studies suggest that injury of neurons by CTL is a peptide–MHC I-dependent process.

Direct engagement of neurons is a very straightforward means for CTL to injure the CNS and is reliant on the ability of neurons to present at least some peptide–MHC complexes. However, an alternative route to neuronal injury could occur indirectly through CTL targeting of the glial cells that support neurons. *In vitro*, CTL targeting of astrocytes was shown to occur in a MHC I-dependent manner (Cabarrocas et al. 2003). Visualization of $[Ca^{2+}]i$ flux in astrocytes exposed to CTL damage revealed that perforin was the primary pathway responsible for CTL-mediated injury of astrocytes (Medana et al. 2001a, 2000) followed by the Fas/FasL pathway if the former mechanism was impaired (Medana et al. 2001a). Confocal studies on tissue samples from Rasmussen's encephalitis patients have also captured images of granzyme B polarization towards astrocytes, suggesting that CTL targeting of astrocyte also occurs *in vivo* (Bauer et al. 2007).

In certain regions of the brain, astrocyte targeting can directly impact neuronal health. This was shown in a model where transferred CTL targeted astrocytes with transgenic expression of the β-gal antigen. Because antigen expression was restricted to astrocytes and not neurons, the bystander killing of neurons observed in the cerebellum was a result of CTL–astrocyte targeting (McPherson et al. 2006). While this phenomenon was observed in the cerebellum, there was little evidence of neuronal death with β-gal expression in retinal astrocytes (McPherson et al. 2006). Therefore, destruction of antigen-expressing astrocytes can occur in the absence of bystander damage. Accordingly, in the white matter of the brain, transgenic hemagglutinin (HA) expression in astrocytes resulted in specific elimination of these cells, without bystander effects, upon transfer of antigen-specific CTL (Cabarrocas et al. 2003). Therefore, impairment of astrocyte support can indirectly result in neuronal injury, but may be depend on the types of neurons located in the anatomical site that is targeted within the CNS.

2.4 *Autoimmunity in the CNS (Experimental Autoimmune Encephalomyelitis)*

Multiple sclerosis (MS) is a complex inflammatory disease of the CNS. Pathologically, MS patients have demyelination and axonal/neuronal damage in the CNS. Although the cause of MS remains unknown, MS is generally thought to be caused by autoimmune targeting of CNS components, including adaptive immune responses to myelin basic protein (MBP), myelin oligodendrocyte glycoprotein (MOG) and proteolipid protein (PLP) (Bernard and de Rosbo 1991; Gold et al. 2006; Kerlero de Rosbo et al. 1993; Lassmann et al. 2007; Siffrin et al. 2007; Sun et al. 1991; Wucherpfennig et al. 1990). Several animal models of experimental autoimmune encephalomyelitis (EAE) have been utilized to study pathological mechanisms associated with CNS autoimmunity. EAE can be induced through several strategies, including immunization with myelin proteins or peptides (e.g., PLP139–151, MOG35–55) in conjunction with adjuvants and pertussis toxin, or, alternatively, through adoptive transfer of myelin-reactive T cells. Recently, a spontaneous model of EAE-like disease was generated using double transgenic mice that have MOG reactivity in both the T and B cell compartments (Krishnamoorthy et al. 2006). Depending on the model used, EAE can vary in: (1) chronicity (monophasic versus relapsing–remitting), (2) pathology (amount of demyelination and axonal/neuronal damage), and (3) components of the adaptive response involved ($CD4^+$ T cells, $CD8^+$ T cells, and B cells) (Gold et al. 2006; Siffrin et al. 2007). Therefore, investigators should proceed cautiously when attempting to link their findings observed in one model to the pathogenesis of human MS.

Myelin-specific $CD4^+$ T cells are typically thought to be involved in the induction of EAE. Autoradiographic analysis of [^{14}C] thymidine-labeled MBP-primed cells after transfer demonstrated homing of these cells into the CNS prior to and during EAE. Using this technique, the majority of cells appeared to be restricted to the perivascular space with little parenchymal infiltration (Cross et al. 1990). Chronological localization of [^{14}C] thymidine-labeled MBP T cells showed that myelin-specific T cells were localized in newer lesions (Cross et al. 1993). Additionally, parenchymal detection of unlabeled infiltrates coincided with clinical symptom kinetics, implicating a role for immune-mediated pathology during disease. More recently, immunohistochemical analysis of T cells in the CNS after MOG35–55 immunization revealed infiltration in the choroid plexus, meninges, and subventricular zones during early phases of disease followed later by additional T cell accumulation around the perivascular regions (Brown and Sawchenko 2007). This suggests that there are two phases of T cell entry during EAE: an early recruitment through the CSF and choroid plexus, followed by a secondary wave of vascular recruitment.

Retroviral transduction of a GFP vector into MBP-specific CD4$^+$ T cells has been utilized to further monitor phenotypes and migratory patterns from the periphery into the CNS during monophasic adoptive EAE progression (Flugel et al. 1999). Initially, *in vitro* activated MBP-specific GFP$^+$ cells trafficked primarily to the parathymic lymph nodes, followed by entry into the blood and spleen by 60 h post transfer. Recruitment into the CNS occurred after 60 h post transfer and was associated with upregulated chemokine receptor expression on splenic MBP-specific GFP$^+$ T cells and depletion of the cells from peripheral sites, including the spleen, lymph nodes, and blood. The kinetics of splenic entry and egress of MBP-specific GFP$^+$ T cells were also confirmed with epifluorescence microscopy (Flugel et al. 2001). Confocal analysis of MBP-specific cells within the spinal cord revealed localization to the meninges, perivascular space, and parenchyma, whereas CD8$^+$ T cells and B cells were mostly confined to the meningeal space. Recruitment of myelin-specific CD4$^+$ T cells into the perivascular space and parenchyma was also observed after transfer of GFP-labeled MOG-specific T cells (Yura et al. 2001), suggesting that this distribution is common to myelin-reactive cells of different specificities.

Once in the CNS, myelin-specific T cells reactivate and upregulate OX-40, IL-2R, and cytokine expression (Flugel et al. 2001). Similar to MBP-specific GFP$^+$ CD4$^+$ T cells, macrophages and reactive microglia (labeled with anti ED1 antibodies) localized to the meninges and parenchyma during EAE (Flugel et al. 2001). Parenchymal entry of macrophages has been associated with myelin-specific CD4$^+$ T cell reactivation in the CNS and subsequent disease severity. Tracking of GFP-labeled autoreactive effector T cells with high or low pathogenic potential showed equal capacities to enter the parenchyma, indicating that CNS entry is not a limiting factor for inducing maximal disease. However, only highly encephalitogenic CD4$^+$ T cells reactivated in the CNS, induced MCP-1 and MIP-1α production, and resulted in ED1$^+$ macrophage infiltration into the parenchyma (Kawakami et al. 2004).

Immunohistochemical and immunofluorescent staining for CD3$^+$ T cells in the CNS of CXCR3$^{-/-}$ mice revealed an altered localization, with a more diffuse parenchymal distribution of both T cells and activated microglia/macrophages (Muller et al. 2007). These data imply that CXCR3 is important in the perivascular retention of T cells during EAE that prevents extensive dispersion throughout the white matter. Because T cells contribute to the secondary recruitment of pathogenic macrophages into the CNS, disruption of T cell localization through blockade of chemotactic signals has the potential to greatly alter the course of disease. By tracking myelin-specific T cells through imaging and flow cytometric methods, many novel insights into EAE disease pathogenesis have been obtained, including the pathogenic link between $CD4^+$ T cells and the secondary recruitment of innate immune cells. With the application of intravital imaging approaches, it now should be possible to further refine the exact contribution of T cells and innate immune cells to disease onset and progression, and to evaluate the efficacy of therapeutics (e.g., antichemotactic agents) in real time.

$CD4^+$ T cells can be subdivided into distinct subsets based on cytokine profiles (Kaiko et al. 2008), and the type of helper cell (e.g., Th1, Th2, or Th17) generated can have a significant impact on subsequent disease progression. Previously, IFNγ-producing, Th1-polarized $CD4^+$ T cells were thought to be essential for EAE. This was based on disease induction following transfer of Th1 polarized cells, IFNγ expression found in EAE lesions, and IFNγ production by T cells isolated from the CNS (Kuchroo et al. 2002). However, mice deficient in molecules associated with the Th1 pathway, including IL-12p35, IL-12Rβ2, IFNγ, IFNγR, or STAT1, had normal or increased disease severity making the role for Th1 cells in EAE controversial (Becher et al. 2002; Bettelli et al. 2004; Cua et al. 2003; Ferber et al. 1996; Gran et al. 2002; Willenborg et al. 1996; Zhang et al. 2003). In recent years, IL-23 was linked with CNS inflammation during EAE. It was observed that EAE could be induced in IL-12 $p35^{-/-}$, but not IL-12 $p40^{-/-}$ (which also acts a subunit of IL-23; Oppmann et al., 2000) or IL-23 $p19^{-/-}$ mice (Cua et al. 2003). IL-23 has been associated with Th17 polarized cells (Bettelli et al. 2007; Langrish et al. 2005; Volpe et al. 2008), defined by IL-17 production, and provides an explanation for the induction of EAE in the absence of Th1 related genes. Similar to what was initially observed with IFNγ, IL-17^+ myelin-specific T cells detected in the CNS by ELISPOT are associated with disease (Hofstetter et al. 2005, 2007). IL-17 is now thought to play a significant role in EAE due to the ability of: (1) IL-23 cultured IL-17 producing cells to induce EAE upon transfer (Langrish et al. 2005), (2) attenuated disease in IL-$17^{-/-}$ mice (Komiyama et al. 2006) or in wild type PLP immunized SJL mice depleted of IL-17 (Langrish et al. 2005) or IL-23 (Chen et al. 2006), and (3) lack of disease in $ROR\alpha^{-/-}ROR\gamma^{-/-}$ mice which lack Th17 cells (Yang et al. 2008). Increased IL-17 expression in MS lesions observed with fluorescent confocal microscopy (Kebir et al. 2007) and gene arrays (Lock et al. 2002) suggest that Th17 cells are also prevalent in human disease.

In vitro, both IL-17 and IL-22 were shown to increase the permeability of brain-derived endothelial cells and reduce the levels of tight junction proteins (Kebir et al. 2007). Electron microscopy revealed the existence of intimate associations with MBP-specific T cells and CNS endothelial cells during extravasation (Raine et al. 1990), creating an opportunity for T cells to release cytokines in close proximity to endothelium. Importantly, confocal microscopy of tissues from MS, but not control, patients demonstrated an increase in IL-17R and IL-22R on CNS endothelium (Kebir et al. 2007), suggesting that these cytokines play a potential role in permeabilization of the BBB during MS. These data have all led to a paradigm shift in the EAE field which suggests that Th17 rather than Th1 polarized T cells are more important for disease pathogenesis; however, IL-12 and Th1 cells may still play a significant role in autoimmune disease and thus their contribution should not be completely ignored. For example, it was recently shown that IL-12 and IL-23 polarized T cells induce different types of EAE that varied in the recruitment of macrophages versus neutrophils (Kroenke et al. 2008). Histological analyses hematoxylin- and eosin-stained CNS sections also revealed that infiltrates localized differently. Transfer of IL-12 polarized cells resulted in meningeal and subpial white matter leukocyte infiltration, whereas IL-23 polarized cells infiltrated more deeply into the CNS parenchyma. Further visualization studies to define (1) the specific distribution of the injected T cells and other cellular subsets, (2) the infiltrating or CNS resident cells that are engaged in situ, and (3) the pathogenic outcome (e.g., apoptosis) of such interactions need to be conducted to define the divergence between IL-12 versus IL-23 pathology. These appear to be two different pathways that lead to indistinguishable clinical disease.

Epitope spreading is a critical process for the induction of relapsing forms of EAE. This phenomenon is characterized by the generation of reactivity to epitopes other than the epitope involved in disease initiation and has been observed during both MS and EAE (Lehmann et al. 1992; McRae et al. 1995; Tuohy et al. 1999; Yu et al. 1996). In the SJL mouse model of EAE, evidence for both intra- and intermolecular epitope spreading exists (McRae et al. 1995). Following PLP 139–151 immunization, spreading to other noncross-reactive PLP epitopes (intramolecular spreading) was observed. PLP reactivity was also observed in mice that received MBP-specific $CD4^+$ T cells, which is an example of intermolecular spreading. Diversification of antigens recognized during the course of EAE contributes significantly to disease recurrence. Evidence supporting this concept includes the association between the timing of epitope spreading and relapse, the ability to transfer disease into naive mice using $CD4^+$ T cells that recognize secondary determinants, and the blockade of relapses by tolerance induction to novel epitopes that only emerge during the epitope spreading phase (Vanderlugt and Miller 2002; Yu et al. 1996). Immunization of SJL mice with PLP 178–191 was followed by immunoreactivity to the PLP 139–151 epitope, which was detected by CFSE dilution of transferred naive PLP 139–151-specific 5B6 $CD4^+$ TCR transgenic cells (McMahon et al. 2005). Interestingly, the diversification of peptide specificity was initiated in the CNS compartment prior to any detectable epitope spreading in the periphery. Isolation of potential APCs localized in the CNS during inflammation, including macrophages, microglia. and DCs, revealed that all three had the capacity to stimulate proliferation in the presence of exogenous antigen. However, only DCs were able to induce naive T cell proliferation with endogenously-derived antigen (McMahon et al. 2005). Visualization of DCs in close proximity to $CD4^+$ T cells and the presence of myelin droplets in CNS DCs (detected by confocal imaging) substantiated the idea that naive myelin-reactive $CD4^+$ T cells were stimulated by DCs in the CNS during EAE (McMahon et al. 2005). DC myelin uptake and juxtaposition to T cells was also noted by immunofluorescence and immunohistology performed on tissues from MS patients, suggesting that DC presentation of myelin antigens to T cells is likely to be relevant in human autoimmune disease (Serafini et al. 2006).

Characterization of CNS DCs during EAE revealed the presence of myeloid, plasmacytoid, and $CD8^+$ DCs (Bailey et al. 2007). Static imagery of the brain and spinal cord demonstrated accumulation of myeloid DCs (mDCs) in areas of inflammation and demyelination where PLP was also detected. In contrast, plasmacytoid DCs were distributed sparsely in uninflamed regions of the CNS. Confocal analyses also revealed close juxtaposition of mDCs with $CD4^+$ TCR transgenic cells specific for PLP 139–151, which become activated during the epitope spreading that follows disease induction with PLP 178–191 (Bailey et al. 2007). Myeloid DCs were shown to induce proliferation of PLP 139–151-reactive $CD4^+$ T cells and cytokine profiles reflective of Th17 lineage cells. Additionally, mDCs isolated from the CNS expressed TGFβ, IL-6 and IL-23 (Bailey et al. 2007), cytokines known to aid in Th17 development (McGeachy et al. 2007), further supporting the role of these DCs in polarizing myelin-reactive $CD4^+$ T cells toward the Th17 lineage. Collectively, these data suggest that the processing of damaged CNS tissue and subsequent presentation by local APCs results in the generation of myelin-reactive Th17 T cells directed against other self-determinants, which then contribute to EAE relapses.

3 Noninvasive Imaging: MRI, mPET, and Bioluminescence

Recently, new technologies for *in vivo* biological research, some which are commonly used for human clinical applications, have been harnessed to monitor alterations that occur over time in animal models of disease. These applications include magnetic resonance imaging (MRI), bioluminescence, and micropositron emission tomography (PET). The major advantage of these systems is the ability to temporally monitor anatomical or biological changes within the same animal in a noninvasive manner. Additionally, the spatial resolution of structures is on the order of millimeters to tens of microns (Pautler 2004; Poeppel and Krause 2008), and both MRI and PET have no real depth limitations (Nair-Gill et al. 2008). These techniques were recently used by immunologists to achieve a better spatiotemporal understanding of tumor growth and immune cell or antibody localization during disease progression. However, it should be noted that we are only at the inception of applying these techniques to basic immunological research, and while we can visualize some aspects of immunity, the ability to examine fine interactions and dynamics of individual cells is currently not feasible due to the limitations in resolution.

MRI is based on the idea that charged particles rotate and have a small magnetic charge associated with their movement. The basic concepts are reviewed in Pautler (2004) and Pirko et al. (2005) and will be only briefly introduced here. MRI is commonly based on the usage of hydrogen (^1H) signals, which has a positively charged proton. Generally, the protons will rotate in random directions, thus canceling out any net movement or magnetic charge. However, when an external magnetic field (B_0) is applied, these spins align themselves with this field. Due to the fairly uniform directional rotation created by the B_0, a net magnetic vector (NMV) or moment is created. Application of a radio frequency perpendicular to the B_0 can displace the protons from the axis in which it was aligned with B_0, generating a new NMV that is at an angle to B_0. Subsequent release of the radio frequency allows for the protons, and thus the NMV, to realign themselves with the B_0. During this process, energy is emitted and exchanged as the nuclei return to alignment with B_0. Because different tissues have different rates of recovery to the position B_0, these discrepancies in energy can be measured and used to resolve anatomical structures. Contrast agents, such as paramagnetic complexes, have the ability to alter the process of B_0 realignment, generating an alteration in the intensity of signals.

MRI has been used to detect immune cells that were labeled *in vitro* prior to injection *in vivo* (Weissleder et al. 1997; Yeh et al. 1993, 1995). More recently, MRI was utilized to track immune cells in two murine models of CNS demyelination: MBP-induced EAE and infection with Theiler's murine encephalomyelitis virus (TMEV) (Pirko et al. 2003, 2004). A novel *in vivo* leukocyte-labeling technique was used in these studies for MRI visualizations. Specifically, antibodies that recognize leukocytic antigens (e.g., CD4 and CD8) were conjugated to magnetic beads (referred to as superparamagnetic antibodies) and then injected into animals prior to MRI analyses. In both systems, these antibodies were successfully used to detect infiltration of leukocytes into the CNS (Pirko et al. 2003, 2004). Therefore, MRI can be used to track *in vivo* labeled cells within the same host over many time points, affording an opportunity to capture noninvasive snapshots of the topography of immune cells in the CNS as a disease progresses. Novel approaches are currently under development to improve the sensitivity of MRI (Zabow et al. 2008); however, current MRI studies of CNS immunity are still confined to yielding information about cells at the population level rather than individual cellular dynamics.

Alternatives to *in vivo* imaging by MRI include bioluminescence or PET. Both depend on the emission of photons that are subsequently detected by the apparatus. In bioluminescence, a reporter gene expressing a light-emitting enzyme, such as luciferase, is expressed in the cells or tissue of interest (Welsh and Kay 2005). Prior to imaging, the substrate is injected into the animal, and this enzymatic reaction results in the release of photons that are subsequently detected by the machine (Negrin and Contag 2006). Similarly, tracking of immune responses by PET also involves expression of PET probes in the cells of interest, either by direct labeling or expression of protein targets for the PET probe. Positrons emitted from the PET isotope-labeled probe interact with tissue electrons and form photons, which can be detected using a microPET machine (Nair-Gill et al. 2008). Quantification of these reporters in both bioluminescence and PET allow for comparisons of the relative amounts of signal accumulated at given time points in controls versus experimental groups. Additionally, using PET it has been suggested that the level of signal intensity correlates with the number of labeled cells present within a region of interest (Su et al. 2004). Application of bioluminescence and PET for tracking leukocytes has mostly been limited to examination of peripheral immunity, although $CD4^+$ T cell trafficking during EAE has been examined using a bioluminescence approach (Costa et al. 2001). Like MRI, these techniques have only recently been applied to tracking leukocyte responses, and in general provide an even lower spatial resolution than MRI (Nair-Gill et al. 2008). Another important aspect to consider when utilizing either PET or bioluminescence is that there is a limitation to the sensitivity of these assays. A certain signal threshold must be achieved prior to detection. Overall, these noninvasive techniques provide an excellent means to examine the general positioning of immune cells in the CNS over time, but further technological advances are required to refine spatial resolution.

4 Dynamic Imaging by Two-Photon Laser Scanning Microscopy

4.1 Introduction/Technical Advances

Neuroscientists have pioneered the usage of multiphoton-based microscopic imaging to assess biological occurrences in living tissue and animals (Denk et al. 1990; Kerr and Denk 2008). Although *in vitro* studies using live cells and microscopy on fixed tissue have yielded important observations, these studies are limited due to removal of cells from their natural environment and the inability to observe dynamic processes, respectively. These technical limitations have been overcome with the inception of two-photon laser scanning microscopy (TPLSM). Based on the principle that simultaneous absorption of multiple low energy photons can combine to reach a high excitation state (equivalent to what is achieved with a single high energy photon), multiphoton-based microscopy has revolutionized the field of *in vivo* imaging. TPLSM is advantageous for tissue microscopy because usage of low energy photons allows for an increased depth penetration and decreased photoxicity, while the imaging field is a focal point, which circumvents the background issues associated with conventional confocal microscopy (Helmchen and Denk 2005; Rocheleau and Piston 2003).

Another advance that has made *in vivo* imaging with TPLSM more feasible is the expression of fluorescent proteins in the cells of interest. Neuroscientists have taken advantage of GFP (Stearns 1995), and other spectral variations, that can be expressed transgenically in mice. For example, thy1 driven expression of XFP (e.g., yellow fluorescent protein, GFP, etc.) labels the entire neuron, which facilitates imaging of everything from axons to dendritic spines (Feng et al. 2000). Recently, immunologists have also begun to utilize TPLSM (Bajenoff and Germain 2007) after labeling immune cells of interest ex vivo and then transferring them *in vivo* to be imaged. A variety of transgenic mice have also been generated that express fluorescent proteins under lineage-specific promoters, which facilitates the tracking of innate and adaptive immune cells. Although initial studies using TPLSM to image the CNS were conducted in the early 1990s (Denk et al. 1994, 1990; Kerr and Denk 2008), the majority of *in vivo* immunological imaging studies conducted to date have focused on the periphery rather than the CNS. Because neuroscientists have paved the way for *in vivo* microscopy in the CNS, we will first briefly review some of the advances made in the field of neuroscience using TPLSM prior to discussing the pioneering studies that have examined CNS immunity.

4.1.1 Neuronal Dendritic Spines

The capacity of the brain to react and rearrange neuronal circuits during development and following experience-driven or environmental changes has been a topic of great interest. To assess the malleability of these connections, researchers have evaluated the stability of dendritic spines (which are protrusions from the dendritic shaft of neurons) as a measure of neuronal plasticity. These structures function as the postsynaptic sites of axons and increase the density of synapses between dendrites and axons (Sorra and Harris 2000). Recently, TPLSM has yielded new insights into the dynamics of spine formation and elimination, the overall stability of the structures, and alterations occurring during experience-dependent manipulations of synaptic plasticity (Alvarez and Sabatini 2007; Pan and Gan 2008).

In vivo imaging of dendritic spine dynamics in normal tissue has demonstrated developmental alterations in neuronal plasticity associated with age. Initial TPLSM studies on dendritic spine stability of apical portions of layer 5 cortical neurons using transgenic mice that express fluorescent proteins in neuronal subsets (Feng et al. 2000) revealed that spines are stable over a period of weeks to months (Grutzendler et al. 2002; Trachtenberg et al. 2002). However, the percentage of stable dendritic spines in young adolescent mice (1–2 months of age) was variable between studies with approximately 50–55% stability for ≥8 days in the somatosensory cortex (Holtmaat et al. 2005; Trachtenberg et al. 2002) and ~73–82% stability over a 1-month or 2-week (Grutzendler et al. 2002) imaging observation period, respectively, in the visual cortex. An increase in overall dendritic spine stability was observed as mice develop and transition into adulthood with approximately 70% stability over 8 days in the somatosensory cortex (Holtmaat et al. 2005) and ~96% over 1 month in the visual cortex (Grutzendler et al. 2002) of ≥4-month-old mice.

Decreased spine elimination contributes to the increased stability of dendritic spines that occurs over time. TPLSM imaging of fluorescent neurons in mice of different ages through thinned skull preparations showed ~17–23% spine elimination over a 1-month period in young mice (1-month-old) that was significantly higher than the ~10% elimination observed in 2-month-old mice, and the ~5–6% elimination observation in mice >4 months of age (Zuo et al. 2005a, b). While spine elimination rates decrease over time, spine formation appears to remain stable (Zuo et al. 2005a, b). TPLSM combined with electron microscopy has shown that generation of new stable spines precedes and results in synaptogenesis, creating new functional neuronal connections (Knott et al. 2006; Nagerl et al. 2007). The rate of spine formation, ~5%, is lower than spine elimination in young mice resulting in a reduction of overall dendritic spines. These studies have shown that as neural networks mature and the number of spines is reduced, the rates of elimination and formation become more equivalent and thus contribute to the long-term stability of dendritic spines with age.

Although there is a general consensus that dendritic spine stability increases over time, the percentage of spines that reach this state is inconsistent between studies. Variability in spine stability may depend on the brain region examined. Differences in dendritic spine motility have been observed in various sensory cortical areas, with lower spine motility observed in neurons located in the visual cortex compared to auditory and somatosensory regions (Majewska et al. 2006), suggesting that localization can impact dendritic spine dynamics. However, the more likely explanation for the disparity between degrees of stability is a technical one based on the marked alterations that occur when the skull is opened to generate a viewing window for *in vivo* imaging. An elegant study by Xu et al. has clearly demonstrated that open skull craniotomy windows result in significantly higher spine turnover in the barrel cortex (~34%) when compared to a thinned skull surgical approach (~6%), which results in less injury to the surface of the brain (Xu et al. 2007). Increased microglial and astrocyte reactivity was observed in tissue sections obtained from mice that underwent craniotomies, but not thinned skull preparations, demonstrating that increased glial activation is associated with more invasive surgeries (Xu et al. 2007). Therefore, the type of cranial window used *in vivo* imaging can greatly impact the behavior of CNS resident cells and must be carefully considered when assessing CNS intravital imaging data.

4.1.2 Microglial Activation

Microglia are CNS-resident innate immune cells that function to monitor and respond to the CNS in both quiescent and pathological states (Garden and Moller 2006; Hanisch and Kettenmann 2007). Previously, the behavior of microglia during normal physiological conditions could not be assessed due to the disruptive nature of the tissue removal process which activated these cells. However, recent studies using TPLSM have provided novel insights into "resting" microglial surveillance of the CNS. Transcranial visualization through a thinned skull of resident microglial-expressing eGFP driven by the fractalkine receptor (CX3CR1) locus (Jung et al. 2000) revealed a homogeneous tissue distribution with approximately 50–60 μm separating individual cells (Nimmerjahn et al. 2005). Time-lapse microscopy revealed that the small somata of microglial generally remained stationary whereas the numerous thin and ramified processes extending from the cell body were highly motile, with continuous random extensions and retractions occurring at similar velocities of ~ 1.5 $\mu m\ min^{-1}$. Due to balanced rates of branch generation and elimination, the overall number of processes remained stable. Filipodia like microglial protrusions, which typically terminated in bulbous ends, were also observed; however, after initial extension, the protrusions were shown to stall for several minutes prior to retraction, suggestive of removal of debris in the healthy brain. Microglia in CX3CR1$^{GFP/+}$ mice were also shown by TPLSM to contact neurons, astrocytes, and blood vessels, demonstrating the high connectivity of microglia with other CNS-resident cells (Nimmerjahn et al. 2005). Based on calculations of protrusion motility rates and the estimated volume of the extracellular space, resting microglia were found to survey the entirety of the brain parenchyma once every few hours.

Microglial dynamics have also been assessed using TPLSM during normal brain development in zebrafish with GFP-tagged microglia (Peri and Nusslein-Volhard 2008). Like mammalian microglia, zebrafish microglia branches underwent extensions and retractions under normal conditions. Apoptotic annexin V^+ DsRed-labeled neurons were phagocytosed by microglia, indicating that microglial neuronal degradation is an essential step in clearance of neuronal debris during CNS development. Further analysis using knockdown studies and TPLSM demonstrated that *in vivo* microglial phagosome to lysosome fusion was dependent on the v-ATPase proton pump, in particular the a1 subunit of the V0 section of this protein (Peri and Nusslein-Volhard 2008), further defining critical elements involved in digestion after microglial engulfment. Thus, under quiescent conditions, "resting" microglia are highly dynamic cells in close contact with other CNS-resident cells and continually monitor the CNS parenchyma and eliminate potentially harmful debris.

In addition to dynamic surveillance of the resting brain, microglia are poised to react quickly to CNS injury and contain damaged tissue. TPLSM time-lapse imaging during acute brain trauma induced by two-photon laser ablation or mechanical injury in the parenchyma of $CX3CR1^{GFP/+}$ mice revealed rapid microglial responses with directed extension of thickened processes towards the injury site that occurred within minutes (Davalos et al. 2005; Kim and Dustin 2006; Nimmerjahn et al. 2005). During this initial phase, minutes to hours postinjury, it was noted that repositioning of microglial processes, but not the somata, to within 90 μm of the microlesion is used to surround and contain the damaged tissue (Davalos et al. 2005; Kim and Dustin 2006; Nimmerjahn et al. 2005). Bulbous termini (Davalos et al. 2005) and inclusions within the microglia, averaging around 4.6 μm in size (Nimmerjahn et al. 2005), suggest that microglial were actively engulfing and degrading tissue debris. Later during injury responses (1–3 days after laser ablation), the somata of activated, ameboid-shaped microglia also converge onto the injury site (Kim and Dustin 2006).

Time-lapse imaging has demonstrated that destruction of an ATP gradient or inhibition of P2Y G protein-coupled metabotropic purinergic receptors, which respond to ATP, is sufficient to impede the reorientation of microglial processes towards the microlesion (Davalos et al. 2005). A potential receptor for this ATP response, the P2Y12 receptor, was shown to be expressed highly on ramified resting microglia and was decreased hours after activation and conversion of microglia to an ameboid state (Haynes et al. 2006). Microglia from $P2Y12^{-/-}$ $CX3CR1^{GFP/+}$ mice have a significantly decreased ability to redirect processes minutes but not hours after laser ablation (Haynes et al. 2006). Therefore, P2Y12 responses are most likely involved in early but not late reorientation of microglial processes after CNS trauma.

After injury, astrocytes, visualized using TPLSM in transgenic mice that express GFP under the control of the glial fibrillary acid protein promoter, polarize toward the lesion in an ATP- and Ca^{2+}-dependent manner. Blockade of connexin hemichannels, which disrupts astrocyte ATP release, prevents redirection of both astrocytes and microglial towards the injury site (Davalos et al. 2005; Kim and Dustin 2006). Because astrocyte responses precede microglial responses (Kim and Dustin 2006), and astrocyte connexin hemichannels are essential for microglial polarization, ATP-induced ATP release from astrocytes after injury is a likely means by which astrocytes and microglia communicate and coordinate their injury responses in the brain parenchyma.

4.1.3 Alzheimer's Disease

Alzheimer's disease (AD), the most common type of dementia, is characterized by the presence of beta amyloid (Aβ) plaques. Amyloid precursor protein (APP), a transmembrane protein expressed by neurons, is cleaved by β-secretase and γ-secretase to form the Aβ polypeptide that is the source of amyloid found in senile plaques (Gotz and Ittner 2008; Kang et al. 1987). The γ-secretase dictates the length of the subsequent Aβ polypeptide, and altered APP metabolism due to mutations in the presenilin components of the γ-secretase has been linked with disease (Scheuner et al. 1996). Cleavage of APP into the longer $A\beta_{42}$ peptide results in the generation of extracellular toxic oligomers that can subsequently aggregate to form plaques (Gotz and Ittner 2008). One of the predominant theories for AD etiology is the amyloid hypothesis, which dictates that imbalanced Aβ production and accumulation initiates and drives pathology (Hardy et al. 1998; Hardy and Selkoe 2002). According to this hypothesis, other hallmark features of AD including neurofibrillary tangles, characterized by intracellular aggregates of the protein tau (Ballatore et al. 2007), are subsequent manifestations of neuronal abnormalities induced by Aβ dysregulation. In recent years, the toxicity of amyloid β oligomers (rather than fibrillar amyloid plaque deposits) has also been investigated as the factor responsible for disease pathogenesis, providing a modified view of the amyloid hypothesis in which soluble mediators become important (Castellani et al. 2008; Klein et al. 2001).

Recently, a series of TPLSM studies were conducted to assess the development of amyloid plaques and their role in the dynamics of neuronal damage in models of AD. Longitudinal *in vivo* imaging of APPswe/PS1d9xYFP mice (B6C3-YFP) (Jankowsky et al. 2001), which have neuronal expression of human APP, presenilin 1, and YFP, revealed that amyloid plaques (marked by methoxy-X04 labeling; Klunk et al. 2002) could form rapidly (Meyer-Luehmann et al. 2008). In order to trace occurrences within the same exact region of the brain over time, vascular and neuronal landmarks, that were distant from affected sites, were used to pinpoint the same location. Daily images of these mice revealed that plaques could be detected in a region previously devoid of plaques 24 h earlier (Meyer-Luehmann et al. 2008). These data indicate that spontaneous plaque formation is a remarkably rapid process that can occur within days rather than slowly over the course of months to years as was previously surmised. Importantly, sequential imaging of mice with fluorescently-tagged neurons (using B6C3-YFP transgenic mice) and microglia (using $PDAPP^{+/-}$ x CX3CR1 $^{GFP/+}$ transgenic mice) demonstrated that both neurite dystrophy and microglial activation occurred after plaque formation (Meyer-Luehmann et al. 2008) and are, therefore, reactionary events rather than initiators of Aβ deposition.

In vivo imaging of established plaques and fluorescent protein-labeled neurons showed significant abnormalities in neurites and dendritic spines in close proximity to the amyloid plaques (D'Amore et al. 2003; Spires-Jones et al. 2007; Spires et al. 2005; Tsai et al. 2004). Although there is a consensus that a "halo" of toxicity surrounds the plaque, the maximal distance of neuronal alterations from the perimeter of the fibrillar deposit was about 15–50 μm (D'Amore et al. 2003; Spires et al. 2005; Tsai et al. 2004). Neuronal dystrophy, disrupted neurite trajectories, and decreased dendritic spine densities were all observed in close proximity to established plaques. Significant axonal swelling, indicative of neuronal dystrophy, was detected in 18% of hippocampal axons located near plaques (Tsai et al. 2004). Disappearance of up to ~36% of neurites was noted over a 4- to 5-week observation period (Tsai et al. 2004); however, others have suggested that neurites are not eliminated rather their trajectories near a plaque are significantly altered giving the appearance of disruption (Spires et al. 2005). Additional acquisition of the z-plane during TPLSM has revealed that neurites curve around plaques rather than penetrate through them (D'Amore et al. 2003; Spires et al. 2005). Dendritic spine densities within 15 μm of the deposit was ~38% lower than regions distal to this point, and in comparison to control animals, there was a ~54% decrease overall (Spires et al. 2005). Temporal *in vivo* imaging TPLSM studies revealed that reduction in the overall number of dendritic spines was due to an increase in the rate of spine elimination that outcompete the slower kinetics of spine formation (Spires-Jones et al. 2007; Tsai et al. 2004). This results in an overall decrease of spine stability in regions near the plaque. Since dendritic spines serve as postsynaptic sites for neurons, significant loss of these structures may attenuate neuronal function during AD.

The mediators that give rise to the zone of toxicity surrounding amyloid plaques are currently unknown, but the distance from the plaque suggests a role for diffusible substances, such as Aβ peptides. Another possibility is reactive oxygen species. TPLSM *in vivo* imaging of oxidation, using fluorescent reporter compounds, demonstrated that reactive oxygen species are associated with thioflavine S positive amyloid plaques, providing one potential mechanism for plaque toxicity (McLellan et al. 2003). The overall loss of dendritic spine stability, significant axonal dystrophy, and altered neurite projections all reflect the types of cellular alterations that may contribute to neuronal dysfunction. Further TPLSM imaging studies are required to more precisely define the toxic mediators that give rise neuronal pathology and subsequent dysfunction.

4.2 Immunity and Autoimmunity in CNS

4.2.1 Dynamics of CNS Autoimmunity

Although many aspects of EAE have been elucidated using flow cytometry, confocal microscopy, and assessment of neurological dysfunction, these methodologies cannot provide any insights into the *in vivo* behavior of immune cells as they contribute to autoimmune disease in the CNS. Recently, TPLSM was used to examine immune cell dynamics for the first time during EAE. To determine the motility of encephalitogenic T cells *in vivo*, MBP-specific $CD4^+$ T cells expressing GFP ($T_{MBP-GFP}$) were transferred into hosts to induce EAE. *In vivo* imaging of spinal cord explants revealed that $T_{MBP-GFP}$ cells fell into two distinct categories of motility: 65% of cells were highly motile, with velocities averaging 6 μm min^{-1}, and the other 35% were stationary. In contrast, GFP^+ OVA-specific T cells cotransferred with unlabeled

MBP-specific T cells showed little confinement with the vast majority (~95%) exhibiting a highly motile phenotype (Kawakami et al. 2005). Confocal microscopy demonstrated the presence of synapse-like structures (marked by TCR/LFA-1 polarization) on $T_{MBP-GFP}$ cells, suggesting a role for cognate peptide–MHC II interactions in the CNS. Consistent with this idea, MHC II blockade resulted in a significant decrease in the percentage of stationary $T_{MBP-GFP}$ cells measured by TPLSM (Kawakami et al. 2005). Similarly, when TPLSM was used to examine a MOG model of EAE, T cell interactions with MHC II$^+$ cells were observed. In this study, a significant portion of cells were found to be stationary (~94%) during the peak of disease (Elyaman et al. 2008). The discrepancy in the number of arrested T cells observed in the two models of EAE suggests motility is influenced by the amount of antigen available. Infusion of soluble MBP during acute disease resulted in a significant increase in the percentage of stationary $T_{MBP-GFP}$ cells, with the majority of cells arresting in the parenchyma within hours of MBP administration (Odoardi et al. 2007). This suggests that additional peptide availability generated in the MOG model by immunization may partially account for the high percentages of stationary cells.

Regardless of antigen availability, MHC II$^+$ cells appear to play a role in both models of EAE. In the MBP model, CD45^{hi} and CD45^{lo} MHC II$^+$ cells were shown to process antigen, presumably allowing for T cell reactivation in the CNS to occur on infiltrating cells or microglia, respectively (Odoardi et al. 2007). Additionally, in the MOG model, meningeal CD11b$^+$ or CD11c$^+$ cells were able to stimulate CD4$^+$ T cells *in vitro*, confirming the stimulatory potential of these APCs (Elyaman et al. 2008). Altogether, these data clearly demonstrate that interactions between encephalitogenic T cells and MHC II$^+$ APCs occur in the CNS, which impact T cell motility and behavior within this compartment. Peptide–MHC II stimulation of encephalitogenic CD4$^+$ T cells within EAE lesions likely results in the reactivation, proliferation, chemokine production, and cytokine release, all culminating in the exacerbation of disease.

Further study of immune cell dynamics and EAE disease pathogenesis using TPLSM will yield additional novel insights into CNS autoimmunity. In the context of EAE, examination of 400-μm vibratome brain sections by TPLSM has facilitated the acquisition of an initial glimpse into the dynamics of T cell–neuronal interactions in a living tissue. Previously, *in vitro* cell culture studies have shown that T cells preferentially interact with neurites and can induce neuronal Ca^{2+} flux (Medana et al. 2000,2001a, b). Consistent with these data, CMTMR-labeled T cell lines applied to brain slices penetrated into the tissue and mostly contacted neurites rather than the neuronal cell body. T cell interactions with neurons also resulted in intracellular neuronal Ca^{2+} oscillations that were presumed to be indicative of cell death. These results suggest that the *in vitro* studies on T cell–neuronal targeting reflect what is observed in ex vivo tissue preparations (Nitsch et al. 2004). However, continued examination of this and other aspects of autoimmune disease pathogenesis will need to be further validated *in vivo* using thinned skull surgical preparations (Xu et al. 2007) or comparable noninjurious *in vivo* approaches. Given the impact that a surgical preparation can have on the dynamics of cells under examination

(Xu et al. 2007), it is important whenever possible to study biological processes in their natural environment. Rapid advances in noninvasive (or minimally invasive) imaging techniques should make this possible.

4.2.2 Antiviral Immunity and Pathogenesis

Understanding the dynamics of CNS antipathogen immunity and the subsequent outcomes of these responses is a fundamental step in the development of more efficacious therapeutics to treat infectious diseases of the CNS. Real time observation of CNS viral infections is particularly challenging due to the size and complexity of the CNS as well as limitations in intravital imaging approaches. Because TPLSM combined with the thinned skull surgical preparation provides minimally invasive means to examine CNS dynamics under natural conditions, we recently set out to capture the first glimpses of a virus-induced disease occurring on the surface of the brain (or meninges). Intracerebral inoculation of adult mice with LCMV induces acute meningitis that results in fatal convulsive seizures (Kang and McGavern 2008). Virus-specific CTL are known to be absolutely required for disease (McGavern and Truong 2004), but the mechanism by which they do so is unknown. To gain novel insights into the pathogenesis of virus-induced meningitis, we recently used TPLSM to first examine the dynamics of GFP-tagged virus-specific CTL in symptomatic mice at day 6 postinfection – a time point when all infected mice present with convulsive seizures (Camenga et al. 1977; Walker et al. 1977). To generate a population of traceable virus-specific CTL to study, we transferred naive $D^bGP_{33–41}$ $CD8^+$ TCR transgenic T cells (P14 GFP^+) intravenously into mice 1 day prior to intracerebral inoculation. This ensured that the naive precursors had the opportunity to become primed with the endogenous repertoire, transition into effector T cells, and migrate into the CNS (and other peripheral tissues) only after receiving the host-derived signals to do so.

TPSLM imaging through a thinned skull viewing window at day 5 postinfection (a time point when all mice were completely asymptomatic) revealed very few CTL surveying the infected meninges and that the meningeal blood vessels were intact (Kim et al. 2008). Meningeal blood vessels were visualized by injecting quantum dots into the blood supply 10 min prior to imaging. In stark contrast to day 5 postinfection, symptomatic mice at day 6 postinfection had a massive number of highly motile CTL throughout the meninges that infrequently penetrated the brain parenchyma. 3D renderings of P14 GFP^+ cells, blood vessels, and skull showed that virus-specific CTL tended to aggregate along meningeal blood vessels, which coincides with the pattern of LCMV infection in the meninges. Meningeal ER-TR7$^+$ stromal cells were found to be the main cell population infected by LCMV on the surface of the brain (Fig. 1).

When the dynamics of P14 CTL were quantified in the LCMV-infected meninges, surprisingly, a highly motile pattern of activity was observed (Kim et al. 2008). Meningeal CTL had an average velocity of 3.4 μm min^{-1}, and only a minority of CTL arrested for very long, suggesting that few cells within the CNS form stable synapses during disease pathogenesis. However, direct injection of anti-MHC I

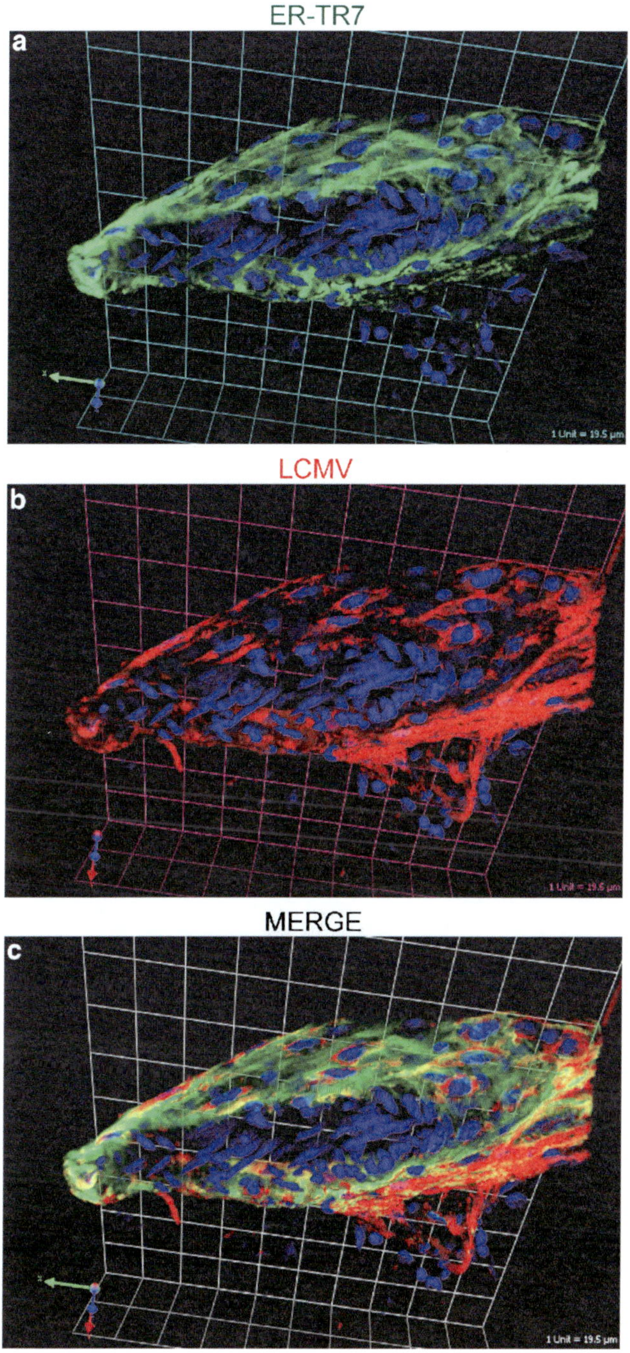

Fig. 1 Localization of LCMV in the meninges following intracerebral inoculation. Sagittal brain vibratome sections obtained from day 6 LCMV-infected mice were stained with antibodies directed against ER-TR7 (*green*) and LCMV (*red*). ER-TR7 is a stromal cell marker that labels fibroblast-like cells. A representative 3D reconstruction of a meningeal blood vessel cross section is shown to illustrate the degree to which LCMV infects fibroblast-like cells that completely surround meningeal blood vessels. (*Grid scale* = 19.5 μm). Cell nuclei are shown in *blue*

antibody into the subarachnoid space revealed that most CTL interacted with MHC I during a 30-min observation period, even if only for a short time. The properties of the LCMV-infected target cell in the meninges (i.e., ER-TR7$^+$ stromal cells) could explain paucity of CTL arrest. Stromal cells may provide strong chemotactic signals that override synapse-forming stop signals (Bromley et al. 2000), which is a possibility that is currently under investigation.

Although CTL are required for LCMV-induced meningitis, the low frequency of stable P14 CTL synapses and the inability of single deficiencies in major CTL effector mechanisms to prevent mortality (Kang and McGavern 2008; Kim et al. 2008; Nansen et al. 1998; Storm et al. 2006; Zajac et al. 2003) suggested an alternative role for CTL in pathogenesis. Analysis of other cellular infiltrates over time revealed a coordinated influx of myelomonocytic cells (i.e., monocytes and neutrophils) only on day 6 postinfection when CTL were present within the CNS (Kim et al. 2008), and these cells localized to the lining of the brain (Fig. 2). Interestingly, TPSLM imaging of myelomonocytic cells in symptomatic mice using lysozyme M-GFP (LysM-GFP) mice (Faust et al. 2000) revealed a synchronized extravasation of myelomonocytic cells from meningeal blood vessels, which resulted in a severe and sustained injury to the vessel walls (Fig. 2). Myelomonocytic cell extravasation correlated positively with sustained leakage of quantum dots from meningeal blood vessels, suggesting that these cells (not CTL) were directly responsible for the onset of fatal convulsive seizures. However, when neutrophils were depleted from mice, no alteration in disease kinetics or the onset of fatal seizures was observed. When TPSLM was again performed in LysM-GFP mice depleted of neutrophils, extravascular GFP$^+$ cells (presumably monocytes/ macrophages) were found to be associated with transient bursts of vascular leakage. Thus, in the absence of neutrophils, monocytes retained the capacity to fatally injure meningeal blood vessels. Only when neutrophils and monocytes were depleted simultaneously was survival extended and acute onset convulsive seizures prevented. These data suggest that neutrophils and monocytes share the capacity to cause vascular leakage during virus-induced meningitis and that neutrophils injure blood vessels through the process of synchronized extravasation, whereas monocytes position themselves extravascularly and mediate vascular damage through another mechanism, possibly adhesion (Ancuta et al. 2004) or chemokine release (Stamatovic et al. 2005).

A role for seizure in mediating rapid onset death in this model is supported by a similar extension in survival observed following anticonvulsant treatment (Walker et al. 1977). It is likely that, once seizure-induced death is averted, infected mice ultimately succumb to another pathogenic mechanism possibly mediated by CTL (Camenga et al. 1977; Walker et al. 1977). To understand precisely how a meningeal disorder could trigger fatal convulsive seizures, it was important to quantify and visualize the degree to which vascular integrity was compromised in LCMV-infected mice. BBB breakdown alone was linked previously to seizure induction in a different model (Marchi et al. 2007). In symptomatic LCMV-infected mice, Evans blue injected intravenously was found leaking from meningeal blood vessels sometimes

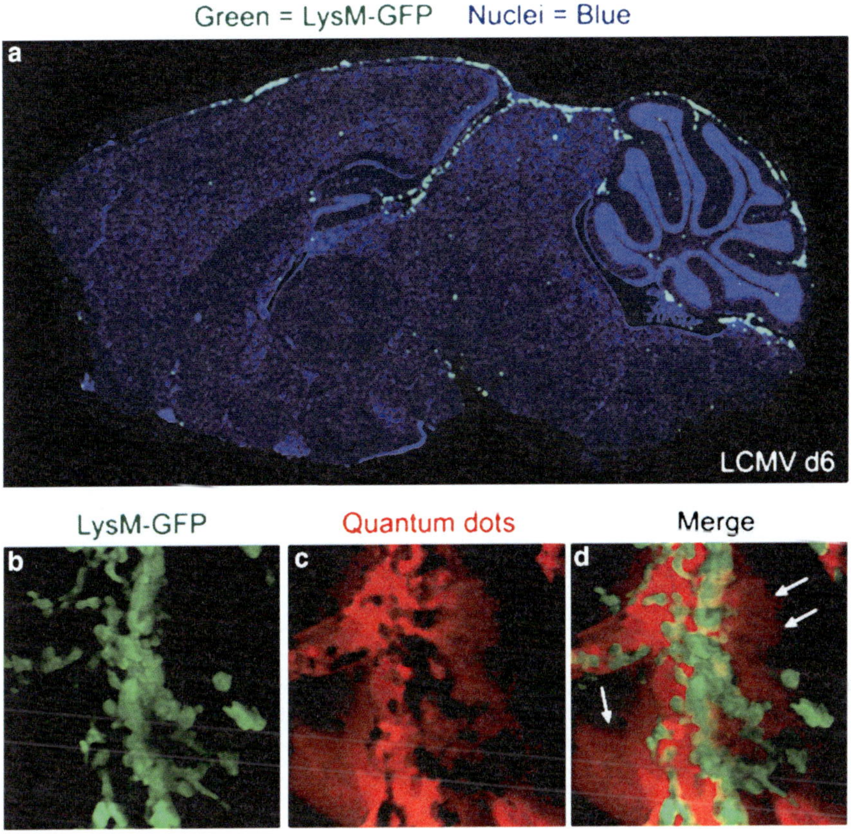

Fig. 2 Injury to meningeal blood vessels mediated by myelomonocytic cells. (**a**) Myelomonocytic cells (i.e., neutrophils and monocytes) were visualized by infecting lysozyme M-GFP transgenic mice intracerebrally with LCMV. The distribution of LysM-GFP$^+$ myelomonocytic cells (*green*) on a sagittal brain reconstruction is shown for a representative symptomatic mouse at day 6 postinfection. Note that LysM-GFP$^+$ cells localize primarily to the meninges and ependyma in LCMV-infected mice. Cell nuclei are shown in *blue*. (**b–d**) A representative 3D reconstruction of two-photon z-stack depicts severe injury to a meningeal blood vessel in a symptomatic LysM-GFP mouse at day 6 postinfection. Quantum dots (*red*) were injected intravenously to visualize vascular integrity. As LysM-GFP$^+$ myelomonocytic cells (*green*) extravasate from meningeal blood vessels, they cause severe vascular leakage. Note the loss in vascular integrity and the leakage of quantum dots into the surrounding meningeal space (denoted with *white arrows*)

hundreds of microns down into the brain parenchyma (Fig. 3). Moreover, when quantified, only the simultaneous depletion of both monocytes and neutrophils resulted in reduced Evans blue leakage. These data suggest that injury to meningeal blood vessels mediated by an overly robust antiviral immune response has the potential to induce fatal convulsive seizures by causing blood-derived material to leak into the brain parenchyma and perturb neuronal homeostasis.

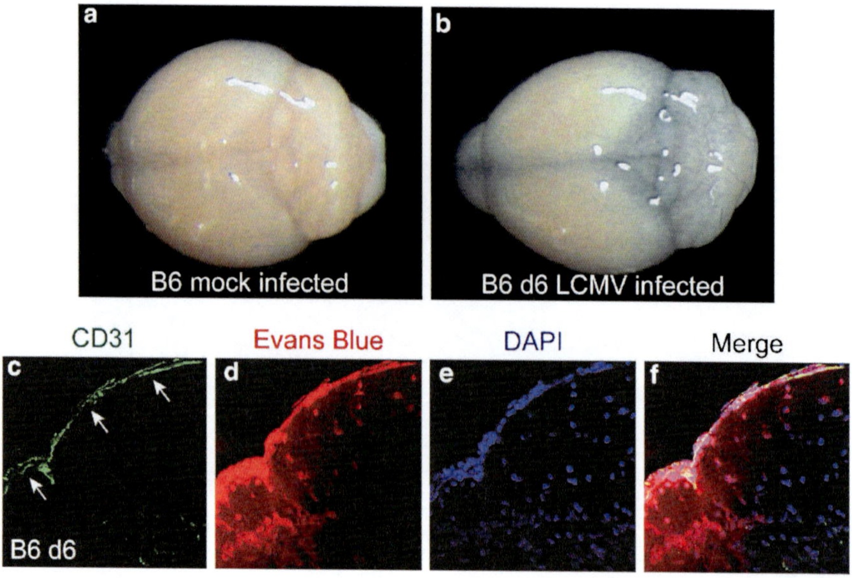

Fig. 3 Leakage of Evans blue into the brain parenchyma of day 6 LCMV-infected mice. (**a**–**b**) To further examine vascular integrity during LCMV-induced meningitis, Evans blue dye was injected intravenously into mice. A representative brain is shown for B6 mock infected (**a**) and a day 6 LCMV-infected mouse (**b**). Note the heavy leakage of Evans blue dye (*blue coloration*) into the brain of the day 6 infected but not the mock-infected mouse. (**c**–**f**) The distribution of Evans blue (*red*) in relation to $CD31^+$ meningeal blood vessels (*green; white arrows*) was examined by confocal microscopy in the brains of mock and LCMV-infected mice at day 6. Minimal to no Evans blue signal was detected in the brains of mock-infected control mice (data not shown). In contrast, severe Evans blue leakage from meningeal blood vessels into brain parenchyma was observed in day 6 LCMV-infected mice. This is indicative of blood brain barrier breakdown

It will be important to understand why $CD8^+$ T cells recruit myelomonocytic cells to a site of viral infection. $CD4^+$ Th17 cells recruit neutrophils in the context of extracellular bacterial infection and gut homeostasis (Weaver et al. 2007). It is unclear why $CD8^+$ T cells would recruit myelomonocytic cells during an acute antiviral response. We observed that LCMV-specific CTL directly produced CCL-3, 4, and 5 (chemokines important in the recruitment of myelomonocytic cells), and late depletion of $CD8^+$ T cells prevented disease onset and significantly reduced the influx of myelomonocytic cells (Kim et al. 2008). It is possible that myelomonocytic cell recruitment is part of a postresponse tissue repair process, but when $CD8^+$ T cells are unable to engage in efficient cytotoxic synapses with ER-$TR7^+$ stromal cells in the meninges, this repair response may itself become proinflammatory and neurotoxic. Genetic defects in selected CTL effector mechanisms result in hemophagocytic syndromes in which myelomonocytic cells engage in widespread and fatal tissue destruction (Menasche et al. 2005). Perhaps ineffective CTL-mediated killing is naturally linked to an aggressive innate response as seen in LCMV-induced meningitis. This may be why loss of additional effector mechanisms by $CD8^+$ T cells in our genetic analyses had no impact on disease (Kim et al. 2008). In any case, recognizing the complementary pathogenic functions of neutrophils and monocytes is critical for devising therapeutic approaches in $CD8^+$ T cell-mediated pathology.

5 Concluding Remarks

To date, visualization of CNS immunity has mostly been limited to static imagery. These studies have yielded novel insights into areas such as CNS peripheral antigen delivery, anatomical localization of leukocyte responses in the CNS, and formation of functionally relevant structures such as the immunological synapse. However, these studies have their own limitations, the most important of which is the inability to examine cellular dynamics in the living tissue. By following the methodological footprints of neuroscientists who have conducted ground-breaking research in the nervous system using intravital imaging approaches, immunologists are now poised to redefine our understanding of CNS immune responses in real time during health and disease. Recent advances in our ability to visualize CNS immunity open innumerable opportunities to study how immune cells influence CNS residents (e.g., glial cells and neurons), how pathogens are handled when they occupy different CNS niches, and how the immune system participates in pathology and repair when the CNS is faced with autoimmunity, infection, and degenerative diseases, etc. Real time imaging enables investigators to gain highly novel insights into CNS processes that could not otherwise be obtained using alternative approaches. With dynamic information in hand, existing hypotheses must often be reformulated to account for actuality. TPLSM and other noninvasive imaging approaches have the potential to provide the most comprehensive understanding to date of two highly evolved and extremely complex mammalian systems: the immune and central nervous systems. This enhanced understanding of CNS immunity should facilitate our ability to more readily purge infectious agents and thwart autoimmune and degenerative disease processes.

Acknowledgments This work was supported by National Institutes of Health grants AI070967-01 and AI075298-01, a grant from The Burroughs Wellcome Fund, and a grant from The Ray Thomas Edwards Foundation (all to D.B.M.) S.S.K. is presently supported by a National Research Service Award (NS061447-01). We would like to thank Drs. Jiyun Kim and Michael Dustin at New York University for providing the image shown in Fig. 2b–d.

References

Alvarez VA, Sabatini BL (2007) Anatomical and physiological plasticity of dendritic spines. Annu Rev Neurosci 30:79–97

Ancuta P, Moses A, Gabuzda D (2004) Transendothelial migration of CD16+ monocytes in response to fractalkine under constitutive and inflammatory conditions. Immunobiology 209:11–20

Bailey SL, Schreiner B, McMahon EJ, Miller SD (2007) CNS myeloid DCs presenting endogenous myelin peptides 'preferentially' polarize CD4+ T(H)-17 cells in relapsing EAE. Nat Immunol 8:172–180

Bajenoff M, Germain RN (2007) Seeing is believing: a focus on the contribution of microscopic imaging to our understanding of immune system function. Eur J Immunol 37(Suppl 1):S18–S33

Ballatore C, Lee VM, Trojanowski JQ (2007) Tau-mediated neurodegeneration in Alzheimer's disease and related disorders. Nat Rev Neurosci 8:663–672

Barcia C, Thomas CE, Curtin JF, King GD, Wawrowsky K, Candolfi M, Xiong WD, Liu C, Kroeger K, Boyer O, Kupiec-Weglinski J, Klatzmann D, Castro MG, Lowenstein PR (2006) In vivo mature immunological synapses forming SMACs mediate clearance of virally infected astrocytes from the brain. J Exp Med 203:2095–2107

Barcia C, Wawrowsky K, Barrett RJ, Liu C, Castro MG, Lowenstein PR (2008) In vivo polarization of IFN-gamma at Kupfer and non-Kupfer immunological synapses during the clearance of virally infected brain cells. J Immunol 180:1344–1352

Bauer J, Elger CE, Hans VH, Schramm J, Urbach H, Lassmann H, Bien CG (2007) Astrocytes are a specific immunological target in Rasmussen's encephalitis. Ann Neurol 62:67–80

Becher B, Durell BG, Noelle RJ (2002) Experimental autoimmune encephalitis and inflammation in the absence of interleukin-12. J Clin Invest 110:493–497

Bernard CC, de Rosbo NK (1991) Immunopathological recognition of autoantigens in multiple sclerosis. Acta Neurol (Napoli) 13:171–178

Bettelli E, Sullivan B, Szabo SJ, Sobel RA, Glimcher LH, Kuchroo VK (2004) Loss of T-bet, but not STAT1, prevents the development of experimental autoimmune encephalomyelitis. J Exp Med 200:79–87

Bettelli E, Korn T, Kuchroo VK (2007) Th17: the third member of the effector T cell trilogy. Curr Opin Immunol 19:652–657

Bien CG, Bauer J, Deckwerth TL, Wiendl H, Deckert M, Wiestler OD, Schramm J, Elger CE, Lassmann H (2002) Destruction of neurons by cytotoxic T cells: a new pathogenic mechanism in Rasmussen's encephalitis. Ann Neurol 51:311–318

Boulanger LM, Shatz CJ (2004) Immune signalling in neural development, synaptic plasticity and disease. Nat Rev Neurosci 5:521–531

Bradbury MW, Cole DF (1980) The role of the lymphatic system in drainage of cerebrospinal fluid and aqueous humour. J Physiol 299:353–365

Bradbury MW, Westrop RJ (1983) Factors influencing exit of substances from cerebrospinal fluid into deep cervical lymph of the rabbit. J Physiol 339:519–534

Bradbury MW, Cserr HF, Westrop RJ (1981) Drainage of cerebral interstitial fluid into deep cervical lymph of the rabbit. Am J Physiol 240:F329–F336

Bromley SK, Peterson DA, Gunn MD, Dustin ML (2000) Cutting edge: hierarchy of chemokine receptor and TCR signals regulating T cell migration and proliferation. J Immunol 165:15–19

Brown DA, Sawchenko PE (2007) Time course and distribution of inflammatory and neurodegenerative events suggest structural bases for the pathogenesis of experimental autoimmune encephalomyelitis. J Comp Neurol 502:236–260

Cabarrocas J, Bauer J, Piaggio E, Liblau R, Lassmann H (2003) Effective and selective immune surveillance of the brain by MHC class I-restricted cytotoxic T lymphocytes. Eur J Immunol 33:1174–1182

Camenga DL, Walker DH, Murphy FA (1977) Anticonvulsant prolongation of survival in adult murine lymphocytic choriomeningitis. I. Drug treatment and virologic studies. J Neuropathol Exp Neurol 36:9–20

Castellani RJ, Lee HG, Zhu X, Perry G, Smith MA (2008) Alzheimer disease pathology as a host response. J Neuropathol Exp Neurol 67:523–531

Ceredig R, Allan JE, Tabi Z, Lynch F, Doherty PC (1987) Phenotypic analysis of the inflammatory exudate in murine lymphocytic choriomeningitis. J Exp Med 165:1539–1551

Chen Y, Langrish CL, McKenzie B, Joyce-Shaikh B, Stumhofer JS, McClanahan T, Blumenschein W, Churakovsa T, Low J, Presta L, Hunter CA, Kastelein RA, Cua DJ (2006) Anti-IL-23 therapy inhibits multiple inflammatory pathways and ameliorates autoimmune encephalomyelitis. J Clin Invest 116:1317–1326

Chua MM, MacNamara KC, San Mateo L, Shen H, Weiss SR (2004) Effects of an epitope-specific CD8+ T-cell response on murine coronavirus central nervous system disease: protection from virus replication and antigen spread and selection of epitope escape mutants. J Virol 78:1150–1159

Costa GL, Sandora MR, Nakajima A, Nguyen EV, Taylor-Edwards C, Slavin AJ, Contag CH, Fathman CG, Benson JM (2001) Adoptive immunotherapy of experimental autoimmune encephalomyelitis via T cell delivery of the IL-12 p40 subunit. J Immunol 167:2379–2387

Cross AH, Cannella B, Brosnan CF, Raine CS (1990) Homing to central nervous system vasculature by antigen-specific lymphocytes. I. Localization of 14C-labeled cells during acute, chronic, and relapsing experimental allergic encephalomyelitis. Lab Invest 63:162–170

Cross AH, O'Mara T, Raine CS (1993) Chronologic localization of myelin-reactive cells in the lesions of relapsing EAE: implications for the study of multiple sclerosis. Neurology 43:1028–1033

Cserr HF, Harling-Berg CJ, Knopf PM (1992) Drainage of brain extracellular fluid into blood and deep cervical lymph and its immunological significance. Brain Pathol 2:269–276

Cua DJ, Sherlock J, Chen Y, Murphy CA, Joyce B, Seymour B, Lucian L, To W, Kwan S, Churakova T, Zurawski S, Wiekowski M, Lira SA, Gorman D, Kastelein RA, Sedgwick JD (2003) Interleukin-23 rather than interleukin-12 is the critical cytokine for autoimmune inflammation of the brain. Nature 421:744–748

D'Amore JD, Kajdasz ST, McLellan ME, Bacskai BJ, Stern EA, Hyman BT (2003) In vivo multiphoton imaging of a transgenic mouse model of Alzheimer disease reveals marked thioflavine-S-associated alterations in neurite trajectories. J Neuropathol Exp Neurol 62:137–145

Davalos D, Grutzendler J, Yang G, Kim JV, Zuo Y, Jung S, Littman DR, Dustin ML, Gan WB (2005) ATP mediates rapid microglial response to local brain injury in vivo. Nat Neurosci 8:752–758

Davis MM, Krogsgaard M, Huse M, Huppa J, Lillemeier BF, Li QJ (2007) T cells as a self-referential, sensory organ. Annu Rev Immunol 25:681–695

Denk W, Strickler JH, Webb WW (1990) Two-photon laser scanning fluorescence microscopy. Science 248:73–76

Denk W, Delaney KR, Gelperin A, Kleinfeld D, Strowbridge BW, Tank DW, Yuste R (1994) Anatomical and functional imaging of neurons using 2-photon laser scanning microscopy. J Neurosci Methods 54:151–162

Dustin ML (2008) T-cell activation through immunological synapses and kinapses. Immunol Rev 221:77–89

Elyaman W, Kivisakk P, Reddy J, Chitnis T, Raddassi K, Imitola J, Bradshaw E, Kuchroo VK, Yagita H, Sayegh MH, Khoury SJ (2008) Distinct functions of autoreactive memory and effector CD4+ T cells in experimental autoimmune encephalomyelitis. Am J Pathol 173:411–422

Faust N, Varas F, Kelly LM, Heck S, Graf T (2000) Insertion of enhanced green fluorescent protein into the lysozyme gene creates mice with green fluorescent granulocytes and macrophages. Blood 96:719–726

Feng G, Mellor RH, Bernstein M, Keller-Peck C, Nguyen QT, Wallace M, Nerbonne JM, Lichtman JW, Sanes JR (2000) Imaging neuronal subsets in transgenic mice expressing multiple spectral variants of GFP. Neuron 28:41–51

Ferber IA, Brocke S, Taylor-Edwards C, Ridgway W, Dinisco C, Steinman L, Dalton D, Fathman CG (1996) Mice with a disrupted IFN-gamma gene are susceptible to the induction of experimental autoimmune encephalomyelitis (EAE). J Immunol 156:5–7

Flugel A, Willem M, Berkowicz T, Wekerle H (1999) Gene transfer into CD4+ T lymphocytes: green fluorescent protein-engineered, encephalitogenic T cells illuminate brain autoimmune responses. Nat Med 5:843–847

Flugel A, Berkowicz T, Ritter T, Labeur M, Jenne DE, Li Z, Ellwart JW, Willem M, Lassmann H, Wekerle H (2001) Migratory activity and functional changes of green fluorescent effector cells before and during experimental autoimmune encephalomyelitis. Immunity 14:547–560

Fung-Leung WP, Kundig TM, Zinkernagel RM, Mak TW (1991) Immune response against lymphocytic choriomeningitis virus infection in mice without CD8 expression. J Exp Med 174:1425–1429

Gairin JE, Mazarguil H, Hudrisier D, Oldstone MB (1995) Optimal lymphocytic choriomeningitis virus sequences restricted by H-2Db major histocompatibility complex class I molecules and presented to cytotoxic T lymphocytes. J Virol 69:2297–2305

Gallimore A, Hombach J, Dumrese T, Rammensee HG, Zinkernagel RM, Hengartner H (1998) A protective cytotoxic T cell response to a subdominant epitope is influenced by the stability of the MHC class I/peptide complex and the overall spectrum of viral peptides generated within infected cells. Eur J Immunol 28:3301–3311

Garden GA, Moller T (2006) Microglia biology in health and disease. J Neuroimmune Pharmacol 1:127–137

Giuliani F, Goodyer CG, Antel JP, Yong VW (2003) Vulnerability of human neurons to T cell-mediated cytotoxicity. J Immunol 171:368–379

Gold R, Linington C, Lassmann H (2006) Understanding pathogenesis and therapy of multiple sclerosis via animal models: 70 years of merits and culprits in experimental autoimmune encephalomyelitis research. Brain 129:1953–1971

Goldmann J, Kwidzinski E, Brandt C, Mahlo J, Richter D, Bechmann I (2006) T cells traffic from brain to cervical lymph nodes via the cribroid plate and the nasal mucosa. J Leukoc Biol 80:797–801

Gotz J, Ittner LM (2008) Animal models of Alzheimer's disease and frontotemporal dementia. Nat Rev Neurosci 9:532–544

Grakoui A, Bromley SK, Sumen C, Davis MM, Shaw AS, Allen PM, Dustin ML (1999) The immunological synapse: a molecular machine controlling T cell activation. Science 285:221–227

Gran B, Zhang GX, Yu S, Li J, Chen XH, Ventura ES, Kamoun M, Rostami A (2002) IL-12p35-deficient mice are susceptible to experimental autoimmune encephalomyelitis: evidence for redundancy in the IL-12 system in the induction of central nervous system autoimmune demyelination. J Immunol 169:7104–7110

Grutzendler J, Kasthuri N, Gan WB (2002) Long-term dendritic spine stability in the adult cortex. Nature 420:812–816

Hanisch UK, Kettenmann H (2007) Microglia: active sensor and versatile effector cells in the normal and pathologic brain. Nat Neurosci 10:1387–1394

Hardy J, Selkoe DJ (2002) The amyloid hypothesis of Alzheimer's disease: progress and problems on the road to therapeutics. Science 297:353–356

Hardy J, Duff K, Hardy KG, Perez-Tur J, Hutton M (1998) Genetic dissection of Alzheimer's disease and related dementias: amyloid and its relationship to tau. Nat Neurosci 1:355–358

Haynes SE, Hollopeter G, Yang G, Kurpius D, Dailey ME, Gan WB, Julius D (2006) The P2Y12 receptor regulates microglial activation by extracellular nucleotides. Nat Neurosci 9:1512–1519

Helmchen F, Denk W (2005) Deep tissue two-photon microscopy. Nat Methods 2:932–940

Herrup K, Yang Y (2007) Cell cycle regulation in the postmitotic neuron: oxymoron or new biology? Nat Rev Neurosci 8:368–378

Hofstetter HH, Targoni OS, Karulin AY, Forsthuber TG, Tary-Lehmann M, Lehmann PV (2005) Does the frequency and avidity spectrum of the neuroantigen-specific T cells in the blood mirror the autoimmune process in the central nervous system of mice undergoing experimental allergic encephalomyelitis? J Immunol 174:4598–4605

Hofstetter HH, Toyka KV, Tary-Lehmann M, Lehmann PV (2007) Kinetics and organ distribution of IL-17-producing CD4 cells in proteolipid protein 139–151 peptide-induced experimental autoimmune encephalomyelitis of SJL mice. J Immunol 178:1372–1378

Holtmaat AJ, Trachtenberg JT, Wilbrecht L, Shepherd GM, Zhang X, Knott GW, Svoboda K (2005) Transient and persistent dendritic spines in the neocortex in vivo. Neuron 45:279–291

Huh GS, Boulanger LM, Du H, Riquelme PA, Brotz TM, Shatz CJ (2000) Functional requirement for class I MHC in CNS development and plasticity. Science 290:2155–2159

Hunter JV, Batchelder KF, Lo EH, Wolf GL (1995) Imaging techniques for in vivo quantitation of extracranial lymphatic drainage of the brain. Neuropathol Appl Neurobiol 21:185–188

Huppa JB, Davis MM (2003) T-cell-antigen recognition and the immunological synapse. Nat Rev Immunol 3:973–983

Huse M, Lillemeier BF, Kuhns MS, Chen DS, Davis MM (2006) T cells use two directionally distinct pathways for cytokine secretion. Nat Immunol 7:247–255

Iannetti P, Nigro G, Imperato C (1991) Cytomegalovirus encephalitis and ganciclovir. Lancet 337:373

Jankowsky JL, Slunt HH, Ratovitski T, Jenkins NA, Copeland NG, Borchelt DR (2001) Co-expression of multiple transgenes in mouse CNS: a comparison of strategies. Biomol Eng 17:157–165

Joly E, Oldstone MB (1992) Neuronal cells are deficient in loading peptides onto MHC class I molecules. Neuron 8:1185–1190

Joly E, Mucke L, Oldstone MB (1991) Viral persistence in neurons explained by lack of major histocompatibility class I expression. Science 253:1283–1285

Jung S, Aliberti J, Graemmel P, Sunshine MJ, Kreutzberg GW, Sher A, Littman DR (2000) Analysis of fractalkine receptor CX(3)CR1 function by targeted deletion and green fluorescent protein reporter gene insertion. Mol Cell Biol 20:4106–4114

Kaiko GE, Horvat JC, Beagley KW, Hansbro PM (2008) Immunological decision-making: how does the immune system decide to mount a helper T-cell response? Immunology 123:326–338

Kang SS, McGavern DB (2008) Lymphocytic choriomeningitis infection of the central nervous system. Front Biosci 13:4529–4543

Kang J, Lemaire HG, Unterbeck A, Salbaum JM, Masters CL, Grzeschik KH, Multhaup G, Beyreuther K, Muller-Hill B (1987) The precursor of Alzheimer's disease amyloid A4 protein resembles a cell-surface receptor. Nature 325:733–736

Karman J, Ling C, Sandor M, Fabry Z (2004a) Dendritic cells in the initiation of immune responses against central nervous system-derived antigens. Immunol Lett 92:107–115

Karman J, Ling C, Sandor M, Fabry Z (2004b) Initiation of immune responses in brain is promoted by local dendritic cells. J Immunol 173:2353–2361

Kawakami N, Lassmann S, Li Z, Odoardi F, Ritter T, Ziemssen T, Klinkert WE, Ellwart JW, Bradl M, Krivacic K, Lassmann H, Ransohoff RM, Volk HD, Wekerle H, Linington C, Flugel A (2004) The activation status of neuroantigen-specific T cells in the target organ determines the clinical outcome of autoimmune encephalomyelitis. J Exp Med 199:185–197

Kawakami N, Nagerl UV, Odoardi F, Bonhoeffer T, Wekerle H, Flugel A (2005) Live imaging of effector cell trafficking and autoantigen recognition within the unfolding autoimmune encephalomyelitis lesion. J Exp Med 201:1805–1814

Kebir H, Kreymborg K, Ifergan I, Dodelet-Devillers A, Cayrol R, Bernard M, Giuliani F, Arbour N, Becher B, Prat A (2007) Human TH17 lymphocytes promote blood-brain barrier disruption and central nervous system inflammation. Nat Med 13:1173–1175

Kerlero de Rosbo N, Milo R, Lees MB, Burger D, Bernard CC, Ben-Nun A (1993) Reactivity to myelin antigens in multiple sclerosis. Peripheral blood lymphocytes respond predominantly to myelin oligodendrocyte glycoprotein. J Clin Invest 92:2602–2608

Kerr JN, Denk W (2008) Imaging in vivo: watching the brain in action. Nat Rev Neurosci 9:195–205

Kida S, Pantazis A, Weller RO (1993) CSF drains directly from the subarachnoid space into nasal lymphatics in the rat. Anatomy, histology and immunological significance. Neuropathol Appl Neurobiol 19:480–488

Kida S, Weller RO, Zhang ET, Phillips MJ, Iannotti F (1995) Anatomical pathways for lymphatic drainage of the brain and their pathological significance. Neuropathol Appl Neurobiol 21:181–184

Kim JV, Dustin ML (2006) Innate response to focal necrotic injury inside the blood-brain barrier. J Immunol 177:5269–5277

Kim JV, Kang SS, Dustin ML, McGavern DB. (2009) Myelomonocytic cell recruitment causes fatal CNS vascular injury during acute viral meningitis. Nature 457:191–195

Klein WL, Krafft GA, Finch CE (2001) Targeting small Abeta oligomers: the solution to an Alzheimer's disease conundrum? Trends Neurosci 24:219–224

Klunk WE, Bacskai BJ, Mathis CA, Kajdasz ST, McLellan ME, Frosch MP, Debnath ML, Holt DP, Wang Y, Hyman BT (2002) Imaging Abeta plaques in living transgenic mice with multi-photon microscopy and methoxy-X04, a systemically administered Congo red derivative. J Neuropathol Exp Neurol 61:797–805

Knott GW, Holtmaat A, Wilbrecht L, Welker E, Svoboda K (2006) Spine growth precedes synapse formation in the adult neocortex in vivo. Nat Neurosci 9:1117–1124

Komiyama Y, Nakae S, Matsuki T, Nambu A, Ishigame H, Kakuta S, Sudo K, Iwakura Y (2006) IL-17 plays an important role in the development of experimental autoimmune encephalomyelitis. J Immunol 177:566–573

Krishnamoorthy G, Lassmann H, Wekerle H, Holz A (2006) Spontaneous opticospinal encephalomyelitis in a double-transgenic mouse model of autoimmune T cell/B cell cooperation. J Clin Invest 116:2385–2392

Kroenke MA, Carlson TJ, Andjelkovic AV, Segal BM (2008) IL-12- and IL-23-modulated T cells induce distinct types of EAE based on histology, CNS chemokine profile, and response to cytokine inhibition. J Exp Med 205:1535–1541

Kuchroo VK, Anderson AC, Waldner H, Munder M, Bettelli E, Nicholson LB (2002) T cell response in experimental autoimmune encephalomyelitis (EAE): role of self and cross-reactive antigens in shaping, tuning, and regulating the autopathogenic T cell repertoire. Annu Rev Immunol 20:101–123

Langrish CL, Chen Y, Blumenschein WM, Mattson J, Basham B, Sedgwick JD, McClanahan T, Kastelein RA, Cua DJ (2005) IL-23 drives a pathogenic T cell population that induces autoimmune inflammation. J Exp Med 201:233–240

Lassmann H, Bruck W, Lucchinetti CF (2007) The immunopathology of multiple sclerosis: an overview. Brain Pathol 17:210–218

Lehmann PV, Forsthuber T, Miller A, Sercarz EE (1992) Spreading of T-cell autoimmunity to cryptic determinants of an autoantigen. Nature 358:155–157

Levite M, Fleidervish IA, Schwarz A, Pelled D, Futerman AH (1999) Autoantibodies to the glutamate receptor kill neurons via activation of the receptor ion channel. J Autoimmun 13:61–72

Ling C, Sandor M, Fabry Z (2003) In situ processing and distribution of intracerebrally injected OVA in the CNS. J Neuroimmunol 141:90–98

Ling C, Sandor M, Suresh M, Fabry Z (2006) Traumatic injury and the presence of antigen differentially contribute to T-cell recruitment in the CNS. J Neurosci 26:731–741

Ling C, Verbny YI, Banks MI, Sandor M, Fabry Z (2008) In situ activation of antigen-specific CD8(+) T cells in the presence of antigen in organotypic brain slices. J Immunol 180:8393–8399

Lledo L, Gegundez MI, Saz JV, Bahamontes N, Beltran M (2003) Lymphocytic choriomeningitis virus infection in a province of Spain: analysis of sera from the general population and wild rodents. J Med Virol 70:273–275

Lock C, Hermans G, Pedotti R, Brendolan A, Schadt E, Garren H, Langer-Gould A, Strober S, Cannella B, Allard J, Klonowski P, Austin A, Lad N, Kaminski N, Galli SJ, Oksenberg JR, Raine CS, Heller R, Steinman L (2002) Gene-microarray analysis of multiple sclerosis lesions yields new targets validated in autoimmune encephalomyelitis. Nat Med 8:500–508

Majewska AK, Newton JR, Sur M (2006) Remodeling of synaptic structure in sensory cortical areas in vivo. J Neurosci 26:3021–3029

Manning PT, Johnson EM, Jr., Wilcox CL, Palmatier MA, Russell JH (1987) MHC-specific cytotoxic T lymphocyte killing of dissociated sympathetic neuronal cultures. Am J Pathol 128:395–409

Marchi N, Angelov L, Masaryk T, Fazio V, Granata T, Hernandez N, Hallene K, Diglaw T, Franic L, Najm I, Janigro D (2007) Seizure-promoting effect of blood-brain barrier disruption. Epilepsia 48:732–742

McGavern DB, Christen U, Oldstone MB (2002a) Molecular anatomy of antigen-specific CD8(+) T cell engagement and synapse formation in vivo. Nat Immunol 3:918–925

McGavern DB, Homann D, Oldstone MB (2002b) T cells in the central nervous system: the delicate balance between viral clearance and disease. J Infect Dis 186(Suppl 2):S145–S151

McGavern DB, Truong P (2004) Rebuilding an immune-mediated central nervous system disease: weighing the pathogenicity of antigen-specific versus bystander T cells. J Immunol 173: 4779–4790

McGeachy MJ, Bak-Jensen KS, Chen Y, Tato CM, Blumenschein W, McClanahan T, Cua DJ (2007) TGF-beta and IL-6 drive the production of IL-17 and IL-10 by T cells and restrain T(H)-17 cell-mediated pathology. Nat Immunol 8:1390–1397

McLellan ME, Kajdasz ST, Hyman BT, Bacskai BJ (2003) In vivo imaging of reactive oxygen species specifically associated with thioflavine S-positive amyloid plaques by multiphoton microscopy. J Neurosci 23:2212–2217
McMahon EJ, Bailey SL, Castenada CV, Waldner H, Miller SD (2005) Epitope spreading initiates in the CNS in two mouse models of multiple sclerosis. Nat Med 11:335–339
McPherson SW, Heuss ND, Roehrich H, Gregerson DS (2006) Bystander killing of neurons by cytotoxic T cells specific for a glial antigen. Glia 53:457–466
McRae BL, Vanderlugt CL, Dal Canto MC, Miller SD (1995) Functional evidence for epitope spreading in the relapsing pathology of experimental autoimmune encephalomyelitis. J Exp Med 182:75–85
Medana IM, Gallimore A, Oxenius A, Martinic MM, Wekerle H, Neumann H (2000) MHC class I-restricted killing of neurons by virus-specific CD8+ T lymphocytes is effected through the Fas/FasL, but not the perforin pathway. Eur J Immunol 30:3623–3633
Medana I, Li Z, Flugel A, Tschopp J, Wekerle H, Neumann H (2001a) Fas ligand (CD95L) protects neurons against perforin-mediated T lymphocyte cytotoxicity. J Immunol 167:674–681
Medana I, Martinic MA, Wekerle H, Neumann H (2001b) Transection of major histocompatibility complex class I-induced neurites by cytotoxic T lymphocytes. Am J Pathol 159:809–815
Menasche G, Feldmann J, Fischer A, de Saint Basile G (2005) Primary hemophagocytic syndromes point to a direct link between lymphocyte cytotoxicity and homeostasis. Immunol Rev 203:165–179
Meyer-Luehmann M, Spires-Jones TL, Prada C, Garcia-Alloza M, de Calignon A, Rozkalne A, Koenigsknecht-Talboo J, Holtzman DM, Bacskai BJ, Hyman BT (2008) Rapid appearance and local toxicity of amyloid-beta plaques in a mouse model of Alzheimer's disease. Nature 451:720–724
Mims CA (1960) Intracerebral injections and the growth of viruses in the mouse brain. Br J Exp Pathol 41:52–59
Monks CR, Freiberg BA, Kupfer H, Sciaky N, Kupfer A (1998) Three-dimensional segregation of supramolecular activation clusters in T cells. Nature 395:82–86
Mora JR, von Andrian UH (2006) T-cell homing specificity and plasticity: new concepts and future challenges. Trends Immunol 27:235–243
Mora JR, Cheng G, Picarella D, Briskin M, Buchanan N, von Andrian UH (2005) Reciprocal and dynamic control of CD8 T cell homing by dendritic cells from skin- and gut-associated lymphoid tissues. J Exp Med 201:303–316
Muller M, Carter SL, Hofer MJ, Manders P, Getts DR, Getts MT, Dreykluft A, Lu B, Gerard C, King NJ, Campbell IL (2007) CXCR3 signaling reduces the severity of experimental autoimmune encephalomyelitis by controlling the parenchymal distribution of effector and regulatory T cells in the central nervous system. J Immunol 179:2774–2786
Murray PD, McGavern DB, Lin X, Njenga MK, Leibowitz J, Pease LR, Rodriguez M (1998a) Perforin-dependent neurologic injury in a viral model of multiple sclerosis. J Neurosci 18:7306–7314
Murray PD, Pavelko KD, Leibowitz J, Lin X, Rodriguez M (1998b) CD4(+) and CD8(+) T cells make discrete contributions to demyelination and neurologic disease in a viral model of multiple sclerosis. J Virol 72:7320–7329
Nagerl UV, Kostinger G, Anderson JC, Martin KA, Bonhoeffer T (2007) Protracted synaptogenesis after activity-dependent spinogenesis in hippocampal neurons. J Neurosci 27:8149–8156
Nair-Gill ED, Shu CJ, Radu CG, Witte ON (2008) Non-invasive imaging of adaptive immunity using positron emission tomography. Immunol Rev 221:214–228
Nansen A, Christensen JP, Ropke C, Marker O, Scheynius A, Thomsen AR (1998) Role of interferon-gamma in the pathogenesis of LCMV-induced meningitis: unimpaired leucocyte recruitment, but deficient macrophage activation in interferon-gamma knock-out mice. J Neuroimmunol 86:202–212
Negrin RS, Contag CH (2006) In vivo imaging using bioluminescence: a tool for probing graft-versus-host disease. Nat Rev Immunol 6:484–490

Neumann H, Cavalie A, Jenne DE, Wekerle H (1995) Induction of MHC class I genes in neurons. Science 269:549–552

Neumann H, Schmidt H, Cavalie A, Jenne D, Wekerle H (1997) Major histocompatibility complex (MHC) class I gene expression in single neurons of the central nervous system: differential regulation by interferon (IFN)-gamma and tumor necrosis factor (TNF)-alpha. J Exp Med 185:305–316

Neumann H, Medana IM, Bauer J, Lassmann H (2002) Cytotoxic T lymphocytes in autoimmune and degenerative CNS diseases. Trends Neurosci 25:313–319

Nimmerjahn A, Kirchhoff F, Helmchen F (2005) Resting microglial cells are highly dynamic surveillants of brain parenchyma in vivo. Science 308:1314–1318

Nitsch R, Pohl EE, Smorodchenko A, Infante-Duarte C, Aktas O, Zipp F (2004) Direct impact of T cells on neurons revealed by two-photon microscopy in living brain tissue. J Neurosci 24:2458–2464

Odoardi F, Kawakami N, Klinkert WE, Wekerle H, Flugel A (2007) Blood-borne soluble protein antigen intensifies T cell activation in autoimmune CNS lesions and exacerbates clinical disease. Proc Natl Acad Sci USA 104:18625–18630

Oliveira AL, Thams S, Lidman O, Piehl F, Hokfelt T, Karre K, Linda H, Cullheim S (2004) A role for MHC class I molecules in synaptic plasticity and regeneration of neurons after axotomy. Proc Natl Acad Sci USA 101:17843–17848

Oppmann B, Lesley R, Blom B, Timans JC, Xu Y, Hunte B, Vega F, Yu N, Wang J, Singh K, Zonin F, Vaisberg E, Churakova T, Liu M, Gorman D, Wagner J, Zurawski S, Liu Y, Abrams JS, Moore KW, Rennick D, de Waal-Malefyt R, Hannum C, Bazan JF, Kastelein RA (2000) Novel p19 protein engages IL-12p40 to form a cytokine, IL-23, with biological activities similar as well as distinct from IL-12. Immunity 13:715–725

Pan F, Gan WB (2008) Two-photon imaging of dendritic spine development in the mouse cortex. Dev Neurobiol 68:771–778

Pautler RG (2004) Mouse MRI: concepts and applications in physiology. Physiology (Bethesda) 19:168–175

Peri F, Nusslein-Volhard C (2008) Live imaging of neuronal degradation by microglia reveals a role for v0-ATPase a1 in phagosomal fusion in vivo. Cell 133:916–927

Pircher H, Burki K, Lang R, Hengartner H, Zinkernagel RM (1989) Tolerance induction in double specific T-cell receptor transgenic mice varies with antigen. Nature 342:559–561

Pirko I, Ciric B, Johnson AJ, Gamez J, Rodriguez M, Macura S (2003) Magnetic resonance imaging of immune cells in inflammation of central nervous system. Croat Med J 44:463–468

Pirko I, Johnson A, Ciric B, Gamez J, Macura SI, Pease LR, Rodriguez M (2004) In vivo magnetic resonance imaging of immune cells in the central nervous system with superparamagnetic antibodies. FASEB J 18:179–182

Pirko I, Fricke ST, Johnson AJ, Rodriguez M, Macura SI (2005) Magnetic resonance imaging, microscopy, and spectroscopy of the central nervous system in experimental animals. NeuroRx 2:250–264

Poeppel TD, Krause BJ (2008) Functional imaging of memory processes in humans: positron emission tomography and functional magnetic resonance imaging. Methods 44:315–328

Power C, Poland SD, Blume WT, Girvin JP, Rice GP (1990) Cytomegalovirus and Rasmussen's encephalitis. Lancet 336:1282–1284

Raine CS, Cannella B, Duijvestijn AM, Cross AH (1990) Homing to central nervous system vasculature by antigen-specific lymphocytes. II. Lymphocyte/endothelial cell adhesion during the initial stages of autoimmune demyelination. Lab Invest 63:476–489

Rall GF, Mucke L, Oldstone MB (1995) Consequences of cytotoxic T lymphocyte interaction with major histocompatibility complex class I-expressing neurons in vivo. J Exp Med 182:1201–1212

Rasmussen T, Olszewski J, Lloydsmith D (1958) Focal seizures due to chronic localized encephalitis. Neurology 8:435–445

Rensing-Ehl A, Malipiero U, Irmler M, Tschopp J, Constam D, Fontana A (1996) Neurons induced to express major histocompatibility complex class I antigen are killed via the perforin and not the Fas (APO-1/CD95) pathway. Eur J Immunol 26:2271–2274

Rivera-Quinones C, McGavern D, Schmelzer JD, Hunter SF, Low PA, Rodriguez M (1998) Absence of neurological deficits following extensive demyelination in a class I-deficient murine model of multiple sclerosis. Nat Med 4:187–193
Rocheleau JV, Piston DW. (2003) Two-photon excitation microscopy for the study of living cells and tissues. Curr Protoc Cell Biol Chap 4:Unit 4.11
Roebroek RM, Postma BH, Dijkstra UJ (1994) Aseptic meningitis caused by the lymphocytic choriomeningitis virus. Clin Neurol Neurosurg 96:178–180
Rogers SW, Andrews PI, Gahring LC, Whisenand T, Cauley K, Crain B, Hughes TE, Heinemann SF, McNamara JO (1994) Autoantibodies to glutamate receptor GluR3 in Rasmussen's encephalitis. Science 265:648–651
Sanchez-Ruiz M, Wilden L, Muller W, Stenzel W, Brunn A, Miletic H, Schluter D, Deckert M (2008) Molecular mimicry between neurons and an intracerebral pathogen induces a CD8 T cell-mediated autoimmune disease. J Immunol 180:8421–8433
Schanen A, Gallou G, Hincky JM, Saron MF (1998) A rash, circulating anticoagulant, then meningitis. Lancet 351:1856
Scheuner D, Eckman C, Jensen M, Song X, Citron M, Suzuki N, Bird TD, Hardy J, Hutton M, Kukull W, Larson E, Levy-Lahad E, Viitanen M, Peskind E, Poorkaj P, Schellenberg G, Tanzi R, Wasco W, Lannfelt L, Selkoe D, Younkin S (1996) Secreted amyloid beta-protein similar to that in the senile plaques of Alzheimer's disease is increased in vivo by the presenilin 1 and 2 and APP mutations linked to familial Alzheimer's disease. Nat Med 2:864–870
Schwendemann G, Lohler J, Lehmann-Grube F (1983) Evidence for cytotoxic T-lymphocyte-target cell interaction in brains of mice infected intracerebrally with lymphocytic choriomeningitis virus. Acta Neuropathol 61:183–195
Serafini B, Rosicarelli B, Magliozzi R, Stigliano E, Capello E, Mancardi GL, Aloisi F (2006) Dendritic cells in multiple sclerosis lesions: maturation stage, myelin uptake, and interaction with proliferating T cells. J Neuropathol Exp Neurol 65:124–141
Shrestha B, Diamond MS (2007) Fas ligand interactions contribute to CD8+ T-cell-mediated control of West Nile virus infection in the central nervous system. J Virol 81:11749–11757
Shrestha B, Samuel MA, Diamond MS (2006) CD8+ T cells require perforin to clear West Nile virus from infected neurons. J Virol 80:119–129
Siffrin V, Brandt AU, Herz J, Zipp F (2007) New insights into adaptive immunity in chronic neuroinflammation. Adv Immunol 96:1–40
Skinner PJ, Haase AT. (2005) In situ staining using MHC class I tetramers. Curr Protoc Immunol Chap 17:Unit 17.4
Skinner PJ, Daniels MA, Schmidt CS, Jameson SC, Haase AT (2000) Cutting edge: in situ tetramer staining of antigen-specific T cells in tissues. J Immunol 165:613–617
Sorra KE, Harris KM (2000) Overview on the structure, composition, function, development, and plasticity of hippocampal dendritic spines. Hippocampus 10:501–511
Spires TL, Meyer-Luehmann M, Stern EA, McLean PJ, Skoch J, Nguyen PT, Bacskai BJ, Hyman BT (2005) Dendritic spine abnormalities in amyloid precursor protein transgenic mice demonstrated by gene transfer and intravital multiphoton microscopy. J Neurosci 25:7278–7287
Spires-Jones TL, Meyer-Luehmann M, Osetek JD, Jones PB, Stern EA, Bacskai BJ, Hyman BT (2007) Impaired spine stability underlies plaque-related spine loss in an Alzheimer's disease mouse model. Am J Pathol 171:1304–1311
Stamatovic SM, Shakui P, Keep RF, Moore BB, Kunkel SL, Van Rooijen N, Andjelkovic AV (2005) Monocyte chemoattractant protein-1 regulation of blood-brain barrier permeability. J Cereb Blood Flow Metab 25:593–606
Stearns T (1995) Green fluorescent protein. The green revolution. Curr Biol 5:262–264
Stevenson PG, Hawke S, Sloan DJ, Bangham CR (1997) The immunogenicity of intracerebral virus infection depends on anatomical site. J Virol 71:145–151
Stinchcombe JC, Majorovits E, Bossi G, Fuller S, Griffiths GM (2006) Centrosome polarization delivers secretory granules to the immunological synapse. Nature 443:462–465

Storm P, Bartholdy C, Sorensen MR, Christensen JP, Thomsen AR (2006) Perforin-deficient CD8+ T cells mediate fatal lymphocytic choriomeningitis despite impaired cytokine production. J Virol 80:1222–1230

Su H, Forbes A, Gambhir SS, Braun J (2004) Quantitation of cell number by a positron emission tomography reporter gene strategy. Mol Imaging Biol 6:139–148

Sun JB, Olsson T, Wang WZ, Xiao BG, Kostulas V, Fredrikson S, Ekre HP, Link H (1991) Autoreactive T and B cells responding to myelin proteolipid protein in multiple sclerosis and controls. Eur J Immunol 21:1461–1468

Szentistvanyi I, Patlak CS, Ellis RA, Cserr HF (1984) Drainage of interstitial fluid from different regions of rat brain. Am J Physiol 246:F835–F844

Trachtenberg JT, Chen BE, Knott GW, Feng G, Sanes JR, Welker E, Svoboda K (2002) Long-term in vivo imaging of experience-dependent synaptic plasticity in adult cortex. Nature 420:788–794

Tsai J, Grutzendler J, Duff K, Gan WB (2004) Fibrillar amyloid deposition leads to local synaptic abnormalities and breakage of neuronal branches. Nat Neurosci 7:1181–1183

Tuohy VK, Yu M, Yin L, Kawczak JA, Kinkel RP (1999) Spontaneous regression of primary autoreactivity during chronic progression of experimental autoimmune encephalomyelitis and multiple sclerosis. J Exp Med 189:1033–1042

Twyman RE, Gahring LC, Spiess J, Rogers SW (1995) Glutamate receptor antibodies activate a subset of receptors and reveal an agonist binding site. Neuron 14:755–762

van der Most RG, Murali-Krishna K, Whitton JL, Oseroff C, Alexander J, Southwood S, Sidney J, Chesnut RW, Sette A, Ahmed R (1998) Identification of Db- and Kb-restricted subdominant cytotoxic T-cell responses in lymphocytic choriomeningitis virus-infected mice. Virology 240:158–167

Vanderlugt CL, Miller SD (2002) Epitope spreading in immune-mediated diseases: implications for immunotherapy. Nat Rev Immunol 2:85–95

Vinters HV, Wang R, Wiley CA (1993) Herpesviruses in chronic encephalitis associated with intractable childhood epilepsy. Hum Pathol 24:871–879

Volpe E, Servant N, Zollinger R, Bogiatzi SI, Hupe P, Barillot E, Soumelis V (2008) A critical function for transforming growth factor-beta, interleukin 23 and proinflammatory cytokines in driving and modulating human T(H)-17 responses. Nat Immunol 9:650–657

Walker DH, Camenga DL, Whitfield S, Murphy FA (1977) Anticonvulsant prolongation of survival in adult murine lymphocytic choriomeningitis. II. Ultrastructural observations of pathogenetic events. J Neuropathol Exp Neurol 36:21–40

Walter L, Albert ML (2007) Cutting edge: cross-presented intracranial antigen primes CD8+ T cells. J Immunol 178:6038–6042

Weaver CT, Hatton RD, Mangan PR, Harrington LE (2007) IL-17 family cytokines and the expanding diversity of effector T cell lineages. Annu Rev Immunol 25:821–852

Weissleder R, Cheng HC, Bogdanova A, Bogdanov A, Jr (1997) Magnetically labeled cells can be detected by MR imaging. J Magn Reson Imaging 7:258–263

Weller RO, Engelhardt B, Phillips MJ (1996) Lymphocyte targeting of the central nervous system: a review of afferent and efferent CNS-immune pathways. Brain Pathol 6:275–288

Welsh DK, Kay SA (2005) Bioluminescence imaging in living organisms. Curr Opin Biotechnol 16:73–78

Whitney KD, McNamara JO (2000) GluR3 autoantibodies destroy neural cells in a complement-dependent manner modulated by complement regulatory proteins. J Neurosci 20:7307–7316

Wiedemann A, Depoil D, Faroudi M, Valitutti S (2006) Cytotoxic T lymphocytes kill multiple targets simultaneously via spatiotemporal uncoupling of lytic and stimulatory synapses. Proc Natl Acad Sci U S A 103:10985–10990

Willenborg DO, Fordham S, Bernard CC, Cowden WB, Ramshaw IA (1996) IFN-gamma plays a critical down-regulatory role in the induction and effector phase of myelin oligodendrocyte glycoprotein-induced autoimmune encephalomyelitis. J Immunol 157:3223–3227

Wucherpfennig KW, Ota K, Endo N, Seidman JG, Rosenzweig A, Weiner HL, Hafler DA (1990) Shared human T cell receptor V beta usage to immunodominant regions of myelin basic protein. Science 248:1016–1019

Xu HT, Pan F, Yang G, Gan WB (2007) Choice of cranial window type for in vivo imaging affects dendritic spine turnover in the cortex. Nat Neurosci 10:549–551

Yang XO, Pappu BP, Nurieva R, Akimzhanov A, Kang HS, Chung Y, Ma L, Shah B, Panopoulos AD, Schluns KS, Watowich SS, Tian Q, Jetten AM, Dong C (2008) T helper 17 lineage differentiation is programmed by orphan nuclear receptors ROR alpha and ROR gamma. Immunity 28:29–39

Yeh TC, Zhang W, Ildstad ST, Ho C (1993) Intracellular labeling of T-cells with superparamagnetic contrast agents. Magn Reson Med 30:617–625

Yeh TC, Zhang W, Ildstad ST, Ho C (1995) In vivo dynamic MRI tracking of rat T-cells labeled with superparamagnetic iron-oxide particles. Magn Reson Med 33:200–208

Yu M, Johnson JM, Tuohy VK (1996) A predictable sequential determinant spreading cascade invariably accompanies progression of experimental autoimmune encephalomyelitis: a basis for peptide-specific therapy after onset of clinical disease. J Exp Med 183:1777–1788

Yura M, Takahashi I, Serada M, Koshio T, Nakagami K, Yuki Y, Kiyono H (2001) Role of MOG-stimulated Th1 type "light up" (GFP+) CD4+ T cells for the development of experimental autoimmune encephalomyelitis (EAE). J Autoimmun 17:17–25

Zabow G, Dodd S, Moreland J, Koretsky A (2008) Micro-engineered local field control for high-sensitivity multispectral MRI. Nature 453:1058–1063

Zajac AJ, Dye JM, Quinn DG (2003) Control of lymphocytic choriomeningitis virus infection in granzyme B deficient mice. Virology 305:1–9

Zhang GX, Gran B, Yu S, Li J, Siglienti I, Chen X, Kamoun M, Rostami A (2003) Induction of experimental autoimmune encephalomyelitis in IL-12 receptor-beta 2-deficient mice: IL-12 responsiveness is not required in the pathogenesis of inflammatory demyelination in the central nervous system. J Immunol 170:2153–2160

Zinkernagel RM, Doherty PC (1973) Cytotoxic thymus-derived lymphocytes in cerebrospinal fluid of mice with lymphocytic choriomeningitis. J Exp Med 138:1266–1269

Zuo Y, Lin A, Chang P, Gan WB (2005a) Development of long-term dendritic spine stability in diverse regions of cerebral cortex. Neuron 46:181–189

Zuo Y, Yang G, Kwon E, Gan WB (2005b) Long-term sensory deprivation prevents dendritic spine loss in primary somatosensory cortex. Nature 436:261–265

Multiphoton Imaging of Cytotoxic T Lymphocyte-Mediated Antitumor Immune Responses

Alexandre Boissonnas, Alix Scholer-Dahire, Luc Fetler, and Sebastian Amigorena

Contents

Abstract The actual contribution of T lymphocytes to protection against tumors is still unclear. *In vitro* imaging experiments show that tumor specific cytotoxic T lymphocytes (CTLs) are competent to kill target cells by conventional cytotoxic pathways. The emergence of multiphoton imaging in the past decade now allows real time *in vivo* imaging of CTLs. New insights are available on the behavior of antitumor T cells during the priming phase, during their traffic within the tumor

A. Boissonnas, A. Scholer-Dahirel, and S. Amigorena (✉)
Institut National de la Santé et de la Recherche Médicale U653, Immunité et Cancer, Pavillon Pasteur, Institut Curie, 26 rue d'Ulm, F-75245, Paris Cedex 05, France

L. Fetler
Centre National de la Recherche Scientifique UMR 168, Laboratoire de Physico-Chimie Curie, Institut Curie, 26 rue d'Ulm, F-75245, Paris Cedex 05, France

D. McGavern and M. Dustin (eds.), *Visualizing Immunity,* Current Topics in Microbiology and Immunology 334,
DOI: 10.1007/978-3-540-93864-4_11,

tissue, and on their interactions with tumor cells during the effector phase. Recent reports suggest that direct killing of tumor cells by CTLs is a slow process, suggesting that the ratio of effector to target cells is determinant, or that additional cytotoxic contribution by other cell types is required to induce efficient tumor rejection. This review will focus on the publications that have imaged antitumor immune responses dynamically and discuss how this new information contributes to understand the implication of CTLs in tumor rejection.

1 Introduction

The immune system evolved to protect the organism against infectious pathogens. To do so, it needs to distinguish between foreign, potentially dangerous factors, and self, innocuous compounds. And it does so quite efficiently. Tumors represent a peculiar situation in this context: tumor cells are self, but dangerous. Not surprisingly, the role of the immune system in our natural defenses against tumors is still a matter of debate. In contrast, there is no doubt today that the immune system has the potential to mediate the rejection of established tumors. Antibodies that recognize tumors are largely used for cancer immunotherapy, including breast cancer and certain lymphomas (Nimmerjahn and Ravetch 2007). Moreover, adoptive transfer of tumor-specific cytotoxic T lymphocytes (CTLs) induces tumor rejection in both mouse models and melanoma patients (Rosenberg et al. 2008).

Although in these protocols of T cell-based passive immunotherapy tumor rejection is often complete and rapid, the mechanisms involved are still a matter of debate. The simplest hypothesis would be that the cytotoxic T cells kill the tumor cells, causing direct disappearance of the tumors. Direct killing is certainly the primary cause of rejection in certain tumor models. Even in this case, the "functional organization" of tumor cell killing and rejection is unclear. We know very little about how T cells invade tumors, in terms of topology and kinetics. We also know very little about the actual killing process of tumor cells by CD8+ T cells *in vivo*, in terms of frequency and duration of the individual killing events. We also understand poorly how the cytotoxic functions of T cells are controlled within the tumor.

Of course, the regulation of T cell functions should be considered in the context of the whole tumor tissue, which includes not only the tumor cells and the infiltrated cytotoxic lymphocytes, but also the number of other cells involved in different aspects of the functional organization of the tumor. These cells include other types of lymphocytes (helper CD4+ T cells, regulatory T cells Tregs, NK cells), endothelial cells (that form the tumor vasculature), myeloid cells (including macrophages, granulocytes and other myeloid populations), and fibroblasts. These cells, which altogether form the tumor stroma, have complex, ill-defined roles, either favoring or reducing tumor growth depending on the cell types and the physiological situations (Mantovani et al. 2008).

In the last 5 years, technological developments in dynamic imaging have enabled us to see, in real time, cells migrating in their physiological environment, interactions between different cell types, and certain types of cell function, such as cytotoxicity (Bousso 2008; Ng et al. 2008). These events can be quantified and analyzed statistically. Multiphoton imaging therefore allows the analysis of the behavior of both individual cells and of cell populations. Most of the published multiphoton studies were performed on lymph nodes (Bousso 2008), whereas only a few reports have imaged tumors (Boissonnas et al. 2007; Breart et al. 2008; Mrass et al. 2006).

Herein, we propose to highlight recent contributions of live imaging to our understanding of how CTLs are involved in the rejection of nascent and established tumors. We will discuss papers analyzing CD8+ T cell fate both in tumor-draining lymph nodes and inside tumors. We will attempt to summarize how dynamic imaging by multiphoton microscopy has contributed to our understanding of the role of cytotoxic T cell responses in tumor rejection.

2 CTL Priming During Antitumor Immune Responses

2.1 *Tumor Antigens and Antitumor T Cell Models: A Technological Challenge*

Tumors express different types of antigens. Tumor-associated antigens (TAAs) are proteins overexpressed in tumors as compared to normal tissue. Tumor-specific antigens (such as mutated antigens or viral proteins in virus-induced tumors) are uniquely expressed in tumors, in contrast to TAAs (Stevanovic 2002). TAAs and tumor-specific antigens constitute the basis for analyzing the induction of antitumor CTL responses. In the different mouse models imaged so far, the antigens used were recombinant model antigens ectopically expressed in tumor cell lines. This represents a real limitation for the interpretation of the results, since these model antigens (usually ovalbumin, OVA) are exogenous antigens expressed at high levels, in contrast to most tumor antigens, which are endogenous proteins expressed at low levels. In addition, the adoptive transfer of tumors is quite apart from any physiological situation. After adoptive transfer, the vast majority of the implanted tumor cells die, creating an important source of antigen available for lymphocyte recognition and response. It is very important not to adoptively transfer T cells too soon after tumor implantation. And even then, the structure of the tumors is not always representative of that of endogenous tumors.

Conversely, the T cells used in most of the models that have been imaged so far are T cell receptor (TCR) transgenic T cells. The TCRs expressed in these T cells are most usually of very high affinity, in contrast to the normal repertoire of antitumor T cells found *in vivo*. The type of T cells used may again represent a possible source of artifacts. In addition, the experimental situations created for two-photon imaging are very different from what could happen during any endogenous immune response against tumors. Indeed, during the development of endogenous tumors, the immune system coexists with tumors while they develop, most often for very long periods of time (months or even years). The sudden adoptive transfer of several millions of tumor-antigen-specific T cells (very high numbers of T cells are required for imaging), in a tumor-bearing mouse is far from any physiological situation. It is unclear today to what extent these experimental systems are representative of endogenous antitumor immune responses. Nevertheless, the experimental settings used for imaging so far can be considered as acceptable models for understanding tumor rejection during tumor immunotherapy by adoptive transfer of T cells.

Dynamic imaging has been performed both intravitally (on anesthetized animals) and on explanted organs (lymph nodes or tumors) in oxygenated medium ex vivo. Although there is no evidence at present of any significant difference between these two modalities of imaging, ideally the results obtained with explanted organs should when possible be confirmed intravitally. Another major impediment of multiphoton imaging is related to the limited penetration depth within the samples. While it is possible to image lymph nodes up to 300–500 μm (this is enough to analyze both B and T cell zones), the imaging penetration in solid tumors is often restricted (usually 100–200 μm at present). Thus, all the imaging performed so far concerns the 1–2% peripheral regions of tumors. It is very important to keep in mind these limitations when analyzing the results that we will discuss here. In any case, these limitations are vastly inherent to the imaging approaches available, which are technically challenging and impose limitations on the sensitivity and the resolution. On the other hand, one should also keep in mind that we are, for the first time, capable of seeing cells in their physiological environment dynamically and in real time.

2.2 *Direct Priming Versus Cross-Priming*

Although the question of how the immune system sees tumors has been the subject of innumerable studies and reviews, the modalities of T cell priming against tumor antigens remains unclear, both in humans and in mice. In carcinomas or sarcomas, it has been shown that antigen dose and tumor localization are determinant for efficient CTL priming (Ochsenbein et al. 2001). Some fibrosarcomas directly prime CTLs, but only if tumor cells can reach secondary lymphoid organs early on and in sufficient numbers (Kundig et al. 1995). If the tumors are transplanted as solid tumor fragments, no T cell priming occurs (Ochsenbein et al. 1999). In contrast, persistent antigen presentation in lymphoid organs can lead to overactivation of the CD8+ T cells and inefficient cytotoxic responses when lymphoma cells are injected in large numbers (Ochsenbein et al. 2001). T cell priming in tumor-draining lymph nodes depends critically on the amount of antigen delivered to lymph nodes, and on the delivery of this antigen by the tumor cells themselves or by antigen-presenting cells migrating from the tumor to the lymph node.

The question of the direct priming versus indirect priming (or cross-priming) in tumor recognition has been a matter of debate for several years (Ochsenbein et al. 2001; Zinkernagel 2002). On the one hand, several studies showed that tumors expressing costimulatory molecules are more efficiently rejected than control tumors (Hellstrom et al. 1995; Maric et al. 1998; Townsend and Allison 1993; Yang et al. 1997), suggesting that direct interactions of tumor cells with T lymphocytes are required. Huang et al, however, suggested that priming of CD8+ T cells with CD80-expressing tumors also depends on host antigen-presenting cells, and that the expression of CD80 and CD86 on tumors increased the amplification of the tumor-specific T cells, rather than priming (Huang et al. 1996a). In several different experimental systems, cross-priming mediates antitumor T cell responses. A seminal study by the group of Levitsky, showed that host mice need to express TAP1/2 in order to mount CTL responses against a tumor (Huang et al. 1996b), indicating that a host-derived antigen-presenting cell is required and that antigen presentation solely by the tumor cells is not sufficient. More recent studies also showed that overexpressing a tumor antigen induced a better cross-presentation in the tumor-draining lymph node, followed by tumor rejection (Spiotto et al. 2002). Recently, an alternative mechanism of priming, called "cross-dressing," was reported wherein DCs acquire MHC class I molecules from dead cells, through direct cell–cell contacts, and prime CD8+ T cells (Dolan et al. 2006a). Optimal priming of CD8+ T cells was obtained *in vivo* when both cross-dressing and cross-priming occur, suggesting that lack of class I molecule expression by tumor cells could reduce the efficacy of the antitumor immune responses. This mechanism has also been observed for antitumor CD4+ T cell activation through MHC class II cross-dressing (Dolan et al. 2006b).

2.3 *Imaging the Priming of Antitumor CD8+ T Lymphocytes in Tumor-Draining Lymph Nodes*

In the last few years, several groups have imaged the priming of CD4+ and CD8+ T cells by DCs. In experimental systems in which antigen or antigen-presenting DCs are adoptively transferred to mice bearing antigen-specific naive T cells, T cell motility evolves according to three distinct phases (Hugues et al. 2004; Mempel et al. 2004; Miller 2004, p 140). (1) The first phase is of reduced motility, without complete arrests (with a mean velocity around 5 μm min^{-1} vs 10–15 μm min^{-1} in the absence of antigen). During this phase, the DC–T cell contacts last for 1–2 min and are very frequent. This first phase can last for several hours, or be much shorter, depending on the experimental settings. In systems where the antigen-presenting cells are loaded with high amounts of peptides, phase one can even be absent (Celli et al. 2007; Garcia et al. 2007). (2) The second phase is characterized by completely arrested T cells (Vmean < 2 μm min^{-1}). During this second phase, T cells interact for several hours with individual DCs. (3) During the third phase, T cells regain motility and move again with a mean velocity around 5 μm min^{-1}.

These studies indicate that T cell–DC interactions are tightly regulated, both in terms of frequency and duration. The frequency is controlled through chemokine–chemokine receptor interactions (Castellino et al. 2006; Hugues et al. 2007), while the duration, especially during phase 2, is determined by a combination of different parameters. First is antigen concentration and the avidity of the TCRs for the peptide–MHC complexes. The time before the complete arrest of the T cell population on the DCs depends on the concentration of antigen presented on the DC surface (Celli et al. 2007; Garcia et al. 2007). Likewise, high affinity peptides induce more pronounced T cell arrests than low affinity peptides (Skokos et al. 2007). Second is the state of DC maturation. Immature DCs, which induce functional T cell tolerance, do not establish stable interactions with T cells, at least with CD8+ T cells (Hugues et al. 2006). Mature DCs, in contrast, interact for several hours with individual T cells. Indeed, we showed that a phase of complete and long lasting T cell arrests is critical for the functional outcome of T cell priming (Hugues et al. 2006). Under conditions of induction of T cell tolerance, the phase of complete T cell arrests is not observed. In addition, DCs, incompetent to induce complete T cell arrests because they lack expression of the adhesion receptor ICAM-1, also failed to induce effective T cell priming (Scholer et al. 2008). In fact, in the absence of long lasting DC–T cell contacts, activated T cells fail to produce *IFNγ* and to survive the contraction phase, which results in defective CD8+ T cell memory.

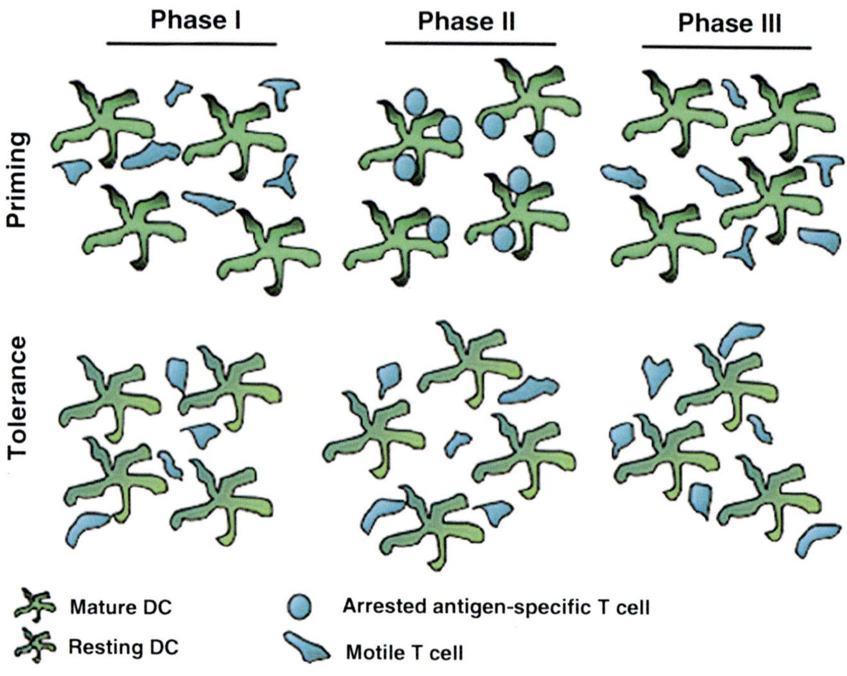

Fig. 1 Priming within a lymph node. The priming phase occurs in three distinct phases. *Phase I:* reduced motility of T cells, without complete arrests. *Phase II:* stable DC–T cell interactions cause complete arrest of T cells. *Phase III:* T cells regain motility. In tolerogenic conditions (immature or not fully functional DCs, low affinity TCRs), DC–T cell interactions during Phase II are not stable

Few studies have attempted to image T cell priming in the context of antitumor T cell responses (in tumor-draining lymph nodes). In a model of OVA-expressing tumors that are rejected after adoptive transfer of CD8+ OVA-specific TCR transgenic T cells (OT1), we showed that OT1 T cells make long lasting arrests in tumor-draining lymph nodes, and that these contacts require the expression of ICAM-1 by the host mice (Scholer et al. 2008). These results suggest that tumor rejection by CD8+ T cells require stable DC–T cell interactions in tumor-draining lymph nodes. More recently, we used a different OVA-expressing tumor cell line, which is not rejected by the adoptive transfer of OT1 T cells, to analyze the role of Tregs on the motility of CD8+ T cells in tumor-draining lymph nodes. This particular tumor is only rejected by OT1 T cells if Tregs are depleted in the mice. Interestingly, Treg depletion induced a strong arrest in the migration of OT1 T cells, causing increased OT1 T cell priming (Boissonnas et al., submitted for publication).

Altogether, imaging of T cell motility during priming by DCs suggests a three-phase model, in which the reduced motility reflects the stability of the DC–T cell interactions, and correlates to the efficiency of the priming (Fig. 1).

3 Migration and Tumor Infiltration by CTLs

3.1 *Selective Inflammation, Imprinting, and Retention by Antigen*

After priming, activated CD8+ T cells egress from lymph nodes and migrate to the tumor site. The ability of CTLs to infiltrate tumors has a major impact on the efficiency of tumor elimination. Although two-photon microscopy permits to investigate certain aspects of T cell entry and migration in tumors, other techniques are better suited for the analysis of tumor infiltration by T cells. Some noninvasive imaging technologies allow the *in vivo* monitoring of adoptively transferred T cells kinetically. Magnetic resonance imaging (MRI), positron emission tomography (PET), and bioluminescence imaging (BLI) allow the analysis of the temporal and spatial distribution of adoptively transferred T cells both in human and mice (Gade et al. 2005; Smirnov et al. 2006; Tumeh et al. 2008). The number of T lymphocytes present within a tumor is the result of several parameters that are not always easy to integrate, including the total number of lymphocytes, their specificity for tumor antigens, the rate of entry into the tumor, their eventual retention within the tumor (time of residence), their eventual proliferation or death within the tumor, and their rate of exit.

Migration of CTLs to peripheral tissues requires selective changes in the expression of homing molecules, such as adhesion and chemokine receptors (Boissonnas et al. 2004). How, then, do T cells acquire the information for their appropriate homing in peripheral tissues? Three possibilities have been proposed: (1) inflammation – activated T cells go to any inflamed tissue, including tumors (when tumors are inflamed), (2) imprinting – depending on their site of priming, activated

T cells will preferentially infiltrate the tissues drained by this lymph node, and (3) retention by antigen – once inside the tissue, the time of residence could depend on the presence of antigen, and thus lead to their selective accumulation.

The role of inflammation in lymphocyte extravasation and in tissue invasion by lymphocytes has been described in great detail (Jain et al. 1996). All the basic rules defined for normal inflamed tissue may, in principle, apply to tumors. The vasculature in tumors, however, is in some cases different from that of normal tissue, and the nature of these modifications seem to vary in different tumors (Jain et al. 2002). The possible relationships between inflammation and the permeability of the tumor vasculature to lymphocytes certainly deserve further attention.

In different animal models, the immunization route determines the subsequent migration of antigen-specific T cells (Butcher et al. 1999). From these observations emerged the idea that priming induces a selective imprinting of activated T cells, depending on the lymph node tissue where priming occurs (Mora et al. 2003). It is unclear, however, if this imprinting is imposed by the location of the lymph node itself, or by the tissue origin of the migrating DCs that present the tissue antigens and induce priming (Calzascia et al. 2005).

A recent report on human breast cancer shows that activated CD4+ T cells accumulate both within the tumors and within the contra-lateral healthy breast tissue (Joncker et al. 2006b). In addition, some of us have shown that, after a brief activation *in vitro*, naive CD8+ T cells infiltrate tumors effectively, even if the tumors do not express the cognate antigen (Boissonnas et al. 2004). In this particular system, we cannot exclude that they secondarily interact with antigen-presenting cells in an antigen-independent manner in lymphoid organs, but it is unlikely that they are able to acquire independently the activation and the homing signal. Jonker et al. (Joncker et al. 2006a) showed that the accumulation of CD4+ T cells in tumors relies on the strength of their priming, suggesting that imprinting is not a prerequisite for homing, but could favor the infiltration of tumors and any other inflamed tissue. In contrast, Calzascia et al. showed, in a model where tumor cells do not express the MHC molecules recognized by the specific CTLs, that antigen cross-presenting cells from the tumor stroma are involved in the retention of T cells inside the tumor and tumor rejection (Calzascia et al. 2003). In the tumor model that we have used, mice were inoculated in a contra-lateral manner with tumor expressing or not the antigen (Boissonnas et al. 2007). We showed a preferential accumulation of antigen-specific CTLs in the tumor expressing the cognate antigen. Therefore, in the presence of antigen, specific T cells are retained in the tissue and may leave or die when antigen is not expressed.

Proliferation within the tumor may also contribute to CTL accumulation. We evaluated that 0.8% of T cells divide within the tumor over 30 min, indicating that up to 40% of the T cells divide every 24 h (Boissonnas et al. 2007). This demonstrates that active proliferation occurs inside tumors (in the absence of antigen, the lower frequencies of infiltration did not allow to estimate the role of antigen in T cell division inside tumors). Altogether, these studies do not support or exclude a major role for retention of T cells by antigen, but rather suggest that different mechanisms contribute to the eventual accumulation of T cells inside tumors.

The antigen specificity of tumor-infiltrating lymphocytes, as well as the mechanisms of targeting and retention of lymphocytes within tumors, are still ill defined.

3.2 *Intratumoral Lymphocyte Trafficking*

So far, the main contribution of two-photon microscopy to the understanding of tumor infiltration relates to the intratumor distribution and migration of lymphocytes. Two independent studies have analyzed the motility and distribution of CD8+ T cells in solid tumors (Boissonnas et al. 2007; Mrass et al. 2006).

The first published study, by the group of Weninger, imaged explanted tumors (Mrass et al. 2006). These authors imaged endogenous T cells infiltrating a TC-1 tumor expressing the E7 antigen after boosting or not the immune system by vaccination with nonreplicative adenovirus expressing E7 antigen. They also used the thymoma cell line (EL4) expressing or not a model antigen (OVA), and imaged adoptively transferred, *in vitro*-activated, antigen-specific T cells. T cells migrated in the tumors with elongated morphology, crawling along collagen fibers, before establishing stable interactions with tumor cells. Interestingly, while T cells stopped and then regained motility in the antigen-expressing tumors, in the absence of antigen expression the motility decreased progressively, indicating that antigen stimulation is required to maintain T cell motility in tumors. Decreased motility, however, was observed 4 days after the adoptive transfer of activated T cells to mice that did not express the cognate antigen, a point when the vitality of the transferred cells may start to decrease. These authors also showed direct contacts in the tumor between T cells and macrophages.

In the second published study, the mice were inoculated in a contra-lateral manner with the same thymoma cell line as above, expressing or not the same model antigen (OVA), before the adoptive transfer of antigen-specific naive CD8+ T cells (Boissonnas et al. 2007). Antigen-specific T cells infiltrated both tumors, but the accumulation was greater in the antigen-expressing tumors. Using both intravital imaging and immuno-histological analysis, we showed that these CTLs make long lasting contacts with antigen-expressing tumor cells, while the tumor cell–T cell contacts in the tumors that do not express antigen are brief and do not cause the arrest of the T cells. Reduced motility and stable contacts with antigen-expressing tumor cells, or other cells from the tumor stroma, may prolong the time of residence, and thereby retain, the T cells within the tumors.

We also found that CD8+ T cells are present in the periphery of the tumors early after adoptive transfer, both in tumors expressing or not the cognate antigen. In antigen-bearing tumors (which are effectively rejected by the adoptively transferred T cells), however, the distribution of the T cells changed rapidly, and deep infiltration of the tumors was observed 6 days after adoptive transfer. In antigen nonexpressing tumors (which are not rejected), CD8+ T cells failed to infiltrate deeply the tumors. Immobile T cells observed early on in antigen-bearing tumors were found in regions of the tumors where the tumor cells were alive. At later time points, T cell motility increased, while most of the tumor cells in those regions were dead. These changes in T cell motility were not observed in tumors that did not express antigen, suggesting that the phase of arrest was related to antigen expression and tumor cell killing by CD8+ T cells.

During the phases of T cell migration, motility occurred along blood vessels or collagen fibers, with the T cells adopting different morphologies. Strikingly, we could find no directionality in the T cell trajectories, suggesting that the mechanism of deep infiltration is not related to vectorized migration. Interestingly, the mRNA levels for several chemokine receptors in effector cells isolated from the tumor was undetectable (Boissonnas et al. 2004), suggesting that, after reaching the tumor, effector cells do not need chemokine receptors to traffic inside the tumor matrix. Mukai et al. (1999) showed that ex vivo activated T cells treated by PTX lose their therapeutic efficacy because they fail to infiltrate tumor parenchyma, but PTX treatment did not alter their cytolytic activity. When T cells are inside the tumor matrix, trafficking and cytotoxic activity may thus be independent of chemokine-receptor expression.

Migration of leucocytes along fibers in a 3D reconstituted collagen matrix is contact-guided and independent of proteolysis or collagen degradation (Wolf et al. 2003). We observed similar results *in vivo* and showed that the tumoral architecture (such as collagen and blood vessels) might influence the trajectories of CTLs (Boissonnas et al. 2007). Mrass et al. proposed that the migration of CTLs in tumors is random rather than guided. Indeed, the level of chemokines and the disorganized infiltration of stromal cells is likely to preclude the presence of distinct chemokine gradients, which could direct CTL migration. In the lymph node, it was previously suggested that the migratory path of T cells was random. It is now known that the migration also relies on the stromal cell architecture (such as reticular fibroblast – FRCs – for T cell zones and follicular dendritic cells – FDCs – for B cell zones) (Bajenoff et al. 2006). Similar motility behavior might happen in the tumor mass. We have imaged the tumor stroma using hematopoietic chimeras of CFP-expressing mice reconstituted with a bone marrow from GFP-expressing mice and observed that, besides the tumor cells, the stroma is very rich in both hematopoietic and nonhematopoietic cells, including elongated cells that look like reticular fibroblasts (Fig. 2). The role of these cells in CTL motility inside tumors is being analyzed.

3.3 A Motility-Based Model for Tumor Infiltration by T Lymphocytes

These results prompted us to propose a working model for tumor rejection, based on successive phases of nondirectional motility followed by phases of cytotoxicity and long arrests in migration (Fig. 3). This model is based on our observation that tumor infiltration is initiated at the periphery of the tumors (Boissonnas et al. 2007). After destruction of the peripheral layer of the tumor, the motility of CTLs

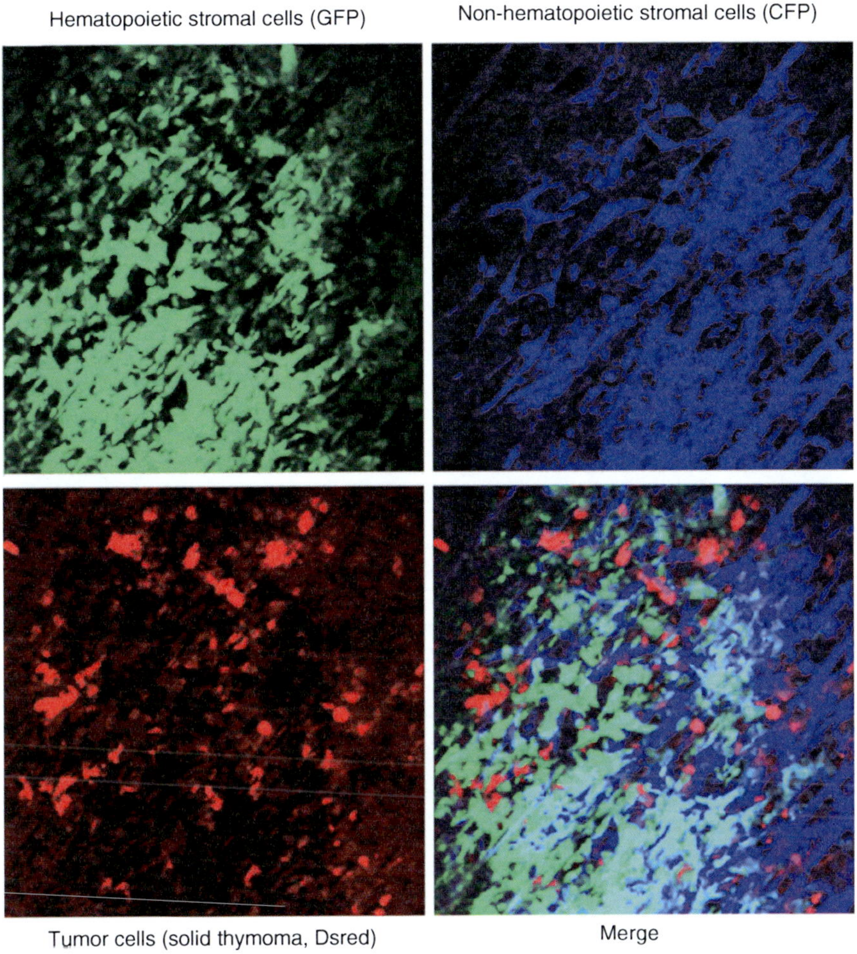

Fig. 2 Imaging the tumor and its stroma. Two-photon imaging of a Ds-red EG7 solid thymoma inoculated subcutaneously in fluorescent hematopoietic chimera mouse. The chimera were generated by reconstitution of an irradiated Tg-UBI-CFP mouse with bone marrow from a Tg-UBI-GFP mouse. Hematopoietic stromal cells are in *green*, nonhematopoietic stromal cells are in *blue*, tumor cells are in *red*

increases, which probably optimizes the search for remaining living tumor cells. Upon encountering antigen-expressing living tumor cells, CD8+ T cells stop again, kill the tumor cells in this new region and regain motility. Since the infiltration starts in the periphery of the tumors, this series of "stop" and "go" phases would result in the progressive directional invasion of the tumors. Consistent with this model, in the nonantigen-expressing tumors, the T cells remained in the periphery and, despite their high motility, they were unable to infiltrate the tumors in depth (Boissonnas et al. 2007).

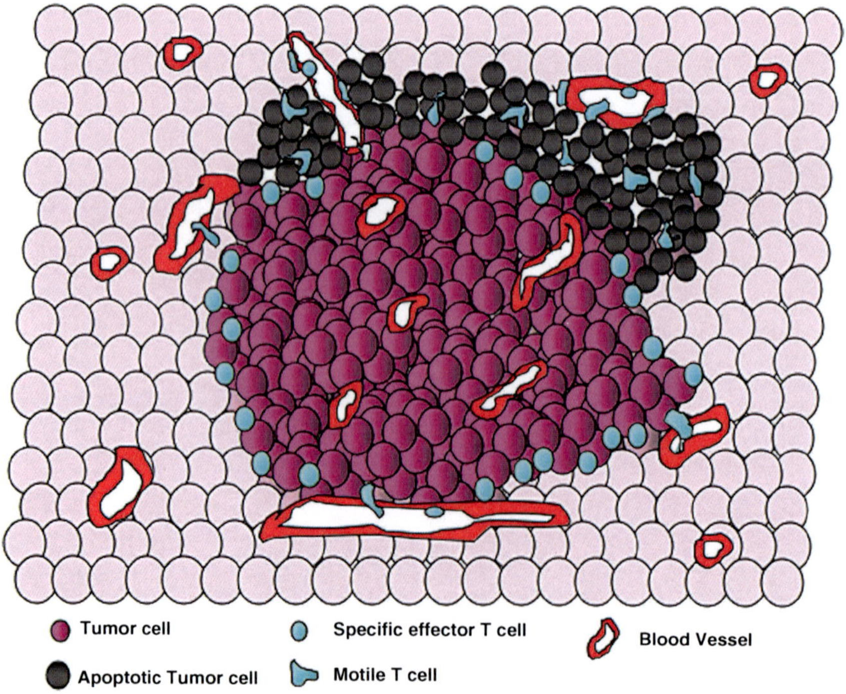

Fig. 3 T cell infiltration into the tumor. Antigen-specific T cells infiltrate the tumor from the blood vessels and make long lasting contacts with tumor cells (*bottom part of the tumor*). After elimination of the tumor cells in that area, T cells regain motility and stop again in healthy tumor cell-containing areas (*upper part of the tumor*). These serial "stop" and "go" phases lead to tumor elimination from the periphery toward the center

This model is based on the assumption that deep infiltration into the tumor occurs by migration of the peripheral T cells, although it is still possible that newly arriving T cells extravasate from deeper vessels, directly in the innermost of the tumor. Several reports demonstrated a defective adhesion of leucocytes in the vicinity of tumors, accounting for a preferential distribution of the T cells in the peritumoral stroma, rather than the parenchyma (Chen et al. 2003; Dirkx et al. 2003). It is therefore most likely, but remains to be addressed directly, that T cells can traffick inside tumors from the periphery towards deeper regions.

Nevertheless, in the course of tumor rejection, the permeability and the functionality of blood vasculature may change, maybe under the influence of increased inflammation caused by tumor rejection (Mrass and Weninger 2006). Indeed, the expression of adhesion molecules may be different according to the origin, the location, or the stage of the tumors. Three-dimensional imaging of T cell trafficking to tumors by MRI showed that, after sequential adoptive transfer, T cell recruitment may occur in different regions of the tumors (Kircher et al. 2003). This observation could also be related to the blood vasculature and the destruction of the tumor in the areas where CTLs infiltrate, resulting in different infiltration sites by the subsequent T cell "waves." Noninvasive imaging should address these issues, by following lymphocytes over time within the whole animal (Dubey et al. 2003).

In conclusion, two-photon dynamic imaging showed active migration of CD8+ T cells in tumors. Antigen expression by the tumor induces specific retention and accumulation of the CTLs. Migration from lymph nodes to the tumor site is dictated by chemokines and adhesion molecules, but once within the tumor, the trafficking occurs along blood vessels or collagen fibers, but the arrests are dictated by the presence of antigen (Fig. 3).

4 Tumor Rejection and Tumor Cell Killing by CTLs

4.1 *Mechanism of Cytotoxicity*

Upon differentiation from resting CD8+ T lymphocytes into CTLs, the effector cells acquire the unique ability to kill target cells. Cytotoxicity is triggered upon TCR engagement by class I MHC molecules expressed on the surface of the target cells (Christinck et al. 1991). The recognition and triggering system is extremely sensitive *in vitro*, but the sensitivity *in vivo* has not so far been accurately measured. Killing by CTLs may occur through several different pathways that may be redundant or synergic, depending on the targets (Shresta et al. 1998). The main pathways are (1) the secretion of cytokines, including TNFα and IFNγ, (2) the release of perforin and granzymes contained in cytolytic granules into the contact zone between the CTLs and their targets, and (3) the engagement of a Fas/Fas ligand pathway that induces caspase-dependent apoptosis in the target cells (Fig. 4).

The cytotoxic activity of CTLs depends on a complex structure that forms at the contact zone with the target cell. This structure, called "immune synapse," represents a platform for both the recognition of the target and the cytotoxic activity of CTLs. The two activities of the synapse are coordinated both in time and space (Stinchcombe et al. 2001). This coordination results in a very high sensitivity of the killing event, at least *in vitro*: a few cognate class I MHC complexes expressed on the surface of the target cells suffice to trigger killing (Christinck et al. 1991), while the secretion of IFNγ requires higher levels of MHC–peptide complexes (Valitutti et al. 1996). The release of the perforin-containing granules causes the formation of pores in the target cell, allowing the release of granzymes into the cytosol of the target cell, thus inducing caspase-dependent cell death (Shresta et al. 1998). The engagement of the Fas/FasL pathway also causes the apoptosis of the target cells through the formation of the death-inducing signaling complex (DISC).

These different pathways are functionally redundant. Deficiencies in the perforin or the Fas pathways rarely show a complete alteration in the cytotoxic activity of the CTLs (Winter et al. 1999). The sensitivity, efficiency, or the frequency of the killing events have all been evaluated *in vitro*, but very little is known about the cytotoxic activity *in vivo*. Understanding how the cytokine pathway integrates the

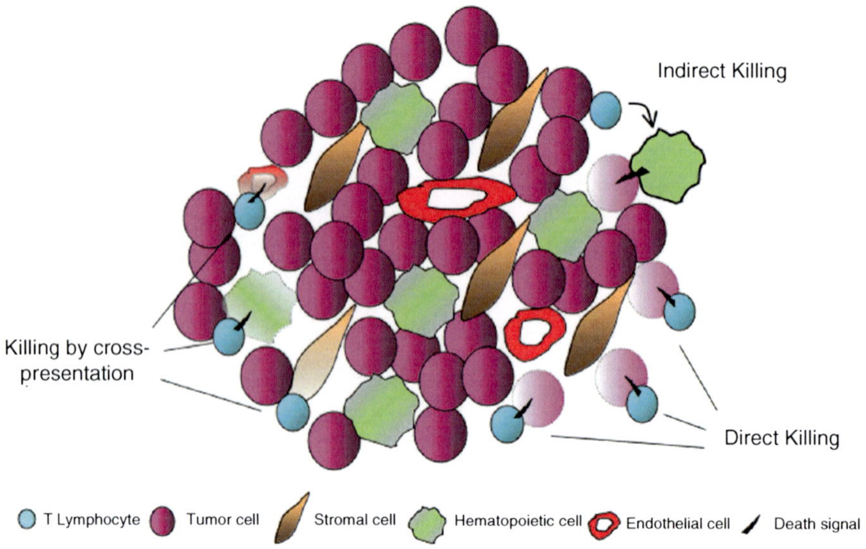

Fig. 4 CTLs targets in the tumor. The different possibilities of CTL cytotoxicity during tumor rejection are represented. *Direct killing*: CTLs induce tumor cell death trough direct interaction and release of cytotoxic factors (perforin,Granzymes FasL, IFNγ, TNFα). *Indirect killing*: CTLs recruit and/or activate hematopoietic cells (trough IFNγ secretion) which eliminates tumor cells. *Killing after cross-presentation*: CTLs target and eliminate stromal cells (fibroblast, hematopoietic antigen-presenting cells, endothelial cells) that disorganize tumor architecture, growth factors, or blood supply

picture is even more complicated. Are the cytokines released by CTLs acting throughout the tumor or only at the vicinity of the activated CTLs, or only on the cells engaged on a direct synapse with the CTL? Can CTLs kill "bystander" cells within the tumor? How does the activation of CTLs, through cytokine secretion, for example, affect other cells within the tumor? Does the killing of tumor cells activate killing by other cells that also infiltrate the tumor (macrophages, NK cells…)? These questions, which may all be addressed using dynamic *in vivo* imaging, are critical when thinking about the ways CTLs induce tumor rejection, since arguments exist supporting both rejection through direct killing of the tumor cells, and indirect rejection through additional effects on the tumor stroma.

4.2 Direct Tumor Cell Killing and Rejection

The first evidence for CTLs killing tumor cells through direct cytotoxicity comes from seminal studies in the 1970s by Klein et al (Klein and Klein 1972) and by Weissman et al (Weissman 1973). These two groups showed, in different experimental settings, that rejection of a solid tumor by CTLs left bystander tumor cells intact. These observations were fully consistent with the exquisite specificity of CTL-mediated killing *in vitro*. Similar results were published recently by the group of Bousso (2008); (Breart et al. 2008), who showed that mixing tumor cells that express or not a model tumor antigen, results in the selective elimination of the antigen-expressing cells, but not of the other tumor cells, even when present at very low numbers. They also showed that the tumor-specific T cells localize specifically to regions of the tumors with high densities of the antigen-expressing tumor cells. In addition, several recent *in vivo* analysis of CTL-tumor cell interactions using two-photon microscopy support direct killing.

The three groups that have so far imaged T cell behavior in solid tumors (Boissonnas et al. 2007; Breart et al. 2008; Mrass et al. 2006) all report that CTLs make long lasting, antigen-dependent, contacts with tumor cells *in vivo*, suggesting that they were inducing cell death through direct contacts. We found a striking correlation in specific regions of the tumor between dead tumor cells (as detected by tunnel) and CTL infiltration, supporting the idea that CTLs are responsible for individual tumor cell death. Mrass et al. observed, in real time, rare events of CTLs in contact with tumor cells undergoing apoptosis (Mrass et al. 2006). The group of Bousso proposed that T cells mediate apoptosis through a direct cell-to-cell contact *in vivo* (Breart et al. 2008). Using a death reporter probe, they imaged in real time, caspase-3 activation in the tumor cells. They reported only 13 events of real time conversion from live cell to dying cell after recording more than 70 h of CTL–tumor cell interactions in fields containing multiple tumor cells. The best evidence for CTLs being the inducers of tumor cell death in this study was that the apoptosis index increased with the number of contacts between a tumor cell and the CTLs. Nevertheless, it is not clear whether, in this quantification, the contacts between T cells and dead tumor cells started before or after the cell death (which could reveal a different affinity of CTL for dying and living cells). Overall, the estimated rate of cell killing was around 1 tumor cell killed every 6 h of contacting each one CTL. This is a quite slow rate of killing, as compared to the rates observed *in vitro*. A prediction from this observation is that the efficiency of tumor rejection will depend critically on the efficiency of CTL infiltration into the tumors. Taking into account the rate of killing and the proportion of tumor cells in contact with one CTL (around 10%), we can estimate that less than 60 h should suffice to contact and kill 100% of the tumor cells, if the rate of tumor cell proliferation is not considered. Altogether, the period of 7 days experimentally observed for complete rejection is in principle consistent with a direct killing-based mechanism for tumor rejection. These calculations, however, do not take into account the distribution of the CTLs within the tumors.

In vivo, it is therefore most likely that the efficiency of tumor infiltration determines the success of tumor rejection. From this point of view, the ability of CTLs to engage sequentially and kill multiple targets is probably critical. Multiple killing by single CTLs has been described *in vitro* by Wiedemann et al. (2006). *In vivo*, we described four distinct behaviors of CTLs in regard to the interaction with tumor cells (Boissonnas et al. 2007): (1) CTLs established stable and long lasting interaction with one tumor cell during all the time of observation, (2) the CTLs made interactions with one tumor and several neighboring cells, this confined interaction may correspond to the ones observed *in vitro* with the ability to polarize lytic granules toward different targets, (3) the CTLs made serial engagements with distant tumor cells, and (4) no significant interactions were detected and the CTLs fleet through the tumor mass. The two first modes were predominant when tumor cell density was high during the early phase of tumor rejection, and the relative proportion of the two latter increased with tumor cell elimination.

Together with the experiments of mixed tumors, these imaging studies provide strong support for the "direct killing" hypothesis. Nevertheless, they do not exclude possible indirect mechanisms. Cytokines and other cell types (such as macrophages, NK cells, or granulocytes), could be involved at later phases of tumor rejection. Indeed, in certain experimental systems, there is solid evidence that rejection is not exclusively due to direct killing of tumor cells by CTLs.

4.3 Indirect Mechanisms of Tumor Rejection

Several lines of evidence indicate that tumor rejection is not the exclusive consequence of the direct killing of tumor cells by CTLs. In various tumor models, the CTLs antitumor activity required IFNγ production (Barth et al. 1991). Dutton's group used adoptive transfer of tumor-specific CTLs deficient for perforin, TNFα, Fas L, or IFNγ (as well as some combinations of these KOs) and showed that only IFNγ-deficiency affected tumor rejection (Hollenbaugh et al. 2004). It must be noted, however, that CTLs defective for either perforin and/or FasL or perforin and/or TNFα may still be cytotoxic, due to the known redundancy of the different pathways in target killing. In addition, in a different tumor model, the same group showed that FasL and TNFα were required for rejection, suggesting that sensitivity of tumor cells to the different cytolytic pathways may vary between different tumors (Dobrzanski et al. 2004). Similarly, Peng et al. showed that perforin and IFNγ requirement for rejection varied with the tumors used (Peng et al. 2000).

The group of Dutton also showed that expression of the receptor for IFNγ on the host cells was required for tumor rejection, even if the tumors expressed the receptor (Hollenbaugh and Dutton 2006). These results demonstrate, at least in this particular tumor model, that the host must respond to IFNγ secreted by the CTLs for the efficient induction of tumor rejection. IFNγ could enhance the cytotoxic activity of the CTLs indirectly (for example, by favoring antigen cross-presentation by the host APCs inside the tumor, or in lymph nodes, or promoting better priming, or remodeling the tumor to allow a more efficient attack of the tumor cells by the CTLs). IFNγ can also enhance cytotoxicity against the tumor cells (through activation of granulocytes or NK cells, for example). It has been shown, indeed, that IFNγ secretion by CTLs induces the recruitment of several other populations such as additional T lymphocytes, macrophages, granulocytes, NK cells, or NK-related cells (such as IKDC) (Bonmort et al. 2007; Hollenbaugh and Dutton 2006), which may all potentially contribute to tumor rejection. In the case of T cells, both reactivation of unresponsive

T cells (Lurquin et al. 2005) and the induction of T cells specific for new tumor determinants ("epitope spreading") have been proposed (Brossart et al. 2000).

One of the better characterized mechanisms of indirect tumor rejection is through destruction of tumor stroma. The group of Shreiber showed that in tumors expressing high levels of antigen, stromal cells cross-present tumor antigens to CTLs, inducing tumor rejection and bystander elimination of antigen loss variants (Spiotto et al. 2004). This observation provides clear evidence that tumor cells are not the unique target for CTLs and raises therapeutic perspectives, such as targeting the tumor stroma to induce rejection (Kammertoens et al. 2005). Blohm et al. showed that. after peripheral CD8+ T cell attack, the center of the tumor was infiltrated by granulocytes destroying the vasculature (Blohm et al. 2006). T cells may also be involved in indirect stromal cell elimination, through macrophages activation (Corthay et al. 2005; Ibe et al. 2001), NK cell activation (Perez-Diez et al. 2007), or endothelial cell destruction, mediated by IFNγ production by CD4+ tumor infiltrating T cells (Qin and Blankenstein 2000).

Altogether, these studies suggest that under certain conditions, and in certain tumor models, antigen-specific T cell responses (mediated by either CD8+ or CD4+ T cells) may function as a trigger for further immune responses. These "secondary" immune responses to the tumor could be mediated by cells of the innate immune system, such as macrophages, NK cells, or granulocytes. The contribution of this antitumor "innate" immune response to tumor rejection, however, remains to be thoroughly investigated.

5 Regulatory T Cells Control Tumor Rejection by CTLs

5.1 Treg-Mediated Suppression of Antitumor T Cell Priming

Even if multiple mechanisms for tumor escape from CTLs have been reported (and extensively reviewed elsewhere) (Gajewski et al. 2006; Nagaraj and Gabrilovich 2008; Pawelec et al. 2006), several recent reports have analyzed the mechanisms of Treg suppression of antitumor T cell responses. It is clear that the frequency of Tregs is increased both in tumor-bearing mice and in cancer patients. In certain cases, prognosis may be linked to the ratio of CD8+ T to Treg cells (Sato et al. 2005). However, the mechanisms of Treg suppression of antitumor immune responses remain unclear. Many reports showed that Tregs may inhibit T cell responses at both the priming and the effector levels (Vignali et al. 2008). The expansion of antitumor CD8+ T cells is inhibited by Tregs, but these effects were mainly observed after *in vitro* restimulation (Somasundaram et al. 2002; Woo et al. 2002) or indirectly by the increase of the frequency of antitumor CTLs after depletion of Tregs (Ercolini et al. 2005; Sutmuller et al. 2001).

Two independent studies used intravital two-photon imaging to analyze the Treg suppression mechanisms of autoimmune CD4+ T cell priming. Both groups showed that Tregs interfere with the stability of the interactions between CD4+ T cells and antigen-presenting cells during T cell priming (Tadokoro et al. 2006; Tang et al. 2006). Using adoptive transfer of naive antigen-specific CD8+ T cells to tumor-bearing mice, we find that Tregs control the priming of CD8+ antitumor T cells in a similar manner, i.e., through the regulation of T cell motility in tumor-draining lymph nodes (Boissonnas et al., submitted). Further two-photon imaging studies will investigate in the near future different aspects of Treg function during antitumor immune responses.

5.2 Treg-Mediated Suppression of Effector Antitumor T Cell Responses

Most of the evidence published so far, however, point to the important role of Tregs in the control of effector antitumor T cells. An interesting report by Chen et al. showed that cotransfer of naive HA-specific CD8+ T cells together with CD4+ Tregs abrogates the rejection of HA-expressing tumors (Chen et al. 2005). Proliferation and cell death of naive CD8+ T cells in the tumor-draining lymph nodes was not affected by the presence of Tregs, suggesting an inhibition at the effector level (Chen et al. 2005). Similarly, Yu et al. (2005) performed local depletion of CD4+ T cells within tumors and showed that the inhibitory effect occurs during the effector phase of the CTLs response.

Regulatory T cells have also been shown by intravital imaging to reduce the exocytosis of cytotoxic granules of effector CTLs, but did not affect the duration of the CTL–target interactions (Mempel et al. 2006). This inhibitory effect did not require long lasting contacts between Tregs and CTLs, but was dependent on their responsiveness to TGF-β. Consistent with this possibility, we found that Tregs and CTLs establish multiple brief interactions, both in tumors and in tumor-draining lymph nodes, although the functional relevance of these interactions remains unclear (Boissonnas et al., submitted).

6 Conclusion

Even though multiple studies have analyzed antitumor immune responses in the last 30 years, many aspects of the actual role of immunity in tumor development and rejection remain unclear. Indeed, it is very difficult to go beyond the rules unique to each tumor model or model antigen used and extract rules applicable to general situations. Understanding how the results obtained in mice could apply to cancer patients is even more difficult. Part of these difficulties probably come from the use of transplantable tumors, rather than spontaneous tumor models. Injecting subcutaneously thousands of tumor cells that grew in fetal calf serum is obviously very different from a tumor that grows over months or years from a single cell precursor in a given organ. The recent development of spontaneous tumor models in different organs expressing model antigens has already helped understand certain discrepancies between mouse models and human cancers. It is most likely that combining spontaneous tumor models and dynamic multiphoton imaging will address some of the critical questions that remain open in the field of tumor immunology in future years.

Acknowledgment The authors would like to thank Dr. Olivier Lantz for helpful discussion and critically reading the manuscript. A.B. has been supported by la Ligue contre le Cancer. A.S-D. was a fellow of Ministère de l'Education et de la Recherche. This work was supported by funding from Institut National de la Santé et de la Recherche Médicale; Centre National de la Recherche Scientifique; La Ligue Contre le Cancer; Association de la Recherche Contre le Cancer (ARC, Leopold Griffuel award); the Institut Curie; Fondation Bettencourt-Schueller; and Contrat EC DC-Thera N°LSBH-CT-2004-512074 "Dendritic Cells for Novel Immunotherapies" and Contrat EC Cancer Immunotherapy LSHC-CT-2006-518234 "Cancer Immunology and Immunotherapy." The authors have no conflicting financial interests.

References

Bajenoff M, Egen JG, Koo LY, Laugier JP, Brau F, Glaichenhaus N, Germain RN. (2006) Stromal cell networks regulate lymphocyte entry, migration, and territoriality in lymph nodes. Immunity 25:989–1001

Barth RJ Jr, Mule JJ, Spiess PJ, Rosenberg SA. (1991) Interferon gamma and tumor necrosis factor have a role in tumor regressions mediated by murine CD8+ tumor-infiltrating lymphocytes. J Exp Med 173:647–658

Blohm U, Potthoff D, van der Kogel AJ, Pircher H. (2006) Solid tumors "melt" from the inside after successful CD8 T cell attack. Eur J Immunol 36:468–477

Boissonnas A, Combadiere C, Lavergne E, Maho M, Blanc C, Debre P, Combadiere B. (2004) Antigen distribution drives programmed antitumor CD8 cell migration and determines its efficiency. J Immunol 173:222–229

Boissonnas A, Fetler L, Zeelenberg IS, Hugues S, Amigorena S. (2007) In vivo imaging of cytotoxic T cell infiltration and elimination of a solid tumor. J Exp Med 204:345–356

Boissonnas A, Scholer-Dahirel A, Simon-Blancal V, Kissenpfennig A, Sparwasser T, Malissen B, Fetler L, Amigorena S. Regulatory T cells control dendritic cells survival in tumor draining lymph node. (Submitted)

Bonmort M, Ullrich E, Mignot G, Jacobs B, Chaput N, Zitvogel L. (2007) Interferon-gamma is produced by another player of innate immune responses: the interferon-producing killer dendritic cell (IKDC). Biochimie 89:872–877

Bousso P. (2008) T-cell activation by dendritic cells in the lymph node: lessons from the movies. Nat Rev Immunol 9:155–165

Breart B, Lemaitre F, Celli S, Bousso P. (2008) Two-photon imaging of intratumoral CD8+ T cell cytotoxic activity during adoptive T cell therapy in mice. J Clin Invest 118:1390–1397

Brossart P, Wirths S, Stuhler G, Reichardt VL, Kanz L, Brugger W. (2000) Induction of cytotoxic T-lymphocyte responses in vivo after vaccinations with peptide-pulsed dendritic cells. Blood 96:3102–3108

Butcher EC, Williams M, Youngman K, Rott L, Briskin M. (1999) Lymphocyte trafficking and regional immunity. Adv Immunol 72:209–253

Calzascia T, Di Berardino-Besson W, Wilmotte R, Masson F, de Tribolet N, Dietrich PY, Walker PR. (2003) Cutting edge: cross-presentation as a mechanism for efficient recruitment of tumor-specific CTL to the brain. J Immunol 171:2187–2191

Calzascia T, Masson F, Di Berardino-Besson W, Contassot E, Wilmotte R, Aurrand-Lions M, Ruegg C, Dietrich PY, Walker PR. (2005) Homing phenotypes of tumor-specific CD8 T cells are predetermined at the tumor site by crosspresenting APCs. Immunity 22:175–184

Castellino F, Huang AY, Altan-Bonnet G, Stoll S, Scheinecker C, Germain RN. (2006) Chemokines enhance immunity by guiding naive CD8+ T cells to sites of CD4+ T cell-dendritic cell interaction. Nature 440:890–895

Celli S, Lemaitre F, Bousso P. (2007) Real-time manipulation of T cell-dendritic cell interactions in vivo reveals the importance of prolonged contacts for CD4+ T cell activation. Immunity 27:625–634

Chen Q, Wang WC, Evans SS. (2003) Tumor microvasculature as a barrier to antitumor immunity. Cancer Immunol Immunother 52:670–679

Chen ML, Pittet MJ, Gorelik L, Flavell RA, Weissleder R, von Boehmer H, Khazaie K. (2005) Regulatory T cells suppress tumor-specific CD8 T cell cytotoxicity through TGF-beta signals in vivo. Proc Natl Acad Sci U S A 102:419–424

Christinck ER, Luscher MA, Barber BH, Williams DB. (1991) Peptide binding to class I MHC on living cells and quantitation of complexes required for CTL lysis. Nature 352:67–70

Corthay A, Skovseth DK, Lundin KU, Rosjo E, Omholt H, Hofgaard PO, Haraldsen G, Bogen B. (2005) Primary antitumor immune response mediated by CD4+ T cells. Immunity 22:371–383

Dirkx AE, Oude Egbrink MG, Kuijpers MJ, van der Niet ST, Heijnen VV, Bouma-ter Steege JC, Wagstaff J, Griffioen AW. (2003) Tumor angiogenesis modulates leukocyte-vessel wall interactions in vivo by reducing endothelial adhesion molecule expression. Cancer Res 63:2322–2329

Dobrzanski MJ, Reome JB, Hollenbaugh JA, Hylind JC, Dutton RW. (2004) Effector cell-derived lymphotoxin alpha and Fas ligand, but not perforin, promote Tc1 and Tc2 effector cell-mediated tumor therapy in established pulmonary metastases. Cancer Res 64:406–414

Dolan BP, Gibbs KD Jr, Ostrand-Rosenberg S. (2006a) Dendritic cells cross-dressed with peptide MHC class I complexes prime CD8+ T cells. J Immunol 177:6018–6024

Dolan BP, Gibbs KD Jr, Ostrand-Rosenberg S. (2006b) Tumor-specific CD4+ T cells are activated by “cross-dressed” dendritic cells presenting peptide-MHC class II complexes acquired from cell-based cancer vaccines. J Immunol 176:1447–1455

Dubey P, Su H, Adonai N, Du S, Rosato A, Braun J, Gambhir SS, Witte ON. (2003) Quantitative imaging of the T cell antitumor response by positron-emission tomography. Proc Natl Acad Sci U S A 100:1232–1237

Ercolini AM, Ladle BH, Manning EA, Pfannenstiel LW, Armstrong TD, Machiels JP, Bieler JG, Emens LA, Reilly RT, Jaffee EM. (2005) Recruitment of latent pools of high-avidity CD8(+) T cells to the antitumor immune response. J Exp Med 201:1591–1602

Gade TP, Hassen W, Santos E, Gunset G, Saudemont A, Gong MC, Brentjens R, Zhong XS, Stephan M, Stefanski J, et al. (2005) Targeted elimination of prostate cancer by genetically directed human T lymphocytes. Cancer Res 65:9080–9088

Gajewski TF, Meng Y, Harlin H. (2006) Immune suppression in the tumor microenvironment. J Immunother 29:233–240

Garcia Z, Pradelli E, Celli S, Beuneu H, Simon A, Bousso P. (2007) Competition for antigen determines the stability of T cell-dendritic cell interactions during clonal expansion. Proc Natl Acad Sci U S A 104:4553–4558

Hellstrom KE, Hellstrom I, Chen L. (1995) Can co-stimulated tumor immunity be therapeutically efficacious. Immunol Rev 145:123–145

Hollenbaugh JA, Dutton RW. (2006) IFN-gamma regulates donor CD8 T cell expansion, migration, and leads to apoptosis of cells of a solid tumor. J Immunol 177:3004–3011

Hollenbaugh JA, Reome J, Dobrzanski M, and Dutton RW. (2004) The rate of the CD8-dependent initial reduction in tumor volume is not limited by contact-dependent perforin, Fas ligand, or TNF-mediated cytolysis. J Immunol 173:1738–1743

Huang AY, Bruce AT, Pardoll DM, Levitsky HI. (1996a) Does B7–1 expression confer antigen-presenting cell capacity to tumors in vivo. J Exp Med 183:769–776

Huang AY, Bruce AT, Pardoll DM, Levitsky HI. (1996b) In vivo cross-priming of MHC class I-restricted antigens requires the TAP transporter. Immunity 4:349–355
Hugues S, Fetler L, Bonifaz L, Helft J, Amblard F, Amigorena S. (2004) Distinct T cell dynamics in lymph nodes during the induction of tolerance and immunity. Nat Immunol 5:1235–1242
Hugues S, Boissonnas A, Amigorena S, Fetler L. (2006) The dynamics of dendritic cell-T cell interactions in priming and tolerance. Curr Opin Immunol 18:491–495
Hugues S, Scholer A, Boissonnas A, Nussbaum A, Combadiere C, Amigorena S, Fetler L. (2007) Dynamic imaging of chemokine-dependent CD8+ T cell help for CD8+ T cell responses. Nat Immunol 8:921–930
Ibe S, Qin Z, Schuler T, Preiss S, Blankenstein T. (2001) Tumor rejection by disturbing tumor stroma cell interactions. J Exp Med 194:1549–1559
Jain RK, Koenig GC, Dellian M, Fukumura D, Munn LL, Melder RJ. (1996) Leukocyte-endothelial adhesion and angiogenesis in tumors. Cancer Metastasis Rev 15:195–204
Jain RK, Munn LL, Fukumura D. (2002) Dissecting tumour pathophysiology using intravital microscopy. Nat Rev Cancer 2:266–276
Joncker NT, Helft J, Jacquet A, Premel V, Lantz O. (2006a) Intratumor CD4 T-cell accumulation requires stronger priming than for expansion and lymphokine secretion. Cancer Res 66:5443–5451
Joncker NT, Marloie MA, Chernysheva A, Lonchay C, Cuff S, Klijanienko J, Sigal-Zafrani B, Vincent-Salomon A, Sastre X, Lantz O. (2006b) Antigen-independent accumulation of activated effector/memory T lymphocytes into human and murine tumors. Int J Cancer 118:1205–1214
Kammertoens T, Schuler T, Blankenstein T. (2005) Immunotherapy: target the stroma to hit the tumor. Trends Mol Med 11:225–231
Kircher MF, Allport JR, Graves EE, Love V, Josephson L, Lichtman AH, Weissleder R. (2003) In vivo high resolution three-dimensional imaging of antigen-specific cytotoxic T-lymphocyte trafficking to tumors. Cancer Res 63:6838–6846
Klein E, Klein G. (1972) Specificity of homograft rejection in vivo, assessed by inoculation of artificially mixed compatible and imcompatible tumor cells. Cell Immunol 5:201–208
Kundig TM, Bachmann MF, DiPaolo C, Simard JJ, Battegay M, Lother H, Gessner A, Kuhlcke K, Ohashi PS, Hengartner H, et al. (1995) Fibroblasts as efficient antigen-presenting cells in lymphoid organs. Science 268:1343–1347
Lurquin C, Lethe B, De Plaen E, Corbiere V, Theate I, van Baren N, Coulie PG, Boon T. (2005) Contrasting frequencies of antitumor and anti-vaccine T cells in metastases of a melanoma patient vaccinated with a MAGE tumor antigen. J Exp Med 201:249–257
Mantovani A, Allavena P, Sica A, Balkwill F. (2008) Cancer-related inflammation. Nature 454:436–444
Maric M, Zheng P, Sarma S, Guo Y, Liu Y. (1998) Maturation of cytotoxic T lymphocytes against a B7-transfected nonmetastatic tumor: a critical role for costimulation by B7 on both tumor and host antigen-presenting cells. Cancer Res 58:3376–3384
Mempel TR, Henrickson SE, Von Andrian UH. (2004) T-cell priming by dendritic cells in lymph nodes occurs in three distinct phases. Nature 427:154–159
Mempel TR, Pittet MJ, Khazaie K, Weninger W, Weissleder R, von Boehmer H, von Andrian UH. (2006) Regulatory T cells reversibly suppress cytotoxic T cell function independent of effector differentiation. Immunity 25:129–141
Miller MJ, Safrina O, Parker I, Cahalan MD (2004) Imaging the single cell dynamics of CD4+ T cell activation by dendritic cells in lymph nodes. J Exp Med 200:847–856
Mora JR, Bono MR, Manjunath N, Weninger W, Cavanagh LL, Rosemblatt M, Von Andrian UH. (2003) Selective imprinting of gut-homing T cells by Peyer's patch dendritic cells. Nature 424:88–93
Mrass P, Takano H, Ng LG, Daxini S, Lasaro MO, Iparraguirre A, Cavanagh LL, von Andrian UH, Ertl HC, Haydon PG, Weninger W. (2006) Random migration precedes stable target cell interactions of tumor-infiltrating T cells. J Exp Med 203:2749–2761

Mrass P, Weninger W. (2006) Immune cell migration as a means to control immune privilege: lessons from the CNS and tumors. Immunol Rev 213:195–212

Mukai S, Kjaergaard J, Shu S, Plautz GE. (1999) Infiltration of tumors by systemically transferred tumor-reactive T lymphocytes is required for antitumor efficacy. Cancer Res 59:5245–5249

Nagaraj S, Gabrilovich DI. (2008) Tumor escape mechanism governed by myeloid-derived suppressor cells. Cancer Res 68:2561–2563

Ng LG, Mrass P, Kinjyo I, Reiner SL, Weninger W. (2008) Two-photon imaging of effector T-cell behavior: lessons from a tumor model. Immunol Rev 221:147–162

Nimmerjahn F, Ravetch JV. (2007) Antibodies, Fc receptors and cancer. Curr Opin Immunol 19:239–245

Ochsenbein AF, Klenerman P, Karrer U, Ludewig B, Pericin M, Hengartner H, Zinkernagel RM. (1999) Immune surveillance against a solid tumor fails because of immunological ignorance. Proc Natl Acad Sci USA 96:2233–2238

Ochsenbein AF, Sierro S, Odermatt B, Pericin M, Karrer U, Hermans J, Hemmi S, Hengartner H, Zinkernagel RM. (2001) Roles of tumour localization, second signals and cross priming in cytotoxic T-cell induction. Nature 411:1058–1064

Pawelec G, Koch S, Griesemann H, Rehbein A, Hahnel K, Gouttefangeas C. (2006) Immunosenescence, suppression and tumour progression. Cancer Immunol Immunother 55:981–986

Peng L, Krauss JC, Plautz GE, Mukai S, Shu S, Cohen PA. (2000) T cell-mediated tumor rejection displays diverse dependence upon perforin and IFN-gamma mechanisms that cannot be predicted from in vitro T cell characteristics. J Immunol 165:7116–7124

Perez-Diez A, Joncker NT, Choi K, Chan WF, Anderson CC, Lantz O, Matzinger P. (2007) CD4 cells can be more efficient at tumor rejection than CD8 cells. Blood 109:5346–5354

Qin Z, Blankenstein T. (2000) CD4+ T cell--mediated tumor rejection involves inhibition of angiogenesis that is dependent on IFN gamma receptor expression by nonhematopoietic cells. Immunity 12:677–686

Rosenberg SA, Restifo NP, Yang JC, Morgan RA, Dudley ME. (2008) Adoptive cell transfer: a clinical path to effective cancer immunotherapy. Nat Rev Cancer 8:299–308

Sato E, Olson SH, Ahn J, Bundy B, Nishikawa H, Qian F, Jungbluth AA, Frosina D, Gnjatic S, Ambrosone C, et al.. (2005) Intraepithelial CD8+ tumor-infiltrating lymphocytes and a high CD8+/regulatory T cell ratio are associated with favorable prognosis in ovarian cancer. Proc Natl Acad Sci USA 102:18538–18543

Scholer A, Hugues S, Boissonnas A, Fetler L, Amigorena S. (2008) Intercellular Adhesion Molecule-1-Dependent Stable Interactions between T Cells and Dendritic Cells Determine CD8(+) T Cell Memory. Immunity 28:258–270

Shresta S, Pham CT, Thomas DA, Graubert TA, Ley TJ. (1998) How do cytotoxic lymphocytes kill their targets. Curr Opin Immunol 10:581–587

Skokos D, Shakhar G, Varma R, Waite JC, Cameron TO, Lindquist RL, Schwickert T, Nussenzweig MC, Dustin ML. (2007) Peptide-MHC potency governs dynamic interactions between T cells and dendritic cells in lymph nodes. Nat Immunol 8:835–844

Smirnov P, Lavergne E, Gazeau F, Lewin M, Boissonnas A, Doan BT, Gillet B, Combadiere C, Combadiere B, Clement O. (2006) In vivo cellular imaging of lymphocyte trafficking by MRI: a tumor model approach to cell-based anticancer therapy. Magn Reson Med 56:498–508

Somasundaram R, Jacob L, Swoboda R, Caputo L, Song H, Basak S, Monos D, Peritt D, Marincola F, Cai D, et al.. (2002) Inhibition of cytolytic T lymphocyte proliferation by autologous CD4+/CD25+ regulatory T cells in a colorectal carcinoma patient is mediated by transforming growth factor-beta. Cancer Res 62:5267–5272

Spiotto MT, Yu P, Rowley DA, Nishimura MI, Meredith SC, Gajewski TF, Fu YX, Schreiber H. (2002) Increasing tumor antigen expression overcomes "ignorance" to solid tumors via cross-presentation by bone marrow-derived stromal cells. Immunity 17:737–747

Spiotto MT, Rowley DA, Schreiber H. (2004) Bystander elimination of antigen loss variants in established tumors. Nat Med 10:294–298

Stevanovic S. (2002) Identification of tumour-associated T-cell epitopes for vaccine development. Nat Rev Cancer 2:514–520

Stinchcombe JC, Bossi G, Booth S, Griffiths GM. (2001) The immunological synapse of CTL contains a secretory domain and membrane bridges. Immunity 15:751–761

Sutmuller RP, van Duivenvoorde LM, van Elsas A, Schumacher TN, Wildenberg ME, Allison JP, Toes RE, Offringa R, Melief CJ. (2001) Synergism of cytotoxic T lymphocyte-associated antigen 4 blockade and depletion of CD25(+) regulatory T cells in antitumor therapy reveals alternative pathways for suppression of autoreactive cytotoxic T lymphocyte responses. J Exp Med 194:823–832

Tadokoro CE, Shakhar G, Shen S, Ding Y, Lino AC, Maraver A, Lafaille JJ, Dustin ML. (2006) Regulatory T cells inhibit stable contacts between CD4+ T cells and dendritic cells in vivo. J Exp Med 203:505–511

Tang Q, Adams JY, Tooley AJ, Bi M, Fife BT, Serra P, Santamaria P, Locksley RM, Krummel MF, Bluestone JA. (2006) Visualizing regulatory T cell control of autoimmune responses in non-obese diabetic mice. Nat Immunol 7:83–92

Townsend SE, Allison JP. (1993) Tumor rejection after direct costimulation of CD8+ T cells by B7-transfected melanoma cells. Science 259:368–370

Tumeh PC, Radu CG, Ribas A. (2008) PET imaging of cancer immunotherapy. J Nucl Med 49:865–868

Valitutti S, Muller S, Dessing M, Lanzavecchia A. (1996) Different responses are elicited in cytotoxic T lymphocytes by different levels of T cell receptor occupancy. J Exp Med 183:1917–1921

Vignali DA, Collison LW, Workman CJ. (2008) How regulatory T cells work. Nat Rev Immunol 8:523–532

Weissman IL. (1973) Tumor immunity in vivo: evidence that immune destruction of tumor leaves "bystander" cells intact. J Natl Cancer Inst 51:443–448

Wiedemann A, Depoil D, Faroudi M, Valitutti S. (2006) Cytotoxic T lymphocytes kill multiple targets simultaneously via spatiotemporal uncoupling of lytic and stimulatory synapses. Proc Natl Acad Sci USA 103:10985–10990

Winter H, Hu HM, Urba WJ, Fox BA. (1999) Tumor regression after adoptive transfer of effector T cells is independent of perforin or Fas ligand. (APO-1L/CD95L) J Immunol 163:4462–4472

Wolf K, Muller R, Borgmann S, Brocker EB, Friedl P. (2003) Amoeboid shape change and contact guidance: T-lymphocyte crawling through fibrillar collagen is independent of matrix remodeling by MMPs and other proteases. Blood 102:3262–3269

Woo EY, Yeh H, Chu CS, Schlienger K, Carroll RG, Riley JL, Kaiser LR, June CH. (2002) Cutting edge: Regulatory T cells from lung cancer patients directly inhibit autologous T cell proliferation. J Immunol 168:4272–4276

Yang G, Hellstrom KE, Chen L. (1997) The role of B7–2. (CD86) in tumour immunity. Expert Opin Investig Drugs 6:677–684

Yu P, Lee Y, Liu W, Krausz T, Chong A, Schreiber H, Fu YX. (2005) Intratumor depletion of CD4+ cells unmasks tumor immunogenicity leading to the rejection of late-stage tumors. J Exp Med 201:779–791

Zinkernagel RM. (2002) On cross-priming of MHC class I-specific CTL: rule or exception. Eur J Immunol 32:2385–2392

Index